EAI/Springer Innovations in Communication and Computing

Series editor
Imrich Chlamtac, European Alliance for Innovation, Ghent, Belgium

Editor's Note

The impact of information technologies is creating a new world yet not fully understood. The extent and speed of economic, life style and social changes already perceived in everyday life is hard to estimate without understanding the technological driving forces behind it. This series presents contributed volumes featuring the latest research and development in the various information engineering technologies that play a key role in this process.

The range of topics, focusing primarily on communications and computing engineering include, but are not limited to, wireless networks; mobile communication; design and learning; gaming; interaction; e-health and pervasive healthcare; energy management; smart grids; internet of things; cognitive radio networks; computation; cloud computing; ubiquitous connectivity, and in mode general smart living, smart cities, Internet of Things and more. The series publishes a combination of expanded papers selected from hosted and sponsored European Alliance for Innovation (EAI) conferences that present cutting edge, global research as well as provide new perspectives on traditional related engineering fields. This content, complemented with open calls for contribution of book titles and individual chapters, together maintain Springer's and EAI's high standards of academic excellence. The audience for the books consists of researchers, industry professionals, advanced level students as well as practitioners in related fields of activity include information and communication specialists, security experts, economists, urban planners, doctors, and in general representatives in all those walks of life affected ad contributing to the information revolution.

About EAI

EAI is a grassroots member organization initiated through cooperation between businesses, public, private and government organizations to address the global challenges of Europe's future competitiveness and link the European Research community with its counterparts around the globe. EAI reaches out to hundreds of thousands of individual subscribers on all continents and collaborates with an institutional member base including Fortune 500 companies, government organizations, and educational institutions, provide a free research and innovation platform.

Through its open free membership model EAI promotes a new research and innovation culture based on collaboration, connectivity and recognition of excellence by community.

More information about this series at http://www.springer.com/series/15427

Lucia Knapcikova • Michal Balog
Dragan Perakovic • Marko Perisa
Editors

4th EAI International Conference on Management of Manufacturing Systems

MMS 2019

Springer

EAI
RESEARCH MEETS INNOVATION

Editors
Lucia Knapcikova
Department of Industrial Engineering and Informatics
Technical University of Kosice
Presov, Slovakia

Michal Balog
Faculty of Manufacturing Technologies
Technical University of Kosice
Presov, Slovakia

Dragan Perakovic
Faculty of Transport & Traffic Sciences
University of Zagreb
Zagreb, Croatia

Marko Perisa
Faculty of Transport and Traffic Science
University of Zagreb
Zagreb, Croatia

ISSN 2522-8595 ISSN 2522-8609 (electronic)
EAI/Springer Innovations in Communication and Computing
ISBN 978-3-030-34271-5 ISBN 978-3-030-34272-2 (eBook)
https://doi.org/10.1007/978-3-030-34272-2

This Springer imprint is published by the registered company Springer Nature Switzerland AG.
The registered company address is: Gewerbestrasse 11, 6330 Cham, Switzerland

Preface

Dear Friends,

It gives me great pleasure to welcome you to the MMS 2019 – 4th EAI International Conference on Management of Manufacturing Systems. The MMS 2019 Conference took place on 8–10 October 2019 in the Centre of the "Pearl of Polish Spa's" Krynica Zdrój.

This Conference was endorsed and organized by the European Alliance for Innovation (EAI) in cooperation with the Faculty of Manufacturing Technologies, Technical University of Košice (Slovakia), and the Department of Information and Communication Traffic, Faculty of Transport and Traffic Sciences, University of Zagreb (Croatia).

I completely agree with the EAI's vision, which is to foster excellence in research and innovation on the principles of transparency, objectivity, equality, and openness. Our guiding principle is community cooperation to create better research, provide fair recognition of excellence, and transform best ideas into commercial value proposition.

When I was asked to be the General Chair of the EAI International Conference on Management of Manufacturing Systems Conference, I gladly accepted the position. The conference aim is creating synergies of "practice and research" and increasing the potential and commercial viability of research and development in the field of innovative technologies in Industry 4.0, logistics and traffic system, smart manufacturing and materials.

I would like to sincerely thank to my department colleagues, especially to Annamária Behúnová, Michal Balog, Patrik Kaščák and Jozef Husár, for their active work and enthusiasm to bring this conference to fruitfulness. Special thanks go to my co-chair of the Conference, Prof. Dr. Dragan Perakovic from the Faculty of Transport and Traffic Sciences, University of Zagreb (Croatia).

Big thanks to our keynote speakers, authors, reviewers, business participants, sponsors, and publishers for their important contributions to the event. Finally, and most importantly, I would like to thank the participants and members of the EAI.

I believe that the MMS 2019 Conference will enrich us all.

Sincerely yours,

Presov, Slovakia Lucia Knapcikova

Credits for Organizing Committee

Steering Committee

Chair:

Prof. Dr. Imrich Chlamtac – Bruno Kessler Professor, University of Trento, Italy

Member:

Assoc. Prof. Dr. Michal Balog – Faculty of Manufacturing Technologies with a seat in Prešov of TU Košice, Slovakia

Organizing Committee

General Chairs

Assoc. Prof. Dr. Michal Balog – Faculty of Manufacturing Technologies with a seat in Prešov of TU Košice, Slovakia

Assoc. Prof. Dr. Lucia Knapcikova – Faculty of Manufacturing Technologies with a seat in Prešov of TU Košice, Slovakia

Prof. Dr. Dragan Perakovic – University of Zagreb, Croatia

TPC Chair and Co-Chair/Program Chairs

Assoc. Prof. Dr. Michal Balog – Faculty of Manufacturing Technologies with a seat in Prešov of TU Košice, Slovakia

Assoc. Prof. Dr. Lucia Knapcikova – Faculty of Manufacturing Technologies with a seat in Prešov of TU Košice, Slovakia

Dr.h.c. Prof. Dr. Jozef Zajac – Faculty of Manufacturing Technologies, TU Kosice, Slovakia

Prof. Dr. Jan Pitel – Faculty of Manufacturing Technologies, TU Kosice, Slovakia

Prof. Dr. Dragan Perakovic – University of Zagreb, Croatia

Assoc. Prof. Dr. Vitalii Ivanov, Ing. Paed. IGIP – Sumy State University, Ukraine

Assoc. Prof. Dr. Michal Hatala – Faculty of Manufacturing Technologies, TU Kosice, Slovakia

Dr. Svetlana Radchenko – Faculty of Manufacturing Technologies, TU Kosice, Slovakia

Assoc. Prof. Dr. Oleg Zabolotnyi – Lutsk National Technical University, Ukraine

Prof. Dr. Michael Herzog – Technical University of Applied Sciences Wildau, Germany

Prof. Dr. Alessandro Ruggiero – University of Salerno, Italy

Prof. Dr. Janusz Grabara – Czestochowa University of Technology, Poland

Dr. Annamaria Behunova – Faculty of Manufacturing Technologies, TU Kosice, Slovakia

Sponsorship and Exhibit Chair

Assoc. Prof. Dr. Michal Balog – *Faculty of Manufacturing Technologies, TU Kosice, Slovakia*

Local Chair

Assoc. Prof. Dr. Michal Balog – *Faculty of Manufacturing Technologies, TU Kosice, Slovakia*

Workshops Committee Chair

Assoc. Prof. Dr. Michal Balog – *Faculty of Manufacturing Technologies, TU Kosice, Slovakia*

Publicity & Social Media Chair

Assoc. Prof. Dr. Michal Balog – *Faculty of Manufacturing Technologies, TU Kosice, Slovakia*

Publications Chair

Assoc. Prof. Dr. Lucia Knapcikova – *Faculty of Manufacturing Technologies, TU Kosice, Slovakia*
Assoc. Prof. Dr. Michal Balog – *Faculty of Manufacturing Technologies, TU Kosice, Slovakia*
Prof. Dr. Dragan Perakovic – *University of Zagreb, Croatia*
Assist. Prof. Dr. Marko Perisa – *University of Zagreb, Croatia*

Web Chair

Dr. Jozef Husar – *Faculty of Manufacturing Technologies, TU Kosice, Slovakia*

Industry Forum Committee Chair

Prof. Dr. Jan Pitel – *Faculty of Manufacturing Technologies, TU Kosice, Slovakia*

Panels Chair

Assoc. Prof. M.A. Dagmar CAGANOVA, PhD. – *Vice-Dean for International Projects, Faculty of Materials Science and Technology in Trnava STU Bratislava, Slovakia*
Prof. Dr. Dragan Perakovic – *University of Zagreb, Croatia*

Special Session Committee Chair

Assoc. Prof. Dr. Michal Balog – *Faculty of Manufacturing Technologies, TU Kosice, Slovakia*

Assoc. Prof. Dr. Lucia Knapcikova – *Faculty of Manufacturing Technologies, TU Kosice, Slovakia*

Assoc. Prof. M.A. Dagmar CAGANOVA, PhD. – *Vice-Dean for International Projects, Faculty of Materials Science and Technology in Trnava STU Bratislava, Slovakia*

Technical Program Committee Session or Track Leaders

Assoc. Prof. Dr. Lucia Knapcikova – *Faculty of Manufacturing Technologies, TU Kosice, Slovakia*

Assoc. Prof. Dr. Vitalii Ivanov, Ing. Paed. ***IGIP***, – *Sumy State University, Ukraine*

Assist. Prof. Marko Perisa – *University of Zagreb, Croatia*

Technical Program Committee

Assoc. Prof. Dr. Lucia Knapcikova – Faculty of Manufacturing Technologies with a seat in Prešov of TU Košice, Slovakia

Assoc. Prof. Dr. Michal Balog – Faculty of Manufacturing Technologies with a seat in Prešov of TU Košice, Slovakia

Dr.h.c. Prof. Dr. Jozef Zajac – Faculty of Manufacturing Technologies with a seat in Prešov of TU Košice, Slovakia

Assoc. Prof. Dr. Michal Hatala – Faculty of Manufacturing Technologies with a seat in Prešov of TU Košice, Slovakia

Dr. Svetlana Radchenko – Faculty of Manufacturing Technologies with a seat in Prešov of TU Košice, Slovakia

Dr. Annamaria Behunova – Faculty of Manufacturing Technologies with a seat in Prešov of TU Košice, Slovakia

Dr. Jozef Husar – Faculty of Manufacturing Technologies with a seat in Prešov of TU Košice, Slovakia

Assoc. Prof. Dr. Martin Straka – Faculty of Mining, Ecology, Process Control and Geotechnology of TU Košice, Slovakia

Prof. Dr. Alessandro Ruggiero – University of Salerno (Universita degli Studi di Salerno), Italy

Prof. Dr. Alena Dankova – Catholic University in Ružomberok, Slovakia

Prof. Dr. Michael Herzog – Technical University of Applied Sciences Wildau, Germany

Prof. Dr. Janusz Grabara – Czestochowa University of Technology, Poland

Prof. Dr. Jerzy Winczek – Czestochowa University of Technology, Poland

Assoc. Prof. Dr. Dagmar Caganova – Slovak University of Technology, Bratislava, Slovakia

Dr. Petr Travnicek – Mendel University in Brno, Czech Republic

Dr. Katarzyna Huk – University of Zielona Gora, Poland

Dr. Jozef Husar – Faculty of Manufacturing Technologies with a seat in Prešov of TU Košice, Slovakia

Prof. Dr. Dragan Perakovik – University of Zagreb, Croatia

Assist. Prof. Dr. Marko Perisa – University of Zagreb, Croatia

Assoc. Prof. Dr. Vitalii Ivanov – Sumy State University, Ukraine

Assoc. Prof. Dr. Ivan Pavlenko – Sumy State University

Assoc. Prof. Dr. Oleksandr Liaposhchenko – Sumy State University

Assoc. Prof. Dr. Oleg Zabolotnyi – Lutsk National Technical University, Ukraine

Assist. Prof. Dr. Justyna Trojanowska – Poznan University of Technology

Mgr. Eng. Yuliia Tarasevych – AGH University of Science and Technology

Prof. Dr. Jan Pitel - Faculty of Manufacturing Technologies, TU Kosice, Slovakia

Prof. Dr. Krzysztof Witkowski – University of Zielona Gora, Poland

Assoc. Prof. Dr. Anna Saniuk – University of Zielona Gora, Poland

Assoc. Prof. Dr. Sebastian Saniuk – University of Zielona Gora, Poland

Conference Manager

Radka Vasileiadis (Pinčáková) – EAI

Contents

Definition of the IoT Device Classes Based on Network Traffic Flow Features

Ivan Cvitic, Dragan Perakovic, Marko Perisa, and Mate Botica

1 Introduction

According to the forecasts presented in [1] by 2020, approximately 31 billion IoT devices will be available globally. By doing so, 41% or 12.86 million IoT devices will be installed within the smart home (SH) concept [2]. Restrictions on IoT devices, in general, and therefore the SHIoT (Smart Home IoT) devices are described in the research [3], which include hardware constraints, demands for high autonomy, and low production costs, thus reducing the possibility of implementing advanced protection methods and increasing the risk of the many threats shown in [4].

Traffic generated by SHIoT devices or MTC (machine-type communication) traffic differs from traffic generated by conventional HTC (human-type communication) traffic, as shown in the survey [5]. Although SHIoT devices are characterized by heterogeneity, MTC traffic is homogeneous to HTC traffic, meaning that devices of the same or similar purpose behave approximately equally or generate similar traffic [6].

The underlying hypothesis of this research is that SHIoT devices can differentiate by traffic flow characteristics such as the ratio of received and sent data and that such features can be utilized to define the classes of IoT devices.

I. Cvitic · D. Perakovic (✉) · M. Perisa
Faculty of Transport & Traffic Sciences, University of Zagreb, Zagreb, Croatia
e-mail: ivan.cvitic@fpz.hr; dperakovic@fpz.hr; marko.perisa@fpz.hr

M. Botica
OiV Transmitters and Communications Ltd., Zagreb, Croatia
e-mail: mate.botica@oiv.hr

L. Knapcikova et al. (eds.), *4th EAI International Conference on Management of Manufacturing Systems*, EAI/Springer Innovations in Communication and Computing, https://doi.org/10.1007/978-3-030-34272-2_1

1.1 Related Research

Identifying devices in the IoT environment is an important step and basis for activities related to the security environment in which such devices exist such as detecting unauthorized activity, detecting unauthorized devices within the network, detecting malicious code, and so on. Authors in the research [7] use a cluster method for classifying 21 IoT devices, classifying devices separately on the basis of 11 features. Research [8] seeks to detect unauthorized devices connected to the monitored network based on device identification. A total of 11 IoT devices have been used for this purpose, which are classified according to the semantic characteristics of the device, i.e., their purpose (child monitoring devices, motion sensors, refrigerators, safety cameras, smoke sensors, sockets, thermostats, televisions, clocks). A similar way of classification, based on the semantic characteristics of the device, is also shown in the research [9] in which authors use a secondary data set collected in [7]. The research covered a total of 15 devices classified into 4 categories regarding the purpose of each device (hubs, electronic devices, cameras, and sockets). Authors, based on the analysis carried out, underline the importance of the variety of devices involved in the data collection phase over the size of the data set itself (the time of preacquisition and the amount of collected traffic) for classifying the SHIoT device. Specific features of MTC traffic were used to address numerous problems in the communications network. Research [10] monitors the impact of MTC traffic on QoS when integrating with HTC traffic in the LTE communications network. Identification and classification of IoT devices in smart cities and campuses and in smart circles using the characteristics of MTC traffic is shown in [7, 11]. Research [12] seeks to identify new demands and challenges in design and management of the mobile communications network imposed by the generation of MTC traffic.

From the current research, it is evident that approaches to the identification and classification of IoT devices so far are based mainly on the semantic features of the device, meaning that the device classes are defined according to the mode of application of such devices or their primary functionalities. The lack of such an approach to defining a device classes can be viewed from the perspective of the dynamism of a smart home environment. According to the statistical indicators presented in [13], the number of SHIoT devices is constantly increasing, followed by the growth in the number of companies developing new solutions and new SHIoT devices. Therefore, the class of the SHIoT device needs to be defined in a way that will be applicable to upcoming SHIoT devices that will differ from the currently available devices according to their functionality and application.

The aim of this research is to define the class of IoT devices based solely on the characteristics of traffic flows generated by such devices. The so defined will be independent of the semantic characteristics and functionality of the device, which is why its application is also possible on new IoT devices that will be introduced in the future.

1.2 Research Methodology

For the purpose of carrying out this research, primary and secondary data were collected. For the purpose of collecting primary data, a laboratory environment was established in the Laboratory for Security and Forensic Analysis of the Information and Communication System of the Department for Information and Communication Traffic of the Faculty of Transport and Traffic Sciences of the University of Zagreb. Secondary data used in this research were collected for the purpose of research [11, 14, 15]. The data collected represent the traffic generated by the IoT devices covered by this research. For the purpose of defining the classes of the device, mathematical and statistical methods and Stata software were used for data processing and interpretation of the results of the research.

2 Data Collection and Processing

The SHIoT devices covered by this research are presented in Table 1. The MAC (media access control) represents unique identifiers of the SHIoT device in the network, the name of the device, the P/S tag that indicates whether the monitored device used for collecting primary or secondary data and to which group belongs according to the segmentation shown in [16].

SHIoT devices are provided by authorized distributors and dealers of a single device manufacturer and are connected to and connected to the communications network in ways recommended by the manufacturer, and in no way have the devices modified at the software and hardware level. It is therefore assumed that the devices used to collect legitimate traffic within this research work in the way they are designed and in no way compromised in any way.

The network topology as well as the characteristics of a smart home environment can be seen in Fig. 1. Devices are connected directly or indirectly with Wi-Fi communications technology with the Fortinet AP 221C wireless access point, with the exception of the Phillips Hue device that communicates with the rest of the local network via Ethernet communication technology. Some devices, such as the Blink Smart Camera, the Netatmo Smart Thermostat, and the Philips Hue Smart Bulb, use IoT hubs with wireless communication, but with ZigBee technology. The reason for this is the energy efficiency of the device as it uses the battery as the ultimate power source, which gives them advantages from the aspect of mobility and independence of the power supply unit as a source of power. The IoT hub is connected to Wi-Fi (or Ethernet in the case of a Phillips Hue device) with a wireless access point technology. The subject mentioned as an adequate point of collecting traffic that SHIoT devices generate is a certain wireless access point.

Because of the known working methods and characteristics of computer and thus wireless Wi-Fi networks, communication in the communications network cannot be collected directly. There are several methods available for collecting traffic, which is

Table 1 SHIoT devices for the purpose of data collection

No.	MAC address	SHIoT device name	Data aggregation source	Functional category
1	00:17:88:78:0a:cb	Phillips Hue Starter kit 2 × E26	P	CL
2	00:17:88:2b:9a:25	Phillip Hue Starter kit 4 × E26	S	CL
3	a8:bb:50:05:31:f3	WiZ Colors ESP_0531F3	P	CL
4	a8:bb:50:05:06:b0	WiZ Colors ESP_0506B0	P	CL
5	d0:73:d5:01:83:08	Light Bulbs LiFX Smart Bulb	S	CL
6	00:24:e4:20:28:c6	Withings Aura Sleep Tracking Mat	S	CL
7	7c:2e:bd:3d:4f:cb	Google Chromecast	P	M
8	18:b7:9e:02:20:44	Invoxia Triby Speaker	S	M
9	e0:76:d0:33:bb:85	PIX-STAR Photo-frame	S	M
10	fc:65:de:31:69:d6	Amazon Alexa Dot	P	M
11	44:65:0d:56:cc:d3	Amazon Alexa Echo	S	M
12	20:df:b9:21:fd:79	Google Home mini	P	M
13	ac:84:c6:5d:97:bc	TPlink Smart Plug HS110	P	MC
14	50:c7:bf:00:56:39	TPlink Smart Plug HS105	S	MC
15	30:ae:a4:57:2d:54	MyStrom switch	P	MC
16	74:da:da:5f:a8:19	D-link DSP-W245 plug	P	MC
17	74:c6:3b:29:d7:1d	iHome Power Plug	S	MC
18	ec:1a:59:79:f4:89	Belkin Wemo switch	S	MC
19	d0:52:a8:00:67:5e	Samsung Smart Things	S	MC
20	74:6a:89:00:2e:25	Blipcare Blood Pressure meter	S	MC
21	70:88:6b:10:0f:c6	Awair air quality monitor	S	MC
22	40:9f:38:e9:28:08	iRoobot Roomba 896	P	SA
23	80:c5:f2:bb:17:95	iRoobot Roomba 895	P	SA
24	00:24:e4:1b:6f:96	Withings Body	S	SA
25	e8:ab:fa:9b:f0:9e	Smartwares C923IP Camera	P	S
26	00:03:7f:27:2c:c3	Blink XT2 Camera	P	S
27	7c:70:bc:5d:5e:dc	Canary View Camera	S	S
28	70:ee:50:18:34:43	Netatmo Welcome Camera	S	S
29	f4:f2:6d:93:51:f1	TPlink Day Night Cloud NC220 camera	S	S
30	00:16:6c:ab:6b:88	Samsung SmartCam	S	S
31	30:8c:fb:2f:e4:b2	Nest Dropcam	S	S
32	00:24:e4:11:18:a8	Withings Smart Baby Monitor	S	S

(continued)

Table 1 (continued)

No.	MAC address	SHIoT device name	Data aggregation source	Functional category
33	18:b4:30:25:be:e4	NEST Protect Smoke Alarm	S	S
34	88:4a:ea:31:66:9d	Ring Video Doorbell	S	S
35	70:ee:50:0c:14:c2	Netatmo Smart Thermostat	P	EM
36	70:ee:50:03:b8:ac	Netatmo Smart Weather Station	S	EM

P primary, *S* secondary, *CL* comfort and lightning, *M* multimedia, *MC* monitor and connectivity, *S* security, *SA* smart appliances, *EM* energy management

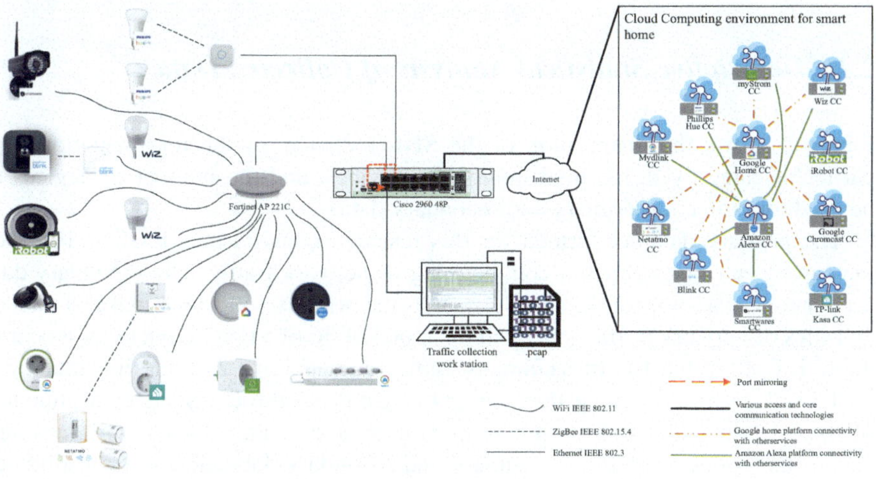

Fig. 1 Smart home laboratory environment

often used to mirror the physical door on the switch. The mentioned method proved to be effective in several studies such as [8, 10, 17, 18] which provides the basis for applying the same method for the purpose of this research.

For the purpose of collecting network traffic, the port mirroring functionality is set up with a software-hardware platform consisting of a wireless access point, the Fortinet AP 221C, the Cisco 2960 Catalyst 48 PoE (Power over Ethernet) switch, and the HP Pavilion dm1 workstation (Microsoft Windows 10 10.0. 17134 build 17134, x64 processor architecture, AMD E-350, 1600 MHz 2 core, 4 GB RAM) with installed Wireshark software tool version 2.6.3.

As shown in Fig. 1, the port mirroring is configured for the physical communication ports (FA0/1 and FA0/3) of the switch with the wireless access point and IoT hub for the Phillips Hue device. The specified ports are configured as source indicating that the overall traffic coming from or to these ports will be mirrored to the destination communication port (FA0/2). There is a traffic collection workstation connected to this port.

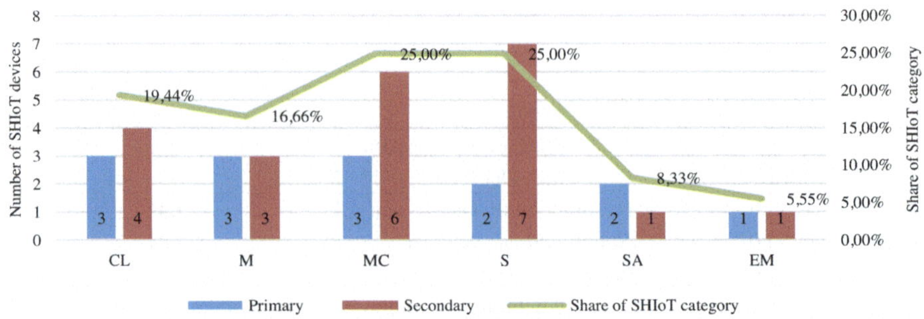

Fig. 2 Distribution of SHIoT devices

2.1 *Descriptive Statistical Analysis of Collected Data*

Figure 2 shows the distribution of the SHIoT device, or the representation of a particular category in the total number of devices, and the number of devices to be used for collecting primary and secondary data.

The primary data set formed for this research consists of a total of 103 files in.pcap format that contain a complete log of network traffic. The secondary data set consists of 41 files of the same format as the primary set, which makes a total of 144 files of network traffic generated by a total of 36 of SHIoT devices as shown in Table 1. Each of the 144 files contains traffic generated in a 24-hour time interval.

Table 2 shows a statistical description of the data set through static measurements of standard deviation and minimum, maximum, and mean values at 24-hour intervals of the collected traffic for primary and secondary data and for a consolidated data set.

The characteristics of the initially collected data are presented in Table 3 and are expressed through the number of collected files containing 24 hours of generated traffic, the number of collected packets, the size of the file, the amount of data collected, and the total time of the data collection. The traffic colleting tool (Wireshark) uses certain metadata that it writes inside the file with the traffic collected, which is the reason for the difference in the size of the file and the amount of data (traffic) that is stored in the file.

For the purpose of defining the SHIoT classes, the process of filtering traffic from a particular.pcap file to the MAC address of the device was performed. The reason for such filtering is to assign an IP address to devices via a DHCP server, which can be changed over time, so it does not represent a reliable feature for precise filtering of traffic to an individual device over a longer period of time.

Table 2 Statistical description of collected data at 24-hour intervals

Statistical measure	Number of packet	File size (Byte)	Amount of collected data (Byte)	Average data rate (Bps)	Average packet rate (packets per second)	Average packet size (Byte)
Primary data						
St. dev.	2613702,8	2503780307	2462270632	28498,13	30,25047	175,4824
Min	1019339	288056862	271747414	3145,21	11,8	252,5
Max	14815959	13562522315	13325466947	154232,8	171,48	899,4
Ar. mean	4428879,6	3416448737	3345586639	38721,71	51,25942	677,782
Secondary data						
St. dev.	1646515,4	804291290,1	765130504,3	12047,91	186,6204	20,69223
Min	527035	89615024	71959664	832,87	136,54	6,1
Max	7720905	3483660828	3322027939	60188,48	910,56	730,47
Ar. mean	2365097,3	986887232,4	908751070	11565,99	330,1257	45,21762
Sum						
St. dev.	2557564,5	2429423670	2396969893	27864,11	165,5721	329,4564
Min	527035	89615024	71959664	832,87	11,8	6,1
Max	14815959	13562522315	13325466947	154232,8	910,56	899,4
Ar. mean	3835384,1	2714894474	2641775718	30881,53	132,6931	492,9182

Table 3 The characteristics of the initial data set

	No. of files	Number of packets	File size (GB)	Collected data size (GB)	Collecting time period (hours)
Primary (sum)	103	456174601	351,89	344,59	2472,01
Secondary (sum)	41	99334088	41,44	38,16	986,45
Total	144	555508689	393,33	382,7554238	3458,47

2.2 *Extraction of Traffic Flows of IoT Devices*

Defining the class of SHIoT devices in this research is based on the statistical characteristics of the traffic flows of a particular device. The traffic flow is defined by a packet with equal source of source IP address, destination IP address, source communication port, destination communication port, and protocol used (TCP or UDP) [19].

The reason for choosing a traffic flow as the level of observation and analysis of the feature is because it represents the aggregated (statistical) data of the packet header for communication between the source and the destination. The packet-level traffic analysis includes more information such as packet content but also requires more computing resources for their storage and processing. An example of the number of traffic flows and the number of packets over a 24-hour time is visible to the Google Chromecast device (covered by this survey) where 11877 individual traffic flows were generated, while the number of packets collected in the same time interval was 2459538. Since most of today's devices and applications use cryptographic methods in communication process, the content of the package cannot be observed and analyzed in an economically, time-consuming, and legally acceptable way. Consequently, observation and analysis of traffic characteristics at the traffic flow level is an acceptable and frequently used approach in many researches.

Extraction of traffic flow features for individual SHIoT device was done using the software tool developed at CICFlowMeter Canadian Institute for Cyber Security at the University of New Brunswick in Canada [20]. This tool allows the extraction of 84 traffic flow characteristics such as source and destination IP addresses, time of the traffic flow, interarrival time of packets, packet number per traffic flow, packet size, amount of data transferred, transfer rate, and so on.

The time period for which the feature extraction is performed is 30 consecutive days, with different number of generated traffic flows depending on the device and its characteristics.

3 Defining SHIoT Device Classes

Classes of SHIoT devices are defined based on traffic flow features. For this purpose, the coefficient of variation of the ratio of the received and sent volume of traffic is used, which represents the index of the predictability level of the IoT's behavior.

3.1 *Determination of Features for Defining SHIoT Device Classes*

The predictability of the IoT devices behavior is a phenomenon that is the result of the communication activities of IoT devices observed in the research [10, 18, 21]. Since IoT devices have a limited number of functionalities, certain devices will behave at approximately the same time as the values of the observed traffic features. Unlike IoT devices, conventional devices (smartphones, desktops, laptops, etc.) support the installation of a large number of applications where the communication activity of such devices depends on end users and how the device is used. According to the above, the IoT device predictability level indices expressed by the coefficient of variation of received and sent data (C_u index) is a measure based on which it is possible to determine the behavior of the IoT device over a given period of time. As the C_u index is closer to 0, the observed device has less deviation of the amount of received and transmitted data, and the level of predictability of such device behavior is considered to be greater than the device whose C_u index is greater than 0.

The C_u index is calculated for mean values of 20 consecutive traffic flows of a single SHIoT device over a time period of 30 days according to expression Eq. (1).

$$C_u = \frac{\sqrt{\frac{1}{N-1}\sum_{i=1}^{N}(x_i - \overline{x})^2}}{\frac{1}{N}\sum_{i=1}^{N} x_i} \tag{1}$$

where:

C_u – index of traffic predictability level for SHIoT device u
N – total number of mean values of the received and sent data volume ratio for 20 consecutive traffic flows in the time period $t = 30$ days
x_i – the mean value of received and sent data volume ratio for 20 consecutive traffic flows

In order to avoid that the mean values to tend 0, which is the problem of the coefficient of variation method application as a normalized dispersion value in the data set, the traffic flows in which the ratio of received and sent data is equal to 0 are removed.

3.2 Defining the IoT Device Classes Based on the Coefficient of Variation

For the purpose of the SHIoT device classes definition based on the C_u index value, the coefficient of variation classification method applied in [22–25] was used. It assumes normal data distribution. Since the distribution of the obtained values (C_u index) are asymmetric in nature (negative skewness), the data have been transformed. The data transformation method was selected using the Ladder of powers (Tukey method) that clearly shows a suitable data transformation function to achieve normal distribution [26].

Figure 3 illustrates the appropriateness of using a logarithmic function for the purpose of data transformation since in this case results in a normal distribution. In addition to the graphical representation, Table 4 shows the chi^2 values of a particular transformation function.

The distribution of the data is closest to the normal chi^2 closer to 0 and P (chi^2) closer to 1. The normal distribution of the data obtained is also confirmed by the Shapiro-Wilk test of normality ($p = 0.7262$)

For the purpose of applying the method of classifying the coefficient of C_u index, it is normalized by the min-max method according to the expression Eq. (2):

$$C_{u(\text{norm})} = \frac{\log(C_u) - \log(C_{u\min})}{\log(C_{u\max}) - \log(C_{u\min})} \tag{2}$$

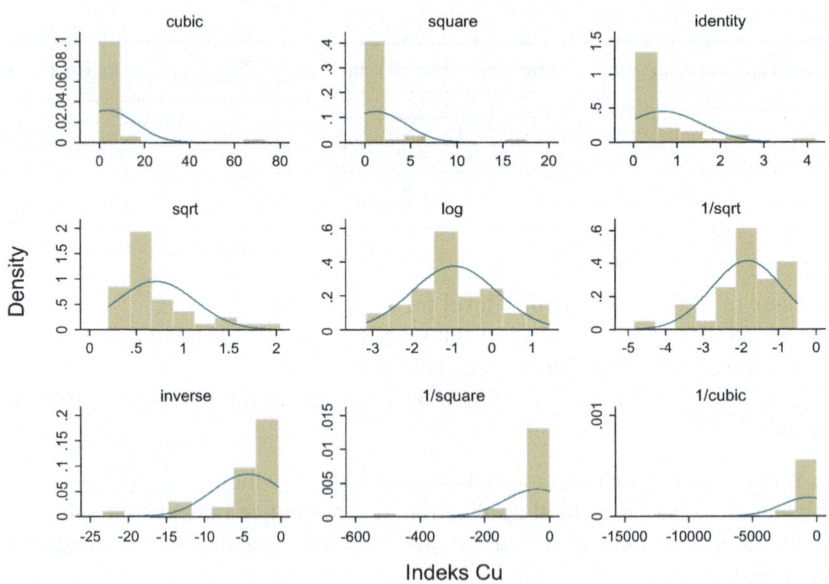

Fig. 3 Histogram of data distribution depending on the transformation function used

Table 4 Ladder of powers results

Transformation	Formula	chi^2	P(chi^2)
Cubic	$C_u{}^3$	50,29	0
Square	$C_u{}^2$	42,42	0
Identity	C_u	25,18	0
Square root	$\sqrt{C_u}$	12,08	0,002
Logarithmic	$\log(C_u)$	**0,44**	**0,804**
1/(square root)	$\sqrt{\frac{1}{C_u}}$	9,48	0,009
Inverse	$\frac{1}{C_u}$	25,07	0
1/square	$\frac{1}{C_u{}^2}$	43,37	0
1/cubic	$\frac{1}{C_u{}^3}$	51,08	0

where:

$C_{u(\mathrm{norm})}$ – the normalized value of the logarithmically transformed C_u index in the interval [0,1]
$\log(C_u)$ – logarithmic value of C_u device u
$\log(C_{u\min})$ – minimum logarithmic value of C_u for all devices
$\log(C_{u\max})$ – maximum logarithmic value C_u for all devices

After the normal distribution of the data was established, the method of classifying the coefficient of variation is applied. It is a result of the mean coefficient variation values and their standard deviations.

The mean value of the coefficient of variation is calculated according to Eq. (3):

$$A_{C_{u(\mathrm{norm})}} = \frac{1}{N}\sum\nolimits_{u=1}^{n} \frac{C_{1(\mathrm{norm})} + C_{2(\mathrm{norm})} + \cdots + C_{n(\mathrm{norm})}}{N} \tag{3}$$

where:

$A_{C_{u(\mathrm{norm})}}$ – the arithmetic mean of the coefficient of variation for all devices
N – number of SHIoT devices
$C_{u(\mathrm{norm})}$ – coefficient of variation of the device u

The standard deviation of the coefficient of variation was calculated according to the formula Eq. (4):

$$\sigma_{C_{u(\mathrm{norm})}} = \sqrt{\frac{1}{N-1}\sum\nolimits_{u=1}^{n}\left(C_{u(\mathrm{norm})} - \overline{C}\right)^2} \tag{4}$$

where:

$\sigma_{C_{u(\mathrm{norm})}}$ – standard deviation of the coefficient of variation for all devices
N – number of SHIoT devices
$C_{u(\mathrm{norm})}$ – coefficient of variation of the device u
$\overline{C}$ – the arithmetic mean of the coefficient of variation for all devices

Based on previously performed data processing, a total of four classes of IoT devices were defined according to the method used in the research [24]. The first class includes devices for which the condition is met $C_{u(\text{norm})} \leq A_{C_{u(\text{norm})}} - \sigma_{C_{u(\text{norm})}}$. The second class includes devices that meet the condition $A_{C_{u(\text{norm})}} - \sigma_{C_{u(\text{norm})}} < C_{u(\text{norm})} \leq \frac{A_{C_u} + \sigma_{C_u}}{2}$. The third class includes devices that meet the condition $\frac{A_{C_u} + \sigma_{C_u}}{2} < C_{u(\text{norm})} \leq A_{C_u} + \sigma_{C_u}$, while the last class includes devices that meet the condition $C_{u(\text{norm})} > A_{C_u} + \sigma_{C_u}$.

4 Research Results and Discussion

Values of C_u index, logarithmically transformed values, and min-max normalized values for each analyzed device are shown in Table 5.

According to the data presented in Table 5, a total of four device classes are defined based on the value of the C_u index. The first class (C1) comprises all devices whose log transformed and normalized value $C_{u(\text{norm})} \leq$ 0,253722. The second class (C2) combines devices that meet the requirement 0,253722 < $C_{u(\text{norm})}$ 0,354866. The third class (C3) includes devices that meet the requirements 0,354866 < $C_{u(\text{norm})} \leq$ 0,709732, while the last class (C4) includes devices that meet the requirements $C_{u(\text{norm})}$ > 0,709732.

Class C1 signifies IoT devices with a very high level of behavior predictability since the coefficient of variation in the ratio of received and sent data is closest to 0. This means that such devices over time behave approximately the same from the aspect of the observed feature. The use of IoT devices of class C1 devices by users, other devices or the environment will not have a significant effect on the change in the value of the C_u index. Class C2 combines devices with a high level of behavior predictability. The use of devices from the specified class by users, other devices or the environment may result in minor changes in the relationship between received and sent data. Devices classed with C3 represent devices with a medium level of behavior predictability. The impact of interaction between users, other devices or the environment on the relationship between received and sent data can be significant. The result of such behavior may be the additional functionalities of devices that at certain times result in a greater amount of data in the incoming or outgoing direction. The last class C4 combines IoT devices with low levels of behavior predictability. The use of such devices and their interaction with the user, other devices, or environment significantly affects the relationship between received and sent data. As a reason, a significantly higher amount of data is received in the incoming direction (download) as a result of the user's request. An example is visible to devices such as Google Chromecast where a user plays video content that requires downloading the same through a Youtube service. This class also includes the Google Home mini device, a smart speaker that can provide different audio content on a user's request, which also causes a greater variation in the relationship between received and sent traffic.

Table 5 Device classes defined by the value of index C_u

No.	SHIoT device	Index C_u	$\log(C_u)$ transformation	Min-max normalization ($C_{u(\text{norm})}$)	Class definition	Class name
1	TPlink Day Night Cloud NC220 camera	0,042916917	−1,367371486	0	$C_{u(\text{norm})} \le A_{C_u} - \sigma_{C_u}$	C1
2	WiZ Colors ESP_0531F3	0,075820416	−1,120213838	0,124242056		
3	TPlink Smart Plug HS105	0,076231674	−1,117864541	0,125423008		
4	WiZ Colors ESP_0506B0	0,08086321	−1,092249024	0,138299504		
5	Samsung Smart Things	0,123562483	−0,908113372	0,230861447		
6	iHome Power Plug	0,148887517	−0,827141714	0,271564558	$A_{C_u} - \sigma_{C_u} < C_{u(\text{norm})} \le \frac{A_{C_u} + \sigma_{C_u}}{2}$	C2
7	Withings Smart Baby Monitor	0,176239975	−0,753895577	0,308384178		
8	NEST Protect Smoke Alarm	0,192606687	−0,715328639	0,327771139		
9	Phillips Hue Starter kit 2 × E26	0,200187894	−0,69856219	0,336199355		
10	Canary View Camera	0,209863653	−0,678062771	0,346504073		
11	Tplink Hs110	0,24742122	−0,606563056	0,382445795	$\frac{A_{C_u} + \sigma_{C_u}}{2} < C_{u(\text{norm})} \le A_{C_u} + \sigma_{C_u}$	C3
12	Belkin Wemo Switch	0,254614637	−0,594116633	0,388702406		
13	Withings Sleep	0,261184872	−0,583051981	0,394264423		
14	D-link DSP-W245 plug	0,27041724	−0,567965624	0,401848085		
15	Netatmo Smart Thermostat	0,290797956	−0,53640865	0,417711253		
16	Amazon Alexa Dot	0,318918293	−0,496320569	0,437862868		
17	Blink XT2 Camera	0,344500361	−0,462810319	0,454707915		
18	Samsung SmartCam	0,34686605	−0,459838205	0,456201948		
19	Light Bulbs LiFX Smart Bulb	0,346886878	−0,459812128	0,456215056		

(continued)

Table 5 (continued)

No.	SHIoT device	Index C_u	$\log(C_u)$ transformation	Min-max normalization ($C_{u(\mathrm{norm})}$)	Class definition	Class name
20	Smartwares C923IP Camera	0,357559305	−0,446651916	0,462830477		
21	iRoobot Roomba 895	0,358681004	−0,445291624	0,463514273		
22	iRoobot Roomba 896	0,379012744	−0,421346187	0,475551248		
23	MyStrom switch	0,432393144	−0,364121201	0,5043173		
24	Blipcare Blood Pressure meter	0,479127026	−0,319549331	0,526722841		
25	Netatmo Smart Weather Station	0,543491131	−0,264807539	0,554240633		
26	Amazon Alexa Echo	0,632948837	−0,198631394	0,587506285		
27	Netatmo Welcome Camera	0,764635407	−0,116545595	0,628769456		
28	Phillip Hue Starter kit 4 × E26	0,791347539	−0,101632744	0,636265899		
29	PIX-STAR Photo-frame	0,958787396	−0,018277684	0,678167108		
30	Withings Body	1,140461786	0,057080738	0,716048538	$C_{u(\mathrm{norm})} > A_{C_u} + \sigma_{C_u}$	C4
31	Google Chromecast	1,267801595	0,103051294	0,739157175		
32	Ring Video Doorbell	1,370122066	0,136759261	0,756101612		
33	Nest Dropcam	1,985562839	0,297883636	0,837096166		
34	Invoxia Triby Speaker	2,468462951	0,392426613	0,884621355		
35	Awair air quality monitor	2,553917945	0,40720694	0,89205118		
36	Google Home mini	4,187473486	0,62195207	1		

This way of classes definition, with respect to classes based on semantic features, allows comprising of IoT devices that are not covered by this research based on their behavior that can be measured by the C_u index. Given the accelerated development and the increasingly frequent application of the IoT concept, classes defined by this research will be able to consolidate IoT devices regardless of their functionalities, purpose, and capabilities.

5 Conclusion

The classification of devices in the IoT concept is a challenging research problem. The initial problem is how to define classes and differentiate IoT devices. Previous researches define the classes based on the semantic characteristics of the device and their purpose and scope of application. Such a class definition method represents a potential problem for new devices whose application and characteristics will differ from the existing ones.

For the purpose of solving this problem, this research has defined classes based on traffic features. To define the class, the coefficient of variation of the received and sent data ratio for an individual device (index C_u) was used. The C_u index was calculated for a total of 36 SHIoTs based on an average of 20 consecutive traffic flows over a 30-day time period and represents the scattering measure of the received and sent data. The data were analyzed, transformed, and normalized by using the Stata tool for statistical analysis and using logarithmic transformation and min-max normalization method. The variation coefficient classification method was applied to define a total of four classes of devices according to the predictability level of their behavior (C1, very high level; C2, high level; C3, middle level; C4, low level).

The class of IoT devices defined in this way provides a framework for further research in the area of classification of the IoT device for the purpose of identifying their behavior and detecting anomalies of network traffic that such devices can generate.

Further research will seek to develop a classification model that will, on the basis of the value of the traffic flow characteristics of a variety of IoT devices, be assigned to the classes defined by this research.

References

1. Statista. (2018). Internet of Things (IoT) connected devices installed base worldwide from 2015 to 2025 (in billions). Retrieved June 24, 2018, from https://www.statista.com/statistics/471264/iot-number-of-connected-devices-worldwide/.
2. Statista. (2018). The Internet of Things (IoT)$_*$ units installed base by category from 2014 to 2020 (in billions). Retrieved June 24, 2018, from https://www.statista.com/statistics/370350/internet-of-things-installed-base-by-category/.

3. Cvitić, I., Vujić, M., & Husnjak, S. (2016). Classification of security risks in the IoT environment. In *26-th Daaam international symposium on intelligent manufacturing and automation* (pp. 0731–0740). https://doi.org/10.2507/26th.daaam.proceedings.102.
4. Ali, B., & Awad, A. (2018). Cyber and physical security vulnerability assessment for IoT-based smart homes. *Sensors, 18*(3), 817. https://doi.org/10.3390/s18030817.
5. Al-Shammari, B. K. J., Al-Aboody, N., & Al-Raweshidy, H. S. (2018). IoT traffic management and integration in the QoS supported network. *IEEE Internet of Things Journal, 5*(1), 352–370. https://doi.org/10.1109/JIOT.2017.2785219.
6. Laner, M., Svoboda, P., Nikaein, N., & Rupp, M. (2013). Traffic models for machine type communications. In *10th IEEE international symposium on wireless communication systems 2013, ISWCS 2013* (Vol. 9, pp. 651–655). Ilmenau, Germany: VDE.
7. Sivanathan, A., Sherratt, D., Gharakheili, H. H., Radford, A., Wijenayake, C., Vishwanath, A., & Sivaraman, V. (2017). Characterizing and classifying IoT traffic in smart cities and campuses. In *2017 IEEE conference on computer communications workshops, INFOCOM WKSHPS 2017* (pp. 559–564). Atlanta, USA: IEEE. https://doi.org/10.1109/INFCOMW.2017.8116438.
8. Meidan, Y., Bohadana, M., Shabtai, A., Ochoa, M., Tippenhauer, N. O., Guarnizo, J. D., & Elovici, Y. (2017). Detection of unauthorized IoT devices using machine learning techniques. *arXiv*. Retrieved from http://arxiv.org/abs/1709.04647.
9. Bai, L., Yao, L., Kanhere, S. S., Wang, X., & Yang, Z. (2018). Automatic device classification from network traffic streams of internet of things. Retrieved from http://arxiv.org/abs/1812.09882
10. Meidan, Y., Bohadana, M., Mathov, Y., Mirsky, Y., Breitenbacher, D., Shabtai, A., & Elovici, Y. (2018). N-BaIoT: Network-based detection of IoT botnet attacks using deep autoencoders. *IEEE Pervasive Computing, 13*(9), 1–8.
11. Sivanathan, A., Habibi Gharakheili, H., Loi, F., Radford, A., Wijenayake, C., Vishwanath, A., & Sivaraman, V. (2018). Classifying IoT devices in smart environments using network traffic characteristics. *IEEE Transactions on Mobile Computing, 18*, 1–1. https://doi.org/10.1109/TMC.2018.2866249.
12. Shafiq, M. Z., Ji, L., Liu, A. X., Pang, J., & Wang, J. (2013). Large-scale measurement and characterization of cellular machine-to-machine traffic. *IEEE/ACM Transactions on Networking, 21*(6), 1960–1973. https://doi.org/10.1109/TNET.2013.2256431.
13. IHS. (2017). The internet of things: A movement, not a market start revolutionizing the competitive landscape. *IHS Markit*. Retrieved from https://ihsmarkit.com/Info/1017/internet-of-things.html.
14. Hamza, A., Ranathunga, D., Gharakheili, H. H., Roughan, M., & Sivaraman, V. (2018). Clear as MUD. In *Proceedings of the 2018 workshop on IoT security and privacy – IoT S&P '18* (pp. 8–14). New York: ACM Press. https://doi.org/10.1145/3229565.3229566.
15. Hamza, A., Gharakheili, H. H., & Sivaraman, V. (2018). Combining MUD policies with SDN for IoT intrusion detection. In *Proceedings of the 2018 workshop on IoT security and privacy – IoT S&P '18* (pp. 1–7). New York: ACM Press. https://doi.org/10.1145/3229565.3229571.
16. Cvitić, I., Peraković, D., Periša, M., & Botica, M. (2018). Smart home IoT traffic characteristics as a basis for DDoS traffic detection. In *Proceedings of the 3rd EAI international conference on management of manufacturing systems* (pp. 1–10). EAI. https://doi.org/10.4108/eai.6-11-2018.2279336.
17. Karimi, A. M., Niyaz, Q., Sun, W., Javaid, A. Y., & Devabhaktuni, V. K. (2016). Distributed network traffic feature extraction for a real-time IDS. *IEEE international conference on electro information technology, 2016–Augus*, 522–526. https://doi.org/10.1109/EIT.2016.7535295.
18. Amar, Y., Haddadi, H., Mortier, R., Brown, A., Colley, J., & Crabtree, A. (2018). An analysis of home IoT network traffic and behaviour. *arXiv:1803.05368*. Retrieved from http://arxiv.org/abs/1803.05368.
19. Aghaei-Foroushani, V., & Zincir-Heywood, A. N. (2015). A proxy identifier based on patterns in traffic flows. In *Proceedings of IEEE international symposium on high assurance systems engineering* (pp. 118–125). IEEE. https://doi.org/10.1109/HASE.2015.26.

20. Habibi Lashkari, A., Draper Gil, G., Mamun, M. S. I., & Ghorbani, A. A. (2017). Characterization of Tor traffic using time based features, (Cic), 253–262. https://doi.org/10.5220/0006105602530262.
21. Doshi, R., Apthorpe, N., & Feamster, N. (2018). Machine learning DDoS detection for consumer internet of things devices. *CoRR, abs/1804.04159*. Retrieved from http://arxiv.org/abs/1804.04159.
22. Vaz, M. A. B., Pacheco, P. S., Seidel, E. J., & Ansuj, A. P. (2017). Classification of the coefficient of variation to variables in beef cattle experiments. *Ciência Rural, 47*(11), 9–12. https://doi.org/10.1590/0103-8478cr20160946.
23. Couto, M. F., Peternelli, L. A., & Barbosa, M. H. P. (2017). Classification of the coefficients of variation for sugarcane crops. *Ciência Rural, 43*(6), 957–961. https://doi.org/10.1590/s0103-84782013000600003.
24. Romano, F. L., Ambrosano, G. M. B., de Araújo Magnani, M. B. B., & Nouer, D. F. (2005). Analysis of the coefficient of variation in shear and tensile bond strength tests. *Journal of applied oral science, 13*(3), 243–246. Retrieved from http://www.ncbi.nlm.nih.gov/pubmed/20878024.
25. Alice, A., Noleto, S., Ferreira, D. C., Rocha, L., Dourado, B., Regia, K., & Sousa, S. (2018). Methods for classifying coefficients of variation in experimentation with poultrys. *Comunicata Scientiae, 9*(December 2017), 565–574. https://doi.org/10.14295/CS.v9i4.2619.
26. Ernst, P. A., Thompson, J. R., & Miao, Y. (2017). Tukey's transformational ladder for portfolio management. *Financial Markets and Portfolio Management, 31*, 317. https://doi.org/10.1007/s11408-017-0292-1.

RFID Monitoring and Accounting System in Health-Care Facilities

Michal Balog, Angelina Iakovets, and Jozef Husar

1 Introduction

The year 2019 is well known for the use of information technologies in every economic sphere. One of them is health-care, which is actively developing. In today's world, there is an urgent need for highly efficient management in hospitals and other health-care institutions. Industry 4.0 dictates rules for every scope of work, thanks to the fact that health-care facilities need to impenetrate this type of technology in its activities [7–9]. Effective health facilities and hospital management are, in both theoretical and practical terms, linked to a country's economic development and it is an important tool in developing modern management systems [4]. Worldwide, the government and health-care institutions are undergoing a process of modernization and reform, which creates new opportunities for innovations in health management [7]. Statistical data show that most hospitals have computerized management for all processes. Hospital management programs eliminate the use of paper documentation and enable quick access to information and speed up information and material flow. Similar software can be used for recording patient health cards, but it is not able to handle the medicine accounting process [4].

Medical facilities are engaged in various activities: health services/patient care, inventory and material flow management, accounting, etc. [15]. By using standard methods of drug registration, a larger number of employees are required (for manual registration, for creating documentation, for entering information onto a computer) [5]. According to the latest data, the European Union (EU) has established trends and guidelines in principles for the introduction and implementation of common

M. Balog · A. Iakovets (✉) · J. Husar
Technical University of Košice, Faculty of Manufacturing Technologies with a Seat in Prešov, Presov, Slovakia
e-mail: angelina.iakovets@tuke.sk

L. Knapcikova et al. (eds.), *4th EAI International Conference on Management of Manufacturing Systems*, EAI/Springer Innovations in Communication and Computing, https://doi.org/10.1007/978-3-030-34272-2_2

standards for identification technologies in the EU. By the end of 2019, all EU countries should introduce a single standard for drug labeling, and then by 2020, biological samples, medical devices, and instruments should be labeled. Therefore, it is necessary to determine the correct identification technology [7]. The use of automatic identification elements to accelerate business processes is gaining popularity and has become a modern trend.

All medical facilities use the similar to traditional principles of managing warehouses, but with changes for specific materials, which are used in this economic field [11, 13, 16]. For the warehouse inventory, elements such as barcodes, radiofrequency identification (RFID) and other types of technologies are used, which are mainly implemented in logistics, where they can be utilized to assign additional data to products and shipment units [3, 6]. Depending on the type of identifier, the amount of data varies. Currently, according to legislation, medical institutions are legally obliged to label all materials and medicines with an European Article Number (EAN) 13-digit code [7]. The statistics show that medicine and medication accounting is one of the main activities of health-care institutions, because it affects a human's life, but appropriate attention is not paid to this. To prevent possible problems with expired medications and to propose an effective method of designation the following research was carried out [4].

These issues were addressed by the European Commission and the Commission Delegated Regulation (EU) 2016/161, and published at the beginning of February 2016, in addition to the European Parliament and Council Directive 2001/83/EC, which sets out detailed rules for the use of secure elements of wrapped medicines for human use. The regulations for the labeling of medicines were supplemented by the following information from February 2019:

- GTIN (Global Trade Identification Number) Product Number
- Shelf life
- Batch number
- Serial number

These data should be presented on the product packaging in a readable form in the GS1 code format. The specified product data are stored in the European Central Database and the National Register. Therefore, when medications are issued, the GS1 Data matrix code is read to confirm the action and verify it with the data of the National Registry. After checking and confirming the genuineness of the medicine, it is given to the patient [4, 7]. The Slovak Republic, like many countries, began to prepare for this transition by implementing and launching the e-health portal. It works on the principle of using a passport with a chip (eID – electronic identification card) by the patient. This chip stores patient data, which will be used for further identification by the general practitioner, nurse or pharmacy. The chip also provides access to the patient's electronic health book (EHB) [7].

2 Literature Overview

To offer the most appropriate technology of designation, the market for the existing identification technologies and principles of moving medicines should be analyzed.

There are two ways of keeping records of medicines delivered to a hospital warehouse:

- Check the availability according to the accompanying documentation, namely, by the number of products actually present
- Electronic registration by code marks on each item [7]

Due to requirements for warehousing in hospitals, mentioned above, it is possible to offer the implementation of RFID identification technology for the warehousing process.

Moving of medications between hospital departments requires the type of drug, its pharmaceutical group, the amount of delivery, and its shelf life to be transmitted as basic information [14]. For accounting records, the following can be used:

- Barcodes (1D codes), especially EAN codes
- QR (quick response) codes
- RFID tags [13]

From the point of view of information about the capacity of the memory capacity, bar codes allow country code, manufacturer code, product code, check digit, product symbol, and license type to be recorded. Most commonly used are:

- EAN-8 (abbreviated)—encoded by 8 digits
- EAN-13 (full) —encoded by 13 digits
- EAN-128—encodes any number of letters and numbers merged into adjustable groups [2]

There are four main types of 2D (QR) codes:

- Digital: from 10 to 3 digits, up to 7,089 digits
- Alphanumeric: supports 10 numbers, letters from A to Z and a few special characters, 11 bits, up to 4,296 characters
- Bit: data in any suitable encoding (ISO 8859-1 standard) up to 2,953 bytes
- Kanji: 13 bits per hieroglyph, up to 1,817 characters [2]

Quick response codes can hold more data than barcodes, but they have one important drawback – they are easily damaged during transfer. 1D and 2D codes require scanning in the visible area, which means that if the image is hard to see and not in the reading direction of the scanner, the code cannot be read [2].

The principle of 1D and 2D code scanning is an example of semi-automatic recording. This type of identification requires several conditions for high-quality reading. The scanner should be able to read the information from the code. The scanner should be placed near the barcode at a certain angle and distance. Barcodes speed up evidence-related processes, but they do not eliminate errors caused by humans [2, 11].

Table 1 Comparison of identification technologies [4]

Parameters	RFID	1D and 2D codes
Placement	Scanning past the visual field	Scanning only in the visual field
Amount of memory	10–10,000 bytes	Up to 100 bytes
Ability to rewrite information and reusing	Yes	No
Scanning distance	Up to 10 m	Up to 2 m
Scan multiple objects at once	Up to 200 tags per second	Only one code
Protection against the influence of the environment: mechanical and chemical damage, temperature, humidity	Enhanced tag protection	Depends on the material
Service life	More than 10 years	Depends on the label's material
Security and anti-fraud protection	Not falsified	Falsified
Readability in case of damage	Unreadable	Difficult
Identification of moving objects	Yes	Difficult
Sensitivity to electric fields	Yes	No

RFID radiofrequency identification

Radiofrequency identification (RFID) is an automatic identification by radiofrequency waves reflected from an RFID tag. The main types of RFID are:

- Passive high-frequency RFID (UHF), which allocates 8 KB of memory
- Active RFID takes up 128 KB [1, 4]

The advantage of this technology lies in the lack of requirement for direct visibility, which allows the device to work over long distances and read faster, despite various obstacles. RFID tags allow information to be recorded (for example, operations performed with goods), are more resistant to environmental conditions, and have a longer service life. To illustrate the main parameters of identification technologies Table 1 was created.

Based on these criteria, RFID technology was selected as appropriate. This technology also meets the requirements of certain groups of drugs, namely protection against counterfeiting. Passive RFID tags are most often used for labeling owing to their cost and ease of use. RFID technology allows identification to be taken to a new level and makes automatic scanning possible without contact with a person. RFID technology involves fully automated identification of medications and automatic scanning of a large number of medicines located near to the RFID scanner [2, 14]. Use of RFID technology in medical facilities was reviewed by a group of scientists Lopez et al. [12]. There is also the possibility of rewriting information (date of manufacture, type of medicine, expiration date, shipment number). This type of system involves the use of a reader equipped with antennas that can receive and record various data. The ability to rewrite data is one of the main differences between RFID and 1D/2D codes. The application of RFID allows:

- Tracking on-line
- Verification of the authenticity along the supply chain
- Automatic generation of accompanying documentation
- Planning of material provision
- Protection again counterfeiting [4].

3 Methodology

To design an effective functional evidential system the right components of the RFID system should be selected that will operate with the evidential program (accounting program or warehousing program) of the health-care facility. For laboratory approbation of the possible RFID accounting system, the main components of the proposed system were selected:

- Computer
- Antennas
- RFID reader
- Passive RFID tags

To produce effective measurements, each medication was assigned by a digital image to its unique code, which was built into the tag. When images and medications were compared, the medications were packed into three types of package. The next step was transporting the packages with medications through the RFID gate. The moment of detection was measured with the help of the special program, Readomer.

The detailed representation of the components is shown in Table 2.

This system was used for testing the best ways of reading, to improve the efficiency of the system. The medications were designated by tags to make medication identification possible. Each tag was a digitally attached photograph of the medication, to identify what kind of medicine was detected by the system. The example of digital attachment is represented in Fig. 1.

Figures 1 and 2 show the drugs in the Alient software and their real appearance.

4 Results

To understand the principle of detecting the medicine and to choose the most effective method of transportation of packages a special test created. This experiment included three basic ways of packaging medications and three basic ways of transporting them. An RFID gate with a width of 2 m was created (Fig. 3).

This width (2 m) was selected owing to the requirement of a wider type of the door in warehouses and hospital wards. Then, various packages were tested for the

Table 2 RFID set [18–22]

Photo	Component	Model of the component
	Reader	Alien ALR-8800
	Antenna (two pieces)	Antenna RHCP, AN480-CR66100WR, 865–956 MHz, 6dB
	RFID tag (ten pieces)	Alien ALL-9540-02 World Tag 860–960
	Computer	Gaben GBTouch 24”
	Software	Alien RFID Gate- way v2.26.03

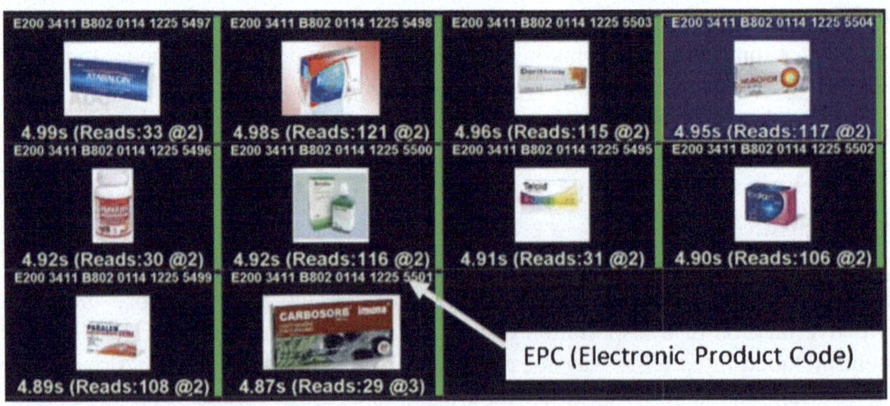

Fig. 1 Medications with radiofrequency identification (RFID) tags

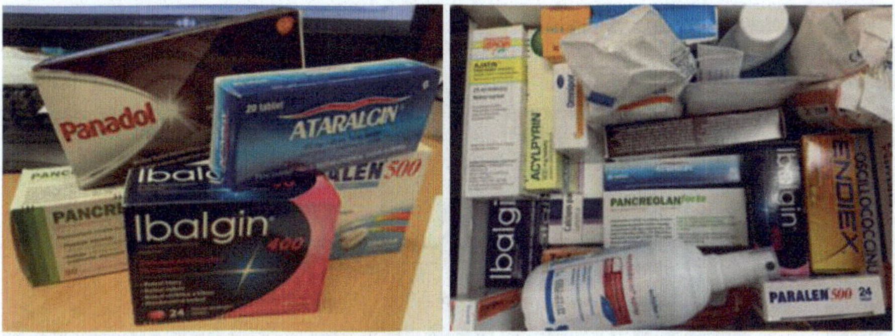

Fig. 2 Medications in original packages

Fig. 3 RFID gate

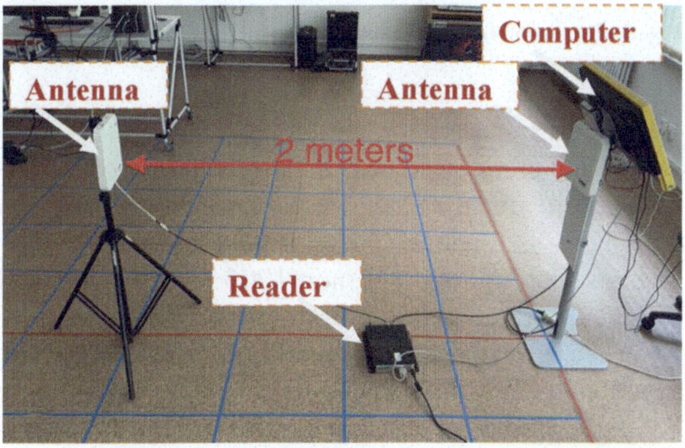

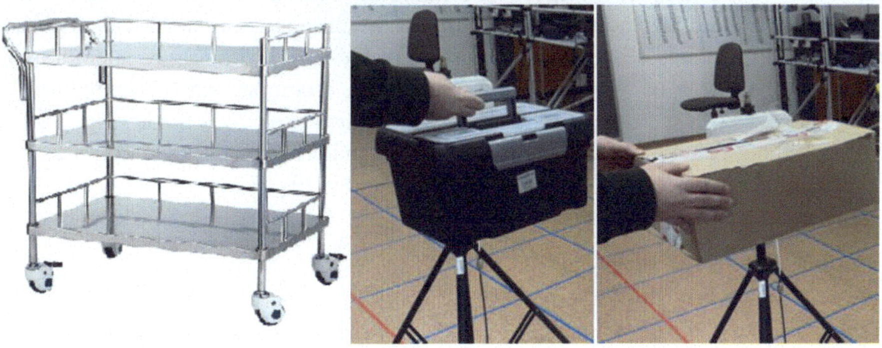

Fig. 4 Components for transportation of medications

transportation of the medicine. To assess the quality of the identification process the Alien RFID Gateway v2.26.03 program and its built-in Readomer were used. Such a function helps to understand how fast the RFID tag is detected. Plastic box, cardboard box, and trolley become tools for testing the reading effectiveness of the tagged medications (Fig. 4).

Fig. 5 Detection of RFID tags in a plastic box

Figures 5 and 6 approve the proposed RFID system as effective, thanks to the yellow and green scale near the digital image of the medicine.

Figure 6 illustrates the data of scanning tags in the cardboard box which is designated as 2 and the metal trolley as 1, where the yellow scales can be noted. These scales indicate the non-100% quality identification of the medicine. The scales have three main colors of quality of identification: green, yellow, and red. Also, the effectiveness of reading depends on the length of the scale. The worst situation is when the image does not appear on the screen. This means that tag was not read. The factor influencing this quality of reading is the iron construction of the trolley. For example, the speed of reading of the Pankreon was 65 reads/s in a cardboard box and 52 reads/s in a metal trolley. Such changes arise with all materials that consist of metals – It is due to principle of the RFID technology operation. To select the best way of transporting the medicine, a comparison of the reading in such packages (plastic box, cardboard box, and plastic bag) was made (Fig. 7). The trolley did not provide quality reading due to its material (metal) (see Fig. 6).

The speed of detection of the tag is represented in Fig. 8.

According to data from Fig. 8:

- A plastic box allows 85 reads/s
- A cardboard box allows 86 reads/s
- A plastic bag allows 87 reads/s

For this reason, it is possible to conclude that the most effective package is the plastic bag. The medical facilities do not have the ability to use only one method of transportation, but any package without metal must be used to provide the most accurate accounting of medication. If the system does not identify any of the medicines, it will lead to a situation where that medication will not get to the accounting system. Plastic packages, as one of the most readable packages, opens up a new possibility of using recycled plastic in the packaging, which will help to save the environment and will enable quality reading of the medications [10].

Fig. 6 Metal trolley and cardboard box

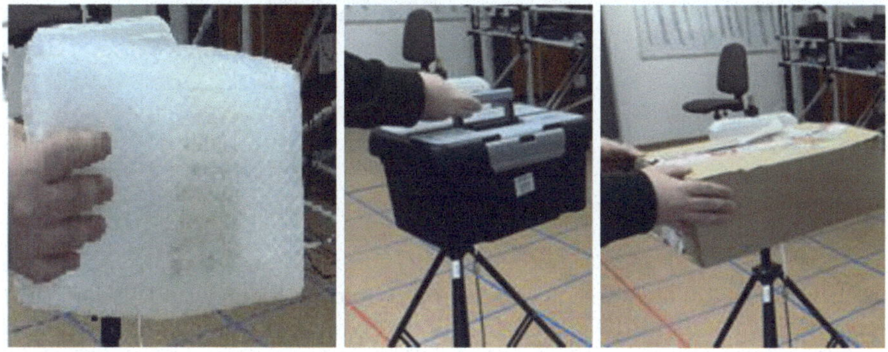

Fig. 7 Packages

To depict a possible variant of the fencing of such a system in a hospital, Fig. 9 was created. This model backs up the principle of the entrance reading, which means that as the hospital employee enters, the medications on the trolley will be read as

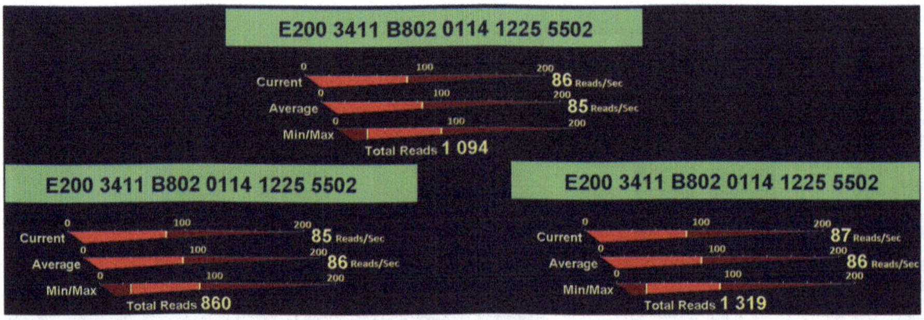

Fig. 8 Speed test by the type of package

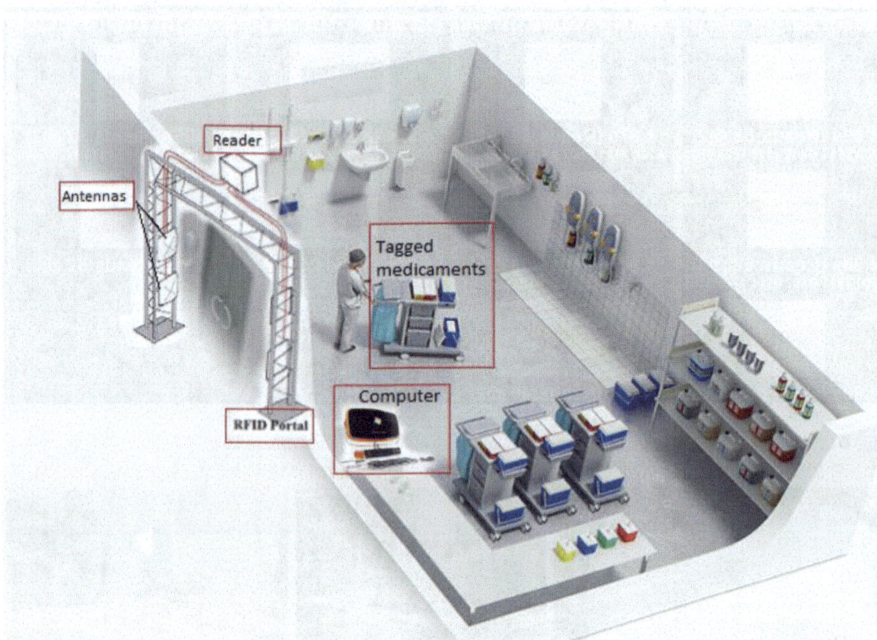

Fig. 9 Options for placement of the RFID system in the health-care facility

products received at the warehouse and shipped from the warehouse. This version of the installation of the RFID components is not the sole one, because this system is flexible to any accounting conditions. Also, it is necessary to mention that there are numerous ways of reading: by handle scanner, by table reader, by wireless antenna, etc. [17–21].

It is important not to forget that each system has its advantages and disadvantages. The RFID medications accounting system has its own (Table 3).

The disadvantages, from Table 2, are not a reason for rejecting such a system, but the reason for further tests of the system under various conditions.

Table 3 Advantages and disadvantages of RFID accounting

Advantages	Disadvantages
User friendliness and simplicity of use	Sensitivity to electric fields
Fast detection of the tagged products	The need for a specialist who will provide technical support for the system
Ability to rewrite information and reuse it	The price of introducing the system is slightly higher than the competition
Scanning distance	
Scanning multiple objects at once	Worst effectiveness in hospital wards with electronic medical devices
Protection against the influence of the environment: mechanical and chemical damage, temperature, humidity	Depends on the material of the package
Service life	Cheap tags require non-metal material of the packages
Security and anti-fraud protection	
Readability in case of damage	
Identification of moving objects	
Fast payback system	

5 Conclusion

Studies have shown the relevance of the experiments. The proposed RFID system was represented as being effective. Such a system can be integrated into existent accounting systems and open up new opportunities for planning and managing material and medication flow, and also to immediately react to the expiration date of the medicine. The RFID accounting system is able to scan a large number of goods at once, thereby speeding up the accounting process and eliminating a large amount of documentation and errors caused by employees. RFID tags satisfy the requirements of the orders established by the European Commission, owing to their ability to hold important information. This type of tag is shielded from external factors and can be reused, which will allow money to be saved. The advantage of such a system is the simplicity of use of the software and the affordability of the equipment. Every health-care institution have the opportunity for lower expenditure on investments by using existing computers. The price of an RFID accounting system is approximately US$ 3098.50. The price comprises:

- Reader Alien ALR-8800: US$ 58,50 [19]
- Antenna RHCP, AN480-CR66100WR, 865–956 Mhz, 6 dB – (two pieces): US$ 500 [21]
- Tags Alien ALL-9540-02 World Tag 860–960 (20,000 tags): US$ 2,540 [20]
- Computer: US$ 0
- Software Alien RFID Gateway v2.26.03:free [17]

Therefore, by reusing the tags, medical facilities can save money. Because the system is flexible in its construction, it opens up the possibility of using alternative cheaper ones, and saving on the process of its implementation. The RFID accounting system opens up new way of management for all medical facilities and Improve the level all their services [5, 9]. The proposed system will provide management in health- care institutions according to the Industry 4.0 trends [8].

Acknowledgement The paper was created within the EUREKA-E project! 11158 U – health – Auto-ID technology and the Internet of things to enhance the quality of health services.

References

1. Erick, C. J., & Christopher, A. C. (2007). *RFID in logistics: A practical introduction,* ISBN: 978–0-84-93-8526-1. CRC Press.
2. William, T. (2011). *Dennis: Parcel and small package delivery industry,* ISBN 13:9781461021544. North Charleston: CreateSpace.
3. Balog, M., Matiskova, D., & Semanco, P. System of Workpiece Automation with RFID Identifier Published Patent Application SK 50063–2016 A/Darina Matisková, Michal Balog, Pavol Semančo – Banská Bystrica: IPO SR – 2017. – 7 sec.
4. Balog, M., Husar, J., & Iakovets, A. *RFID system for management of Supplies*. In Healthcare, automation and management in theory and practice, 13th anniversary conference of universities, high schools and practice, February 2019, Stará Lesná, SR, ISBN 978–80–553-3250-5.
5. Ferreira, D. C., & Marques, R. C. (2019). Do quality and access to hospital services impact on their technical efficiency. *Omega-International Journal of Management Science, 86*, 218–236. https://doi.org/10.1016/j.omega.2018.07.010.
6. Hrehova, S., & Fechova, E. *Design of system for evaluating and sorting products with tools of statistical process control*. In: MMS Conference 2017. – Gent: EAI. 2018. p. 1–8. – ISBN 978-1-63190-158-4.
7. Husar, J., & Iakovets, A. *Proposal for monitoring the movement of biological samples in hospitals*. Automation and management in theory and practice, 13th anniversary conference of universities, high schools and practice, February 2019, Stará Lesná, SR, ISBN 978-80-553-3250-5.
8. Husar, J., & Knapcikova, L. Principles and characteristics of Industry 4.0, Management of manufacturing systems 2018, TUKE, FVT, Dubrovnik, Croatia, November 2018.
9. Husar, J., & Iakovets, A. Fourth industrial revolution technologies, management of manufacturing systems 2018, TUKE, FVT, Dubrovnik, Croatia, November 2018.
10. Knapčíková, L., Rimár, M., & Fedák, M. *New trends in waste management in the selected region of the Slovak Republic, 2018*. In: MMS Conference 2017. – Gent: EAI, 2018. p. 1–9. – ISBN 978–1–63190-158-4.
11. Li, J. Q., et al. (2018). PSOTrack: A RFID-based system for Random moving objects tracking in unconstrained indoor environment. *IEE Internet of Things J, 5*(6), 4632–4641. https://doi.org/10.1109/JIOT.2018.2795893.

12. Lopez, Y. A., Franssen, J., Narciandi, G. A., & Pagnozzi, J. (2018). RFID technology for management and tracking e-health applications. *Sensors (Basel), 18*(8), 2633. https://doi.org/10.3390/s18082663.
13. Ttejesh, B. S. S., & Neeaja, S. (2018). Warehouse inventory management system using IoT and open source framework. *Alexandria Engineering Journal, 57*(4), 3817–3823. https://doi.org/10.1016/IJ.aej.2018.02.003.
14. Tiwari, V., Quinlan, J. J., & Levine, W. C. (2019). Scaling perioperative services across health care systems. *International Anesthesiology Clinics, 57*(1), 1–17. https://doi.org/10.1097/AIA.0000000000000218,WIN.
15. Schuller, K. A., Cronin, C. E., Nicks, S. E., Jing, X., Kingori, C., & Morrone, M. (2019). Development and application of a rubric to compare strategies for improving access to health care in rural communities in the United States. *Evaluation and Program Planning, 74*, 61–68. https://doi.org/10.1016/j.evalprogplan.2019.02.013.
16. Yang, J. Y. *Design and study of intelligent warehousing system based on RFID technology*. IEEE 345 E47thst, NY 10017USA, IEEE (393–396). https://doi.org/10.1109/ICITBS.2019.00103. ICITBS, Chansha Peoplesr China.
17. Gaben homepage. http://www.gaben.cz/cz/stacionarni-terminaly/gbtouch. Last accessed 2 May 2019.
18. AtlasRFIDstore.com, homepage. https://www.atlasrfidstore.com/alien-f800-rfid-development-kit/. Last accessed 3 May 2019.
19. Alien ALR-8800 price. https://www.ebay.com/itm/Alien-ALR-8800-Enterprise-RFID-Reader-/332120063306. Last accessed 3 May 2019.
20. Alien RFID tag. https://www.atlasrfidstore.com/alien-squiggle-rfid-white-wet-inlay-aln-9740-higgs-4/. Last accessed 3 May 2019.
21. RFID antenna. https://www.atlasrfidstore.com/zebra-an480-rhcp-rfid-antenna-865-956-mhz/. Last accessed 2 May 2019.
22. LNCS Homepage. http://www.springer.com/lncs. Last accessed 21 November 2016.

Smart, Smarter, and Smartest City: The Method to Comparison of Cities

Michal Jerabek, Jan Kubat, and Vit Fabera

1 Introduction

The most people have competitive spirit, and they like to compare their results with others. This statement is especially true for municipal representatives. They would like to compare their town/village with other towns/villages. The output of comparison can be a measure of representative's fruitfulness as politicians. Many municipalities attempt to be a "Smart City." One of the "Smart City" definition is introduced in [1]: "A smart city is a designation given to a city that incorporates information and communication technologies (ICT) to enhance the quality and performance of urban services such as energy, transportation and utilities in order to reduce resource consumption, wastage and overall costs. The overarching aim of a smart city is to enhance the quality of living for its citizens through smart technology." Essential characteristic is a focus on "to enhance the quality of living for its citizens." Not only citizens should be the target group, but the aim should be "to enhance the quality of staying for all people in the city." It means such people are considered who come to cities to work and of course visitors/tourists.

If we want to make a comparison of city "smartness," we have to:

1. Find suitable parameters to compare; Sect. 2 discuses parameter selection.

M. Jerabek (✉) · V. Fabera
Czech Technical University in Prague, Faculty of Transportation Sciences, Prague, Czech Republic
e-mail: jerabem1@fd.cvut.cz; fabera@fd.cvut.cz

J. Kubat
AMBIS a.s., Prague, Czech Republic
e-mail: jan.kubat@wo.cz

L. Knapcikova et al. (eds.), *4th EAI International Conference on Management of Manufacturing Systems*, EAI/Springer Innovations in Communication and Computing, https://doi.org/10.1007/978-3-030-34272-2_3

2. Select an appropriate method which is able to process values of found parameters; the DEA method is introduced in Sect. 3.

   Section 4 describes the software solution.

2 Parameters of the "Smart City"

The main criterion how successfully the concept of "Smart City" is implemented is "to enhance the quality of staying for all people in the city." The set of evaluative parameters can be composed by quantitative indicators (e.g., the count of parking places), qualitative indicators (e.g., citizen's safety and e-help), or their combinations. Parameters can be also computed using arithmetical operations applied to other parameters, typically sum, rate, and mean. It is possible to speculate about complexes or components – for example, the Smart Street is complex, but Smart Bench is a component [2]. Some example of parameters are bellow.

Typical parameters from the transportation point of view are:

- The length of communications
- Communication management and maintenance costs
- Traffic infrastructure investment
- The count of crossing with public transport preferences
- Average time to spend in congestion
- Public transport line length
- The count of employees in public transport
- The count of vehicles in public transport
- The count of transported persons
- Public transport costs

Parameters describing demographic aspects:

- Number of the population
- Population structure
- The count of people traveling to work

Parameters describing cadastral territory:

- The area
- The count of buildings
- The count of housing units

3 DEA Method

Charnes et al. [3] firstly introduced the data envelopment analysis (DEA) model in 1978. DEA estimates relative efficiency within the set of given decision-making units (DMUs); DMU is town/village in our application. It is a nonparametric method that combines given sets of inputs and outputs and calculates relative efficiency of each DMU. Since 1978, DEA becomes an important tool for measuring efficiency within various sectors. During time, the original model got various extensions. Cooper et al. [4] came in 1999 with the imprecise data envelopment analysis (IDEA) and initiated the approach of fuzzy DEA [5].

The original DEA model [3] was called by its authors CCR model. It looks for the combination of optimal weights of given inputs and outputs so that the sum of weighted outputs divided by the sum of weighted inputs is maximal. The CCR model can be written as maximization problem for each DMU; see Eq. (1) [6].

$$\text{Max} \quad E = \frac{\mathrm{v}^T y_0}{\mathrm{u}^T x_0} \text{ subject to } \mathrm{v}^T y_j \leq \mathrm{u}^T x_j \; u \geq 0, v \geq 0 \tag{1}$$

where u, v are the vector variables. They represent weights of inputs and outputs of DMU_0. Variables y_0 and x_0 are output and input vectors of the DMU_0. y_j and x_j are vectors of outputs and inputs of the DMU_j. The objective function E represents efficiency. If the objective function equals 1, the DMU is efficient. The original CCR model is input oriented and supposes constant returns to scale (CRS). The CCR model can be written as linear programming problem by adding a condition to have sum of weighted inputs equal to 1, and so it is enough to look for maximum of the nominator only as in Eq. (2).

$$\text{Max} \quad E = \mathrm{v}^T y_0 \text{ subject to } \mathrm{u}^T x_0 = 1 \; \mathrm{v}^T y_j \leq \mathrm{u}^T x_j \; u \geq 0, v \geq 0 \tag{2}$$

For the interpretation purposes, it is convenient to construct the dual problem to the CCR problem in linear form Eq. (3).

$$\text{Min} \quad F = \Theta \text{ subject to } y^T \lambda \geq y_0 \; x^T \lambda \leq \Theta \, x_0 \; \lambda \geq 0 \tag{3}$$

where λ is the vector of variable that represents DMU weights. Θ is the scalar value that shows the necessary reduction of inputs to become efficient for the analyzed DMU (DMU_0). The vector λ also shows the peer DMUs to the analyzed DMU. Peer DMUs are DMUs that show the inefficient DMU the path how to reach efficiency.

To construct the DEA model with variable returns to scale (VRS) instead of constant returns to scale, we add the constraint Eq. (4) to the linear problem Eq. (3).

$$e^T \lambda = 1 \tag{4}$$

where e is the vector of ones. Such model is called BCC model (Barnes, Charnes, Cooper).

By restricting λ to be vector of binary numbers rather than nonnegative real numbers as in Eq. (3), we obtain the nonconvex rather than convex set of feasible solutions. Such problem is called Free Disposal Hull (FDH) [7].

We will illustrate the methods on the example of ten towns. We will compare their efficiency in the public expenses on smart road transport (Table 1).

The three mentioned methods of efficiency calculation are depicted in Fig. 1. It shows efficiency calculation example of one input (roads per employee) and one output (roads per citizen). The lines represent production possibility frontier. The DMUs on the line are efficient, while DMUs under the line are not. The constant returns to scale (CRS) showed only one efficient DMU (D), variable returns to scale (VRS) two efficient DMUs (D and F), and free disposal hull (FDH) four efficient DMUs (D, F, G, J).

If one constructs the DEA models for all four input variables and three output variables as set in Table 1, he gets the results as in Table 2. The efficient towns have efficiency equal to 1 and are depicted with the mark *.

As one may see, there is four efficient towns in the CRS efficiency DEA model, five efficient towns in the VRS DEA model, and eight efficient towns in the FDH model. All efficient towns in CRS DEA were also efficient in VRS DEA and FDH models, and all efficient towns in VRS DEA model are also efficient in the FDH

Table 1 Inputs and outputs

Town	Unit costs of maintenance (CZK/km)	Unit costs of service (CZK/km)	Unit costs of SSZ crossing (CZK/unit)	Roads per employee (km/employee)	SSZ crossings per crossing	Road length per citizen (km/citizen)	Parking places per citizen (unit/citizen)
A	737 563	448 802	527	2.67	0.42	0.0016	0.0250
B	764 358	418 945	519	1.31	0.59	0.0018	0.0428
C	180 937	249 126	185	2.90	0.56	0.0026	0.0328
D	418 650	4 811	208	1.05	0.50	0.0028	0.0381
E	450 404	323 060	1 241	1.60	0.52	0.0022	0.0338
F	221 471	108 701	909	1.86	0.17	0.0045	0.0188
G	367 781	313 417	2 803	1.47	0.11	0.0029	0.0338
H	313 247	7 016	2 281	3.63	0.17	0.0031	0.0341
I	662 547	269 054	1 689	2.44	0.12	0.0021	0.0317
J	430 861	91 407	16 064	1.67	0.08	0.0035	0.0257

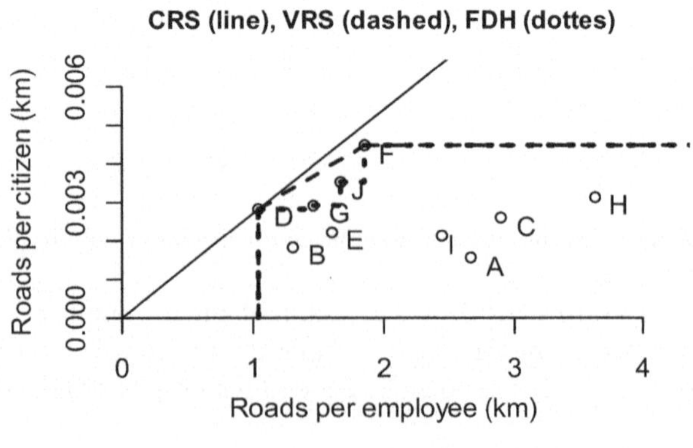

Fig. 1 Efficiency in input-oriented CRS (constant returns to scale), VRS (variable returns to scale), and FDH (free disposal hull) models for one input (roads per employee) and one output (roads per citizen) variables

Table 2 Efficiency of DEA model under VRS, CRS, and FDH options

	Efficiency			Peer towns for inefficient towns in VRS model	Values of input variables to reach efficiency in VRS DEA model			
Town	CRS	VRS	FDH		Unit costs of maintenance (CZK/km)	Unit costs of service (CZK/km)	Unit costs of SSZ crossing (CZK/unit)	Roads per employee (km/employee)
A	0.42	0.50	0.57	C, D, F	372 289	43 918	266.1	1.35
B	0.95	1.00*	1.00*		764 358#	418 945#	518.6#	1.31#
C	1.00*	1.00*	1.00*		180 937#	249 126#	185.0#	2.90#
D	1.00*	1.00*	1.00*		418 650#	4 811#	208.4#	1.05#
E	0.86	0.89	1.00*	B, C, D	401 736	86 711	229.9	1.43
F	1.00*	1.00*	1.00*		221 471#	108 701#	908.7#	1.86#
G	0.94	0.96	1.00*	C, D, F	353 955	52 252	340.8	1.41
H	1.00*	1.00*	1.00*		313 247#	7 016#	2 281.3#	3.63#
I	0.48	0.54	0.63	D, F	354 894	38 402	434.8	1.31
J	0.83	0.83	1.00*	D, F	335 277	48 738	504.5	1.39

model. For the inefficient town, we may obtain so-called peer towns. By product of result vector (matrix) λ and matrix of inputs x from Eq. (3), we can calculate matrix of input variable values to reach efficiency x^e.

$$x^e = \lambda^T x \tag{5}$$

For efficient DMUs, $x^e = x$. The values of VRS DEA problems are presented in Table 2. The already efficient towns have their x^e values marked with # in Table 2.

4 WEBDEAR: Software Solution

The pilot software application WebDEAr is created to analyze municipality efficiency within project "Application of nonparametric methods (DEA, FDH) to analyze and to compare the efficiency of municipalities." This application should serve as a service application to municipalities to support their decision-making. Outputs of software should contribute to improve local politics. The application will be used by governing bodies, especially by the Ministry of the Interior of the Czech Republic that is application guarantee.

The application is designed as three-tier architecture having presentation, application, and database layers. The access to the application will be realized by web pages, and import and export will be possible in Excel format.

The access to the pilot version is planned to be authenticated. The final version will be accessible without authentication. All operations and data about clients (IP address, browser, etc.) will be logged to be able to find clients in the case of troubles and improve the application.

The application will display as static data about towns/villages as dynamically calculated data. User will be able to change stored static data during session temporarily, and the efficiency will be recalculated.

Changes will be discarded after logout, persistent data will be only guaranteed one. Figure 2 shows the architecture of the WebDEAr application. The core is a database system to store data and DEA module programmed in PL/SQL. The module uses R programming language to calculate effectivity by the DEA method. The free package deaR will be applied which was created by Vicente Coll-Serrano, Rafael Benítez, and Vicente J. Bolós at the University of Valencia [8]. Current version 1.0 works with following DEA models:

- Conventional DEA models
- Basic (radial) models (envelopment and multiplier forms)
- Directional distance function model
- (Weighted) Additive model
- Super-efficiency additive model
- Radial super-efficiency model
- (Weighted) Nonradial model

Fig. 2 The architecture of the WebDEAr application

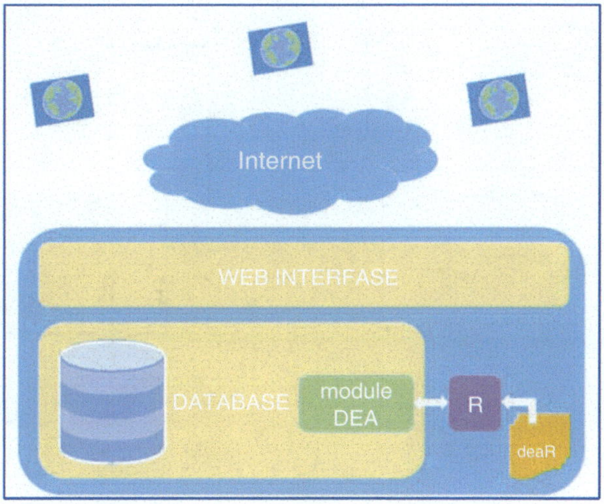

- Preference structure model
- (Weighted) Slack-based model
- (Weighted) Super-efficiency slack-based model
- Cross-efficiency (crs12 and vrs13)
- Bootstrapping (Simar and Wilson algorithm)
- FDH model

Productivity

- Malmquist index

Fuzzy DEA models

- Kao and Liu model
- Possibilistic model
- Guo and Tanaka model
- Fuzzy cross-efficiency

The relational scheme of the database was designed based on requirement analysis. The scheme contains 23 relations (tables), and this is a kernel of data structures to store information (Fig. 3).

The main table is MUNICIPALITY containing basic information about towns/villages; the actual size is 6249 records. Categories of municipalities are stored in the CATEGORY table, and the category for each municipality is assigned by the relationship table CATEGORY_MUNICIPALITY.

Very important information is statistical and financial data. The list of statistical data is stored in the table LIST_STATISTICAL_DATA which contains reference to the table UNIT. Relationship of statistical data to municipalities is stored in the table MUNICIPALITY_STATISTICAL_DATA. Financial information are divided into PARAGRAPHS and ITEMS, and their domains are saved in the

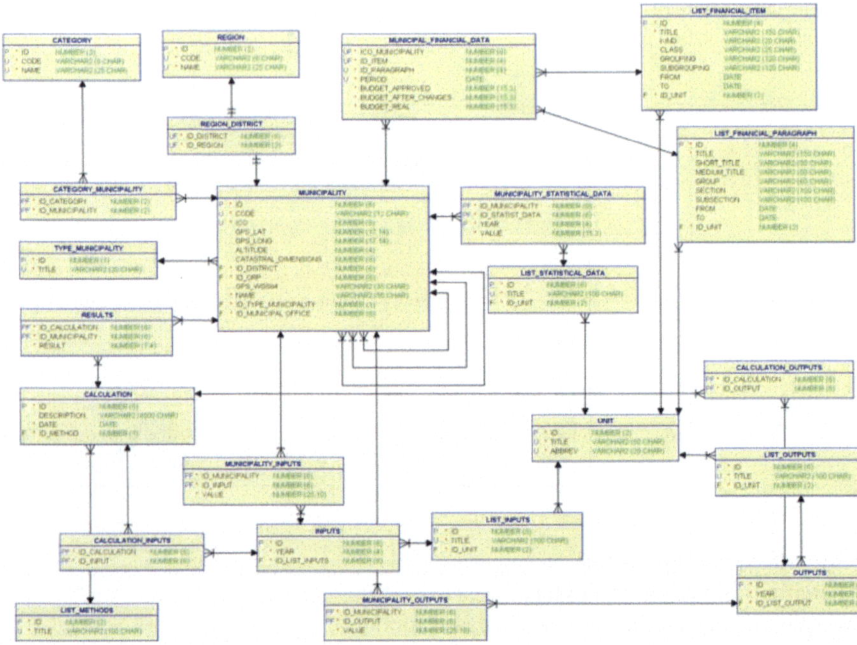

Fig. 3 Relational scheme of the database

LIST_FINANCIAL_PARAGRAPH and LIST_FINANCIAL_ITEM tables. Both tables have bindings to the UNIT table. Financial data about towns/villages are stored together with period in the table MUNICIPAL_FINANCIAL_DATA where three types of information are saved – budget approved, budget after changes, and real budget.

The table MUNICIPAL_FINANCIAL_DATA contains 5 263 845 records related to 524 records in the table LIST_FINANCIAL_PARAGRAPH and to the 527 records in the table LIST_FINANCIAL_ITEM. Financial data is collected from 2012 to 2017.

The DEA method to analyze the efficiency of municipalities uses several inputs and outputs from statistical and financial tables. Tables INPUTS and OUTPUTS serve as repository to store statistical and financial data which were used to compute efficiency. Tables references (foreign key) to LIST_INPUTS and LIST_OUTPUTS. Used data (with regard to possibly later change) are saved in the tables MUNICIPALITY_INPUTS and MUNICIPALITY_OUTPUTS.

The database schema takes into account to keep all results if calculations are repeated. Basic information about each calculation are stored in the table CALCULATION including reference to the used methods (the table LIST_METHODS). Results of each calculation for the selected town/village are stored in the table RESULTS.

The originator of data was Czech Statistical Office (CZSO). There was an inconsistency in financial data among list of municipalities, paragraphs, and items.

5 Conclusion

The analyses of the efficiency of municipalities and their comparison are important tools for municipal politics and governments (ministry, region governments). The application of the nonparametric DEA (data envelopment analysis) method to analyze is one of the ways how to transfer theory to the praxis and to contribute to the improvement of decision-making and control in public sphere. The paper concludes achieved results of project which solves application of DEA method in praxis. The designed software helps to put the method into practice.

Acknowledgments This work is done within the project "Application of nonparametric methods (DEA) to analyze and to compare the efficiency of municipalities" that is supported by TAČR (Technology Agency of the Czech Republic), program Eta (project code TL01000463).

References

1. Definition – What does Smart City mean? [Online]. Available: https://www.techopedia.com/definition/31494/smart-city [2019/02/25].
2. Přibyl P., & Přibyl O. (2015). Definition of a smart street as smart city's building element. Smart cities symposium Prague 2015, ISBN: 978-1-4673-6727-1, 10.1109/SCSP.2015.7181575.
3. Charnes, A., Cooper, W. W., & Rhodes, E. (1978). Measuring the efficiency of decision making units. *European Journal of Operational Research, 2*, 429–444.
4. Cooper, W. W., Park, K. S., & Yu, G. (1999). IDEA and AR-IDEA: Models for dealing with imprecise data in DEA. *Management Science, 45*, 597–607.
5. Kao, C., & Liu, S. T. (2000). Fuzzy efficiency measures in data envelopment analysis. *Fuzzy Sets and Systems, 119*, 149–160.
6. Jablonský, J., & Dlouhý, M. (2004). *Modely hodnocení efektivnosti produkčních jednotek*. Praha: Professional Publishing.
7. Deprins, D., Simar, L., & Tulkens, H. (1984). Measuring labor inefficiency in post offices. In M. Marchand & e.a (Eds.), *The performance of public enterprises: concepts and measurement* (pp. 243–267). Amsterdam: Elsevier.
8. Package dear [Online]. Available: https://www.uv.es/vcoll/deaR.html. [2019/02/25].

Aerial Photogrammetry and Unmanned Aerial Vehicles (UAVs) Like a Smart Technology for Digital As-Built Mapping of Existing Buildings

Matus Tkac, Peter Mesaros, Marcel Behun, and Tomas Mandicak

1 Introduction

It often happens to engineers that they have to record geometric data quickly, and the measurements cannot be performed directly on the object. In this case, photogrammetry provides good solution, which means performing spatial measurements based on 2D photos. Photogrammetry entered in a new era what is due to technological development (computers and cameras) and the evolution of the mathematical model describing the physical model. Computing revolution and the development of digital cameras also had major effect [1]. The input to photogrammetry is photographs, and the output is typically a map, a drawing, a measurement, or a 3D model (e.g., of some existing building) of some real-world object or scene. Photogrammetry can be classified into several ways, but one standard method is to split the field based on camera location during photography. On this basis, we have aerial photogrammetry and terrestrial (or close-range) photogrammetry. In aerial photogrammetry, the camera is mounted in an aircraft and is usually pointed vertically toward the ground. Multiple overlapping photos of the ground are taken as the aircraft flies along a flight path. The aircraft traditionally have been fixed wing manned craft, but

M. Tkac · P. Mesaros (✉) · T. Mandicak
Technical University of Košice, Košice, Slovakia

Faculty of Civil Engineering, Department of Construction Technology and Management, Košice, Slovakia
e-mail: matus.tkac@tuke.sk; peter.mesaros@tuke.sk; tomas.mandicak@tuke.sk

M. Behun
Technical University of Košice, Košice, Slovakia

Faculty of Mining, Ecology, Process Control and Geotechnology, Institute of Earth Resouces, Košice, Slovakia
e-mail: marcel.behun@tuke.sk

L. Knapcikova et al. (eds.), *4th EAI International Conference on Management of Manufacturing Systems*, EAI/Springer Innovations in Communication and Computing, https://doi.org/10.1007/978-3-030-34272-2_4

many projects now are done with drones and UAVs [2]. Unmanned aerial vehicle (UAV) systems as a data acquisition platform and as a measurement instrument are becoming attractive for many surveying applications in civil engineering. For safety and security reasons, commercial UAV use is typically restricted to flights within line-of-sight (LOS) of an operator. Typically, the operator is referred to a pilot. It ensures that a pilot has permanent control and interaction in case of unexpected events, for example, other aircraft(s) operate nearby or changes happen to any of the environmental conditions which might influence the UAV's flight conditions, such as wind parameters [3]. Drones can be fixed wing or rotor. With the inclusion of height sensors, acceleration and tilt sensors, GPS for geo-positioning, flight control software, gimballed camera mounts, and high-quality cameras, modern drones are becoming very powerful mapping tools. There are two styles of drone photography used in photogrammetry: nadir photography and convergent photography. Nadir means pointing straight down, and convergent means the camera is at an angle to the ground [2].

2 Digital As-Built Mapping Using UAV: Case Study

The main purpose of this case study was to measure the exterior of the building using an unmanned aerial vehicle. Looking at Fig. 1, it can be seen that the Faculty of Civil Engineering has a simple layout in the *L* shape with a built-up area of 3150 m^2, a height of 20 meters, and a roof angle of 24°.

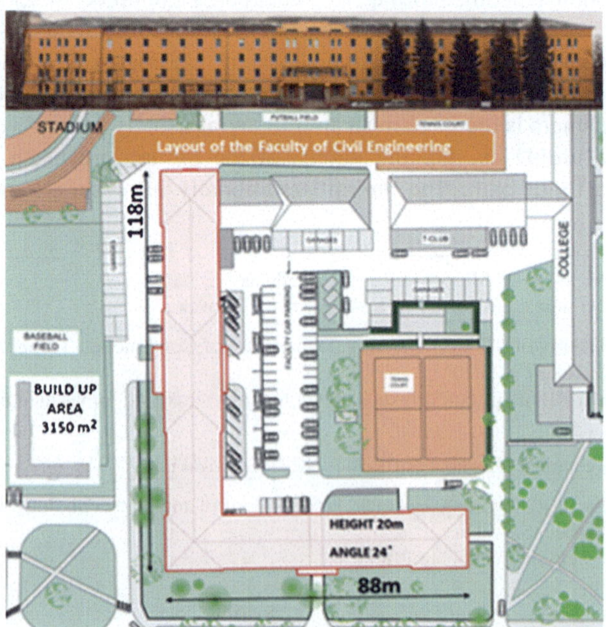

Fig. 1 Layout of the faculty of civil engineering, Slovakia, Košice. (Source: authors)

The drone FALCON 8 was used to create aerial images from the exterior of the Faculty of Civil Engineering. The Falcon 8 is a professional, unmanned aerial vehicle with a patented *V*-shaped design and features best-in-class safety and advanced performance ideal for inspection and close mapping [4]. This type of the drone is possible to classify such as Octocopter (number of rotors 8). The Cockpit Controller is the main control for the Falcon 8 system and features an innovative joystick design for single-hand flight control. The Sony Alpha 7R$_*$ is a full-frame photo camera with an image resolution of 36 MP, which allows it to display even the finest details [5]. What accuracy can be achieved in drone photogrammetry? One of the most important parameters of aerial survey with UAV is the spatial resolution, which in photogrammetry terms is described as Ground Sampling Distance (GSD). In literature, it is defined as the distance between two consecutive pixel centers measured on the ground. In practice, it is simply the size of the pixel in the field [6]. Looking at Fig. 2, it can be seen that for overall aerial mapping of the building, it is necessary to create vertical and oblique aerial images. In this case study, vertical images were focused on the building's roof. As can be seen in Fig. 2, vertical aerial images were created at a height of 70 m on the ground and the value of GSD on the ground was 1 cm and on the roof 7 mm. The GSD depends on the parameters of the camera (mainly camera resolution and focal length) and what is important on

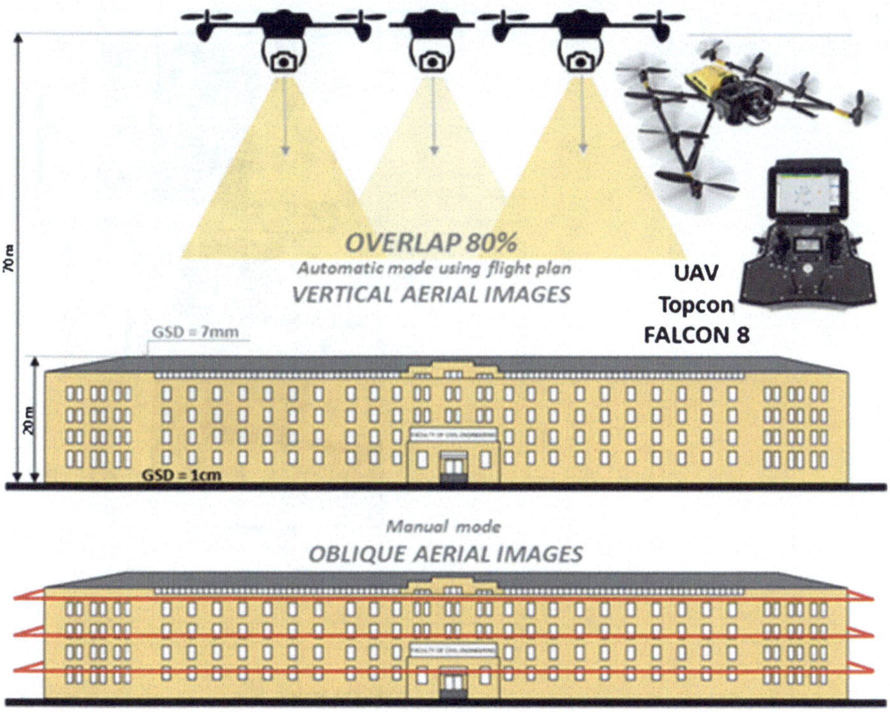

Fig. 2 Principle of creating aerial images using UAV. (Source: authors)

the flight altitude [6]. More simply, less flight altitude means greater accuracy. On the other hand, for digital mapping of the facade of the building, it is necessary to create oblique aerial images around the building in different heights (in this case, three altitude levels have been selected).

2.1 Data Collection

Vertical aerial photographic coverage of the roof is normally taken as a series of overlapping flight strips [6]. Flight strips were created in the software AscTec Navigator with overlap of aerial images 80%. The PC software AscTec Navigator enables planning and performing complex flight patterns for remote surveying applications and the combination of telemetry data with the recorded images [7]. Looking at Fig. 3, it can be seen the technological process of creating aerial images with three selected aerial photos. The overall time for automatic mode was in this case 3 minutes, and 100 aerial images were created. For aerial works, it is necessary to obtain permission from the Ministry of Defense of the Slovak Republic. The waiting period for permission from the Ministry of Defense lasts about 2 or 3 weeks.

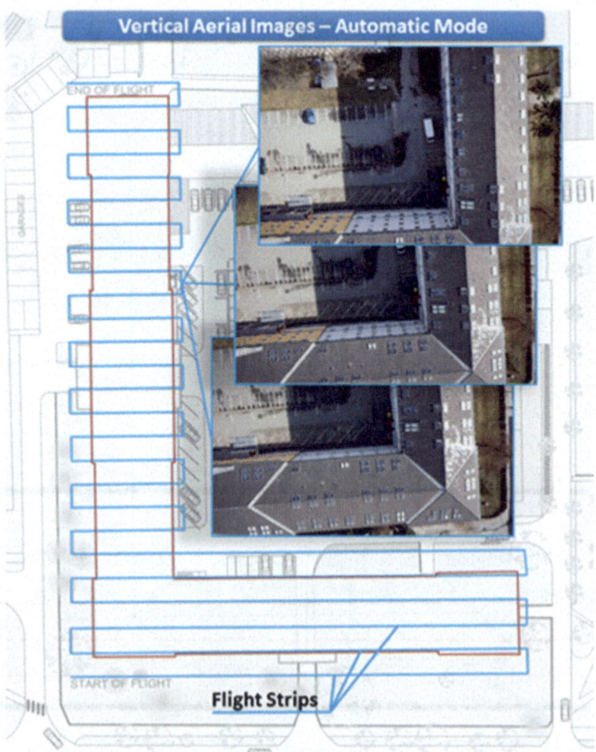

Fig. 3 Technological process of creating vertical aerial images – automatic mode. (Source: authors)

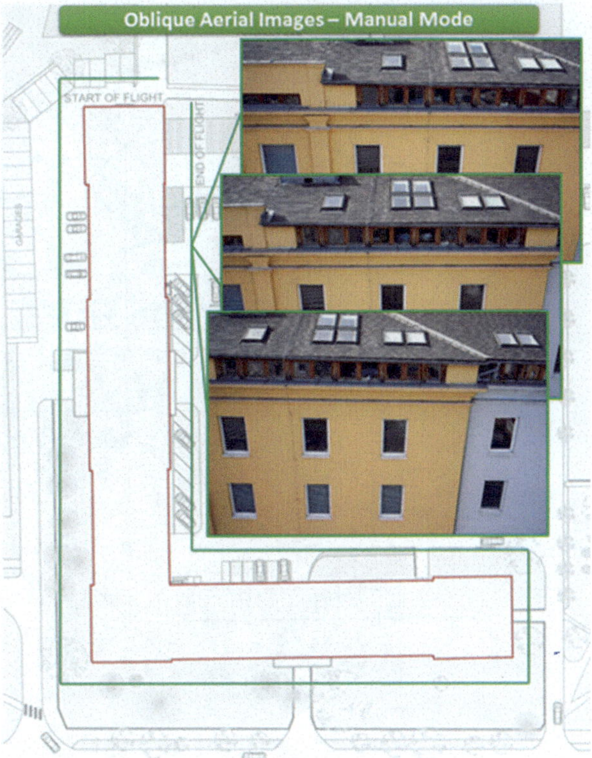

Fig. 4 Technological process of creating oblique aerial images – manual mode. (Source: authors)

After creating aerial images, it was necessary to send aerial images for assessment again to the Ministry of Defense. The waiting period for assessment from the Ministry of Defense lasts about again 2 or 3 weeks.

Looking at Fig. 4, it can be seen the technological process of creating oblique aerial images which were focused on the building's facade. This process was created in the manual mode, and it means that this process is about the pilot's practical and personal experiences. In this mode, 950 aerial images were created in the distance from the building approximately 10–15 m in three altitude levels. The overall data collection time was 2 hours, and 1050 aerial images were created. The disadvantage of this drone (Falcon 8) is its take-off weight (2.8 kg), which has a major impact on battery life. The flight conditions for the data collection were not the most ideal because the flight took place in colder weather (about 5 °C) at the end of February and at a wind speed of about 15 m/s. It was not possible to capture slightly cloudy weather, so there were many shadows on the building's facade, which could not be influenced. These effects meant that one battery lasted only 7 minutes, and since the overall flight time was 2 hours, the overall of 10 batteries had to be replaced throughout the overall flight.

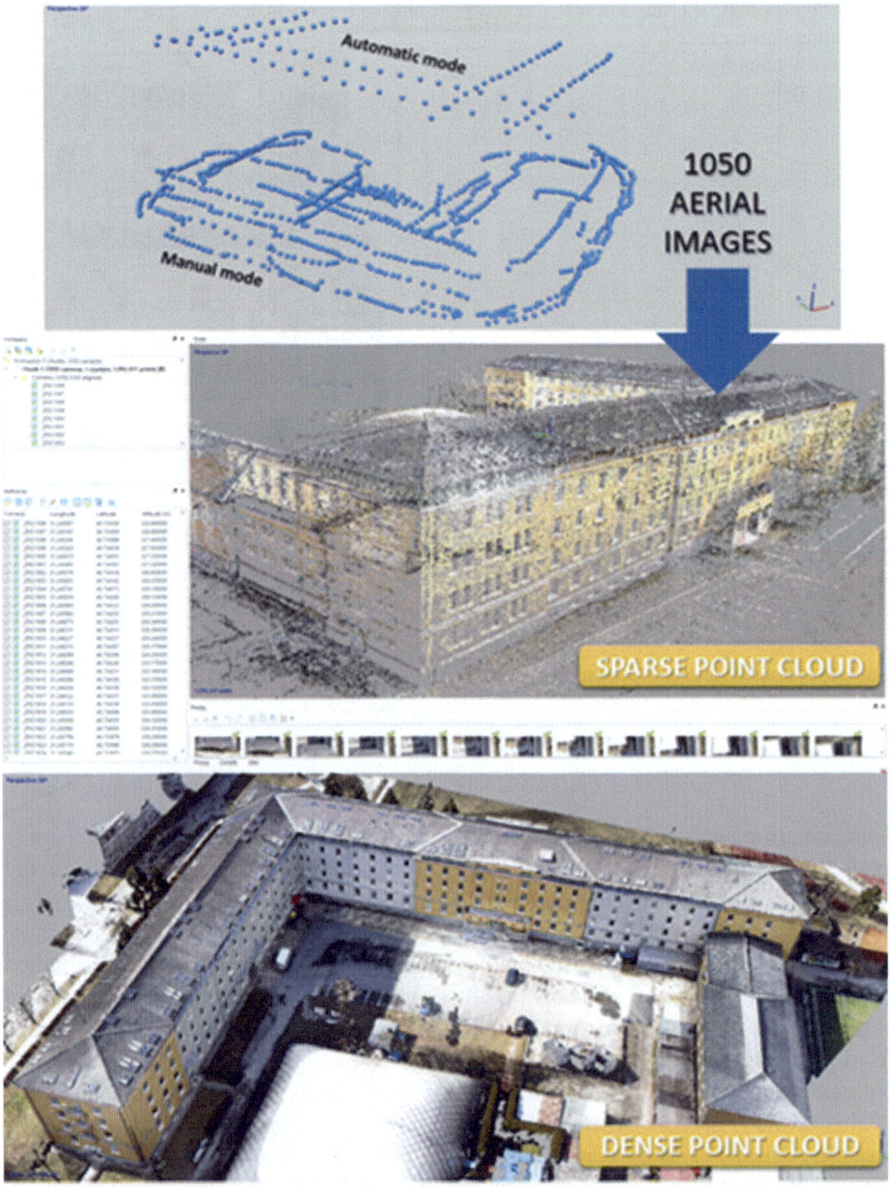

Fig. 5 Point cloud – result of aerial photogrammetry. (Source: authors)

2.2 Data Processing

The data processing phase was performed in the software Agisoft Photoscan Professional. Looking at Fig. 5, it can be seen that the result of the aerial surveying is 3D model in the form of the point cloud. The point is a real as-built 3D model but which consist of several million spatial points and each point has own coordinate, and it, X, Y, and Z. The overall time for creating the point cloud from 1050 aerial images was 33 hours *(1 hour for sparse point cloud and 32 hours for dense point cloud)*. The point cloud was created only in the middle quality because the high and the ultrahigh quality were very time consuming for a computer parameters. Of course, lesser quality means less time, but also the computer's parameters are decisive for the overall time (in this case: processor Intel(R) Core(TM) i7, memory 64 GB, graphic card NVIDIA GeForce GTX 1070).

Looking at Fig. 6, it can be seen different kinds of cutting plane from the point cloud in the software Autodesk ReCap. What is the main task of the point cloud? The main task is to create a new as-built documentation from some existing building where the point cloud can be importing into different kinds of software (such as AutoCad, Revit, Sketchup, ArchiCad, etc.). The point cloud after importing serves as a very accurate template, e.g., for 3D as-built modeling. Figure 6 also

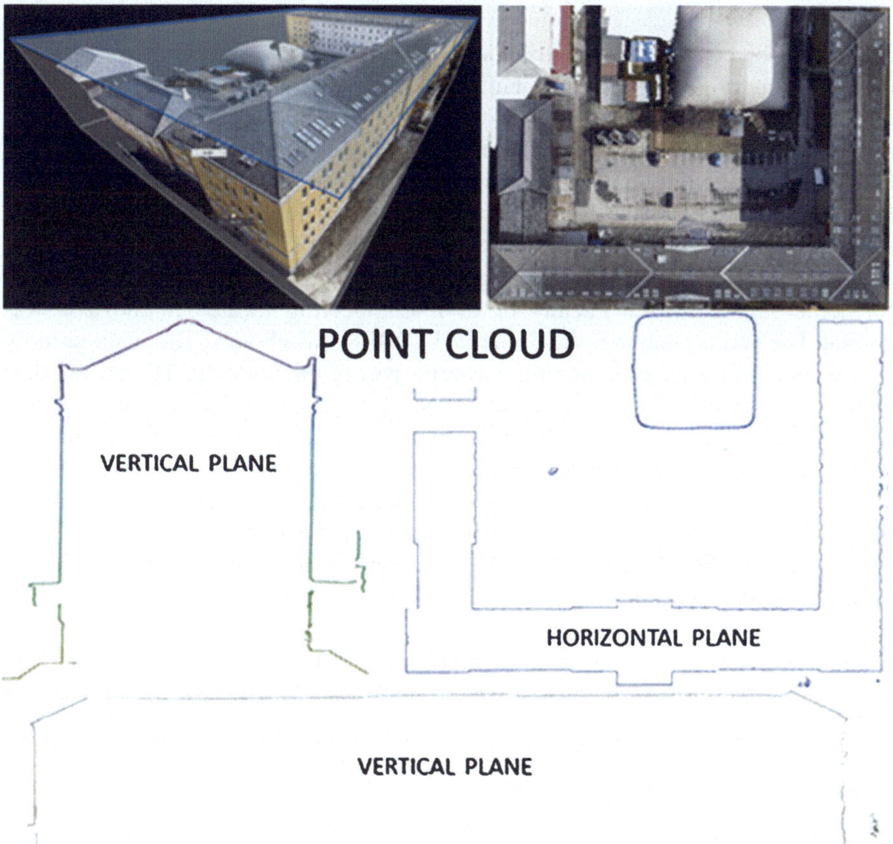

Fig. 6 Point cloud – different kinds of cutting planes.(Source: authors)

shows that the point cloud in the horizontal and vertical planes does not contain spatial information from the interior; it means that point cloud is possible to create only 3D as-built model of the exterior because it contains only spatial information from the exterior part of the building. How to get a spatial information from the interior part of the building in the form of the point cloud? For example, using technology of terrestrial laser scanning. It is a little bit more complicated process, but on the other hand, the combination of the terrestrial laser scanning and aerial photogrammetry can bring effective solution for complete digital as-built surveying of existing buildings. For this case study, it means that with combination of these two technologies, it would be necessary to create the point cloud from the interior of the building using a 3D laser scanner and connect it to the point cloud from the aerial unmanned vehicle.

3 Conclusion

The main objective of this contribution was to describe the technology of aerial photogrammetry for a digital as-built surveying of the existing building using unmanned aerial vehicles (UAVs – drones). Photogrammetry is the science of making measurements from photographs and can be divided into aerial photogrammetry and terrestrial photogrammetry. Placing a camera on a drone naturally led to drone-based photogrammetry – getting measurements, maps, and 3D models from photos taken by a drone. Drone aerial photography is one of the best ways to monitor a construction site, a building, civil engineering projects, etc. In this contribution for aerial surveying, the Faculty of Civil Engineering located in Slovakia was chosen. For aerial mapping, the drone FALCON 8 was chosen. The main purpose of this surveying was to measure the exterior part of the building. The overall data collection time was 2 hours, and the overall data processing time was 33 hours. The main advantage of using the drone is that they can measure the building in a relatively short time with high accuracy (1 cm accuracy can be achieved) and can measure hard-to-reach places. The main disadvantages of UAVs are as follows: more difficult weather conditions, permission for aerial work from the Ministry of Defense of the Slovak Republic (waiting period can be as long as 1 month), and time-consuming data processing phase.

Acknowledgments The paper presents a partial research results of project KEGA no. 059TUKE-4/2019, "M-learning tool for intelligent modeling of building site parameters in a mixed reality environment."

The paper presents partial research results of project APVV-17-0549 "Research of knowledge-based and virtual technologies for intelligent designing and realization of building projects with emphasis on economic efficiency and sustainability."

References

1. Molnár, B. (2010). Developing a web based photogrammetry software using DLT. *Pollack Periodica, 5*(2), 49–56. https://doi.org/10.1556/Pollack.5.2010.2.5.
2. What is Photogrammetry, http://www.photogrammetry.com/, last accessed 2019/05/15.
3. Sieber, S., & Teizer, J. (2014). Mobile 3D mapping for surveying earthwork projects using an unmanned aerial vehicle (UAV) system. *Automation in Construction., 41*, 1–14.
4. The Intel Falcon 8 + System, https://www.intel.com/content/www/us/en/products/drones/falcon-8.html, last accessed 2019/05/15.
5. Buczkowski, A. (2019). How accurate is your drone survey? Everything you need to know. GEO awesomeness, https://geoawesomeness.com/accurate-drone-survey-everything-need-know/, last accessed 2019/05/17.
6. Planning of Aerial Photography – Overlaps, crab, drift, scale, image movement, height accuracy, camera and others, https://photogrammetrydevelopment.blogspot.com/2015/12/planning-of-aerial-photography-overlaps.html, last accessed 2019/05/18.
7. Intel Corporation. (2017). ASCTEC NAVIGATOR MANUAL, 145 pages (2017), https://www.intel.com/content/dam/support/us/en/documents/drones/AscTec-Navigator_User-Manual.pdf, last accessed 2019/05/18.

Lifts as Part of Smart Buildings

Juraj Sinay, Marta Vargova, Juraj Glatz, and Marianna Tomaskova

1 Digital Industry

Compared to previous industrial revolutions, the current fourth revolution is developing at an exponential rather than linear rate. Moreover, it affects almost every industry. The scale of these changes influences the transformation of complex production technologies. The communication of people connected by mobile devices and access to knowledge is almost unlimited. This is due to the development of innovative solutions in areas such as automation, intelligent machine, and machine systems, application of Internet-based technology, autonomous vehicles, 3-D printing, nanotechnology, biotechnology, materials science, energy storage, and quantum computing [1].

Digitization allows for a new dimension in the use of lifts. It is used, for example, for monitoring their efficient operation, defining possible threats to their operation and applications for their control by means of smartphones even from remote locations. Its application allows for the incorporation of newly designed lifts, as an equal object, into smart building systems [1].

The concept of lifts as well as their operation in the use of new telematics and digitization methods will change in the future, similar to the conditions of car traffic. The first activities in the area of cars were based on the use of one sensor. Currently, it is an application of a "complex computer on wheels." In the same way, it can be assumed that lifts will become "hanging computers" characterized by mechatronic principles in the future.

J. Sinay · M. Vargova (✉) · J. Glatz · M. Tomaskova
Technical University of Košice, Faculty of Mechanical Engineering, Department of Safety and Quality of Production, Košice, Slovak Republic
e-mail: marta.vargova@tuke.sk

L. Knapcikova et al. (eds.), *4th EAI International Conference on Management of Manufacturing Systems*, EAI/Springer Innovations in Communication and Computing, https://doi.org/10.1007/978-3-030-34272-2_5

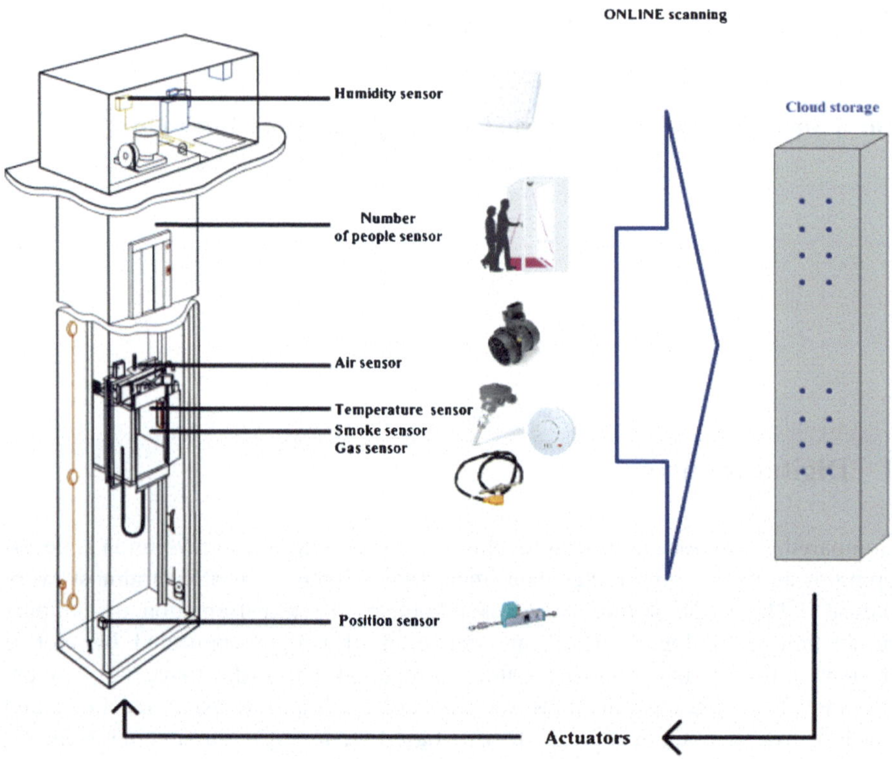

Fig. 1 Example of sensor placement in a lift complex. (Ref. [3])

Information on the operating or technical condition of the lifts is integrated into complex smart building control systems or their complexes by concentrating all operational information in combination with other lifts used in the building or building system in the central control center [2].

Using sensors located in lifts (Fig. 1), which sense relevant information about their operation, for example, speed or temperature, can be diagnosed in real time (online). Subsequently, this information can be transmitted to maintenance professionals by using data mobility. After their processing and evaluation, preconditions for their effective intervention are created. As these systems work continuously, conditions are created to apply predictive maintenance methods to minimize or prevent malfunctions and accidents [1].

2 Smart Buildings

The first mention of smart buildings appeared in the United States in the 1980s in the American Institute for Smart Buildings. The building is characterized as smart

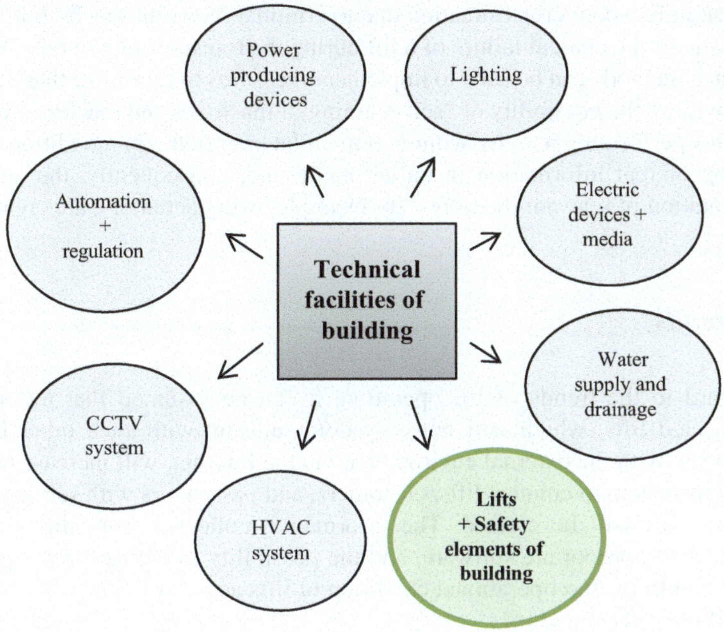

Fig. 2 Structure of technical facilities of building. (Author)

when it provides a productive and cost-effective environment by optimizing four basic elements – building construction, technical facilities, management services, and their mutual relations. The structure of the building's technical facilities is shown in Fig. 2, with lifts being one of the strategic objects also in terms of their safe operation [3].

The safety of smart buildings includes, in particular, hazards due to disturbance of security, fire safety and safety related to the failure of some construction elements of lift systems [4].

In the context of minimizing the impact of human factor on risk management, automatically operating actuators are used in the lift structure or lift system as actuators of mechatronic lift systems that change input information into mechanical intervention in the lift system [5].

2.1 Integration of Lifts into the Complex of Smart Building

When operating lifts as part of a smart building complex, the possibilities for their use for passenger transport as well as means of rescue or evacuation of persons due to the existence of unacceptable hazards must be created. Such operating conditions include mainly the occurrence of fire in the structures of buildings as an investment unit, a dangerous situation due to criminal interventions by third parties or terrorism, or a technical failure of a lift during the transport of persons. Artificial intelligence methods can be used to implement measures to minimize these types of risks, providing the possibility of "self-learning of machines and machine systems." This means performing activity without human intervention – preconditions for the processing of real information in online mode and, consequently, the automatic implementation of relevant measures, for example, using actuators, are created [2].

2.2 Trends

With regard to the trends of lift operation, it can be assumed that the share of interconnected lifts, which will not only communicate with each other but also communicate with the external environment via the Internet, will increase (Fig. 3).

A closed system to connect lifts, customers, and passengers with service centers and technicians will be created. The information collected from this system is analyzed using appropriate software, and the possibility to define the prognosis of technical condition and operational condition of lifts as part of complex systems of smart buildings is created [6].

Contemporary trends are also focused on effective maintenance of lifts and thus on the management of technical and human risks as part of their complex safety with the aim of increasing the comfort of their operation. Digitizing the collection of relevant information on the operation of the lifts allows for their use for other functions and, thus, for the versatility of their deployment as part of smart

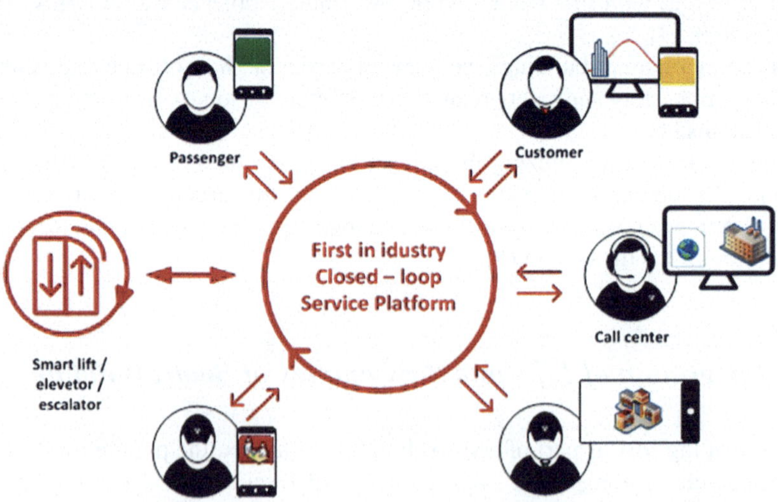

Fig. 3 Smart lift/escalator communication. (Source Fi. Schindler)

buildings or application of the principles of Industry 4.0 Strategy. Currently, there are systems that gather information from more than 100,000 lifts and escalators – of which 20,000 are in Switzerland [6], which are stored in Cloud centers. These data continuously provide information on the technical condition of lifts during their lifetime. It is possible to timely identify malfunctions, accidents, or negative events, and there are efficient online diagnostics to create conditions for their safe operation. These solutions can be applied not only in installations of new types of lifts but also by their implementation into some types of lifts currently in use [6].

When managing information on the operating parameters of lifts, their evaluation in Cloud centers and the subsequent implementation of appropriate interventions in their control systems or their construction groups or modules, it is necessary to define the risks associated with these activities. With regard to active data, information, or mobility, it is important to apply risk management systems for remote control of lifts so as to create conditions for their minimization and thus preventing unauthorized interventions [4].

2.3 The Concept of Smart Lifts

Modern lifts can nowadays use, for example, App management and RFID-based sensors, and they have the ability to remember relevant lift operation information, use applications of artificial intelligence methods to evaluate the information received, and then provide recommendations of relevant measures to the operator or people.

Information on operation and technical condition of lifts are obtained via built-in sensors – for example, in accordance with Fig. 1, these data are subsequently analyzed, thus creating an opportunity to increase the efficiency of monitoring of their technical condition and consequently their maintenance. Each state of movement, vibration, temperature, or acoustic load is measured and evaluated in real time. Information obtained about lift users is considered an "added value" in relation to the economy of operation. Collaboration with leading IT companies is required to deliver and evaluate relevant operational and technical information about lifts. Principles of information processing via Big Data technology as well as principles of artificial intelligence in collaboration with lift designers are applied for the efficiency of this process. The efficiency of all operations in the operation of smart lifts, as part of smart buildings within the Smart Places concept, requires active communication between lift manufacturers, building users, and service technicians with the ultimate goal of controlling acceptable risks, thereby creating conditions for their safe operation [3].

3 Evacuation or Fire Lifts as Part of Smart Buildings

Fires are one of the most common and greatest risks in buildings. The systems for their early identification and subsequent initiation of all measures to reduce the negative consequences of possible fires are part of the management of intelligent building operation. The nature of activity monitoring within these buildings creates the conditions for efficient application of the principles of Industry 4.0 Strategy. For this reason, the area of fire lift management must be part of the integrated security of an intelligent building [1].

Modern high-rise buildings are designed by architects to incorporate several types of lifts, including evacuation or fire lifts. In the case of evacuation of persons, these are used as "express lifts" that cross great distances at high speed.

The criteria that characterize evacuation or fire lifts include the following:

- Separate power supply
- Cabin installed devices for communication with the outside environment in redundant connection
- The lift shaft must be pressurized by air
- Flue gas or smoke sensors, inbuilt devices to release flue gas or smoke
- Cabin sidewall protection or thermal insulation against fire

Safe operation of evacuation/fire lifts requires the following:

- A separate generator for each lift
- Overpressure on staircases
- Independent sensor systems for different physical quantities
- Development of a detailed evacuation plan

Figure 4 illustrates a model of evacuation or fire lift communication in the event of a threat in a smart building or a building complex [7].

To minimize the risks resulting from real threats related to evacuation or fire lift systems, it is necessary to:

- Use noncombustible building materials (which the evacuation or fire lift consists of)
- Design a layout (where lifts are located)
- Choose escape routes (where lifts are located)
- Use equipment with minimal fire potential (e.g., self-extinguishing power cables) that will significantly affect the occurrence and spreading of fire

In the case of buildings with increased risk of fire or with the requirement for higher protection of property or persons, to ensure acceptable level of risk, it is necessary to implement specific systems of active intelligence of EPS, such as CCTV, display systems as part of communication interface [3].

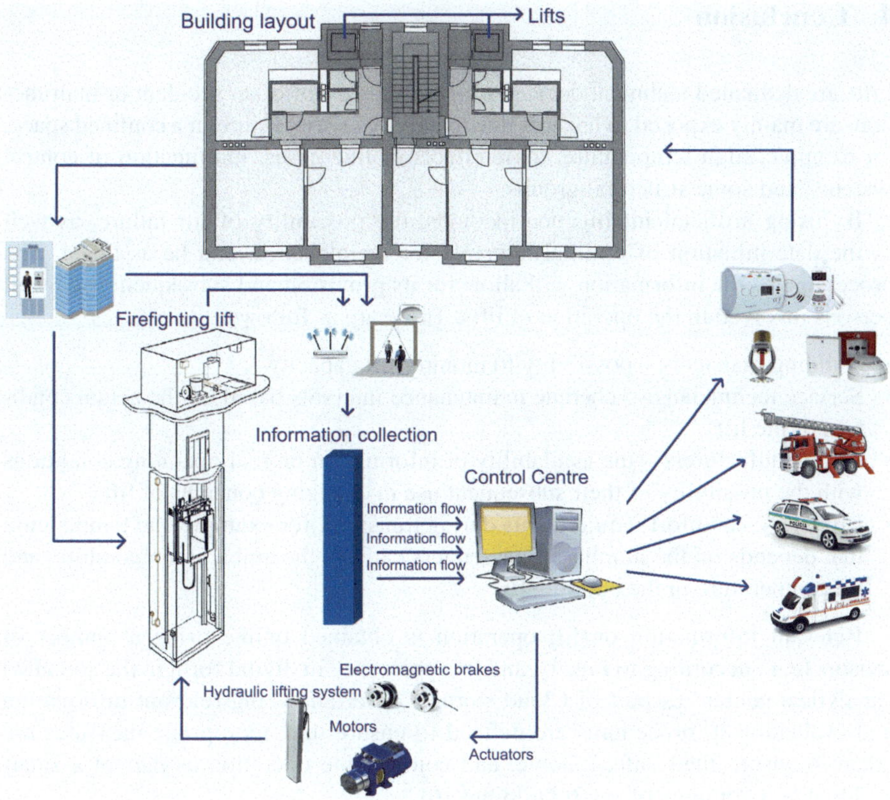

Fig. 4 Lift communication system as part of smart building. (Ref [3])

3.1 Selected Characteristics of Lift Evacuation Process

Some famous high-rise buildings in the world use a combination of diverse types of lifts, for example, according to [7], there are 29 passenger lifts, 3 high-performance freight lifts, and 4 evacuation or fire lifts in one tower.

Due to the characteristics of evacuation or fire lifts, it can be assumed that their use can provide the fastest possible escape from the building [7].

The following factors must be taken into account when a command for evacuation of a building is initiated:

- Where the fire occurred?
- How intense is the fire?
- Is it possible to get the fire under control as soon as possible?

These and similar factors have to be taken into account when defining risks even in the event of a terrorist attack directly on the building or buildings in its immediate vicinity, power cuts, adverse weather conditions, earthquakes, or false alarm [7].

4 Conclusion

Lifts are dedicated technical devices which, in the event of an accident or malfunction, are mainly exposed to hazards due to gravity as well as fire in a confined space, for example, high temperature, toxic effects of flue gases, malfunction of control systems, and some structural groups.

By using artificial intelligence methods, the possibility of lift failure, as well as the determination of optimal intervals for maintenance, can be assumed. The processing of the information will allow for its provision and subsequent use by all persons involved in the operation of lifts. These are as follows:

- Building managers – possibility to monitor lift capacity
- Service technicians – schedule maintenance intervals based on the actual condition of the lift
- Lift manufacturers – the availability of information on real operating conditions with the possibility of their subsequent use in changing concepts of lifts
- Lift users – comfort requirements during transport, for example, the temperature that depends on the number of passengers carried, the ambient temperature, and the temperature in the building [6]

Relevant information on lift operation is obtained online using a number of sensors (e.g., according to Fig. 1), and it is processed in digital form in the so-called "analytical center" as part of Cloud storage. After analyzing relevant information and evaluating it, procedures are defined to ensure that appropriate measures are taken to ensure their safe, reliable, and comfortable operation as part of a smart building or a complex of smart buildings [6].

This article was elaborated within APVV-15-0351 "Development and application of risk management models in the conditions of technological systems in accordance with Industry 4.0 Strategy" project.

References

1. Glatz, J., Sinay, J., Tomašková, M., & Vargová, M. (2019). Lift as subject of risk analysis in the context of smart buildings.
2. Making Smart & Intelligent Buildings a Reality Partha Kundu General Manager, Eco-Buildings Schneider electric., https://www.slideshare.net/seindia/partha-kundusmartcityoct2014. 2014/11/05.
3. Sinay, J., Tomašková, M., & Nagyová, M., (2018). In: Fire Protection 2018: conference proceedings. Ostrava, Czech Republic, Vysoká škola báňská – Technická univerzita Ostrava s. 130–136 [CD-ROM]. ISBN 978-80-248-4273-8
4. Čo tvorí inteligenciu budov?, http://www.stevo.szm.com/papers/2012/ES_3_STEVO-2.pdf. 2019/03/02.
5. Príslušný typ výťahu k príslušnému druhu stavby:, 2019/03/410.
6. Hegglin, R. (2017). Gebäude vernetzen sich mit Menschen..., Luzern: Hochschule.
7. Barry M. Scholes, Managing Director von Evac + Chair International.

The Concept of Implementing NFC Identification Technology for Transporting and Monitoring of Biological Samples

Jozef Husar, Lucia Knapcikova, Patrik Kascak, and Matej Hrabcak

1 Introduction

Healthcare is currently undergoing a transformation process. Under this term, we mean the innovation of healthcare facilities, the construction of new specialized departments and the modernization of the health documentation system. The Ministry of Health of the Slovak Republic has launched a project called e-Health, the goal of which is digitization of healthcare to improve healthcare provision. The main element of this project is to share health documentation between healthcare facilities or individual medics. The project is divided into six units, and each of them is oriented towards one specific part [1].

This article addresses the issue of one of the e-Health project headings, named e-Lab, focusing on results and labelling of laboratory samples. The task was to propose and apply wireless technologies to process of sampling and processing of biological samples. This concept focuses on near-field communication (NFC) technology, which is nowadays implemented in many mobile devices. Its great advantage is that it ensures writing and reading for a short distance. Interactions between NFC devices are carried out smoothly and securely, even without complex connection setup. NFC technology is a popular wireless technology that uses

J. Husar (✉) · L. Knapcikova · P. Kascak
Technical University of Košice, Faculty of Manufacturing Technologies with a Seat in Prešov, Department of Industrial Engineering and Informatics, Presov, Slovakia
e-mail: jozef.husar@tuke.sk; lucia.knapcikova@tuke.sk; patrik.kascak@tuke.sk

M. Hrabcak
Technical University of Košice, Faculty of Manufacturing Technologies with a Seat in Prešov, Department of Technical Systems Design and Monitoring, Prešov, Slovak Republic
e-mail: matej.hrabcak@tuke.sk

L. Knapcikova et al. (eds.), *4th EAI International Conference on Management of Manufacturing Systems*, EAI/Springer Innovations in Communication and Computing, https://doi.org/10.1007/978-3-030-34272-2_6

existing features of contactless card standards. People are nowadays able to simplify and speed up their everyday situations by expanding the NFC [2].

NFC references can be found in the division of radio frequency identification technology in the Working Frequencies subgroup. NFC technology is practically RFID; however, the difference is that its communication distance is shorter in order to ensure transmitting of information. NFC Forum was created by Sony, Philips and Nokia in 2004, the main goal of which was to educate organizations and enforce standards enabling to make use of this technology. Devices that are NFC compatible must normally comply with the standards specified by the NFC Forum. These standards provide the user with smooth communication with any NFC device or NFC tag [3]. The first set of specifications for NFC transponders was created in 2006 by the NFC Forum. The first NFC-compatible mobile phone was the Nokia 6131. For smartphones, it was a device on the Android platform, specifically the Samsung Nexus S. NFC technology has evolved and expanded over time. Nowadays, this technology is expanded to such an extent that almost every single electronic device has NFC technology [4].

1.1 Data Transfer and Security Principle

If we want to implement NFC technology, we need to approach the principle of its operation. This takes place when information is transferred between two inductively coupled coils. All actively powered devices generate an electromagnetic field of 13.56 MHz around themselves. There is interaction and data transfer if the second device is in the vicinity of this electromagnetic field. The greatest distance between these devices can be 20 cm, but in practice, this distance is 10 cm. The device can only be either a receiver or a transmitter when communicating. Such communication is being called to take place in the half-duplex mode. Switching between transmitting and receiving takes place at regular intervals [4]. When tag emulation takes place through a secure element, the tag to be emulated is secured by a device's secure element through an Android-based application. Consequently, if the user holds the device over the terminal, the NFC controller in the device redirects all reader data directly to the secure element. This element itself communicates with the NFC terminal, and there is no application to enter this communication. The Android application may request transaction status information from the secure element to notify the user after the transaction has been concluded [5, 6].

1.2 Blood Sampling Principle

Blood is biological material that is most commonly sampled by doctors not only for preventive examinations but also for specialized examinations to obtain biological analysis and blood count. The goal of blood testing is to get a credible result. It is important to comply with a procedure that eliminates adverse factors affecting the correctness of the result when blood sampling is done. The greatest amount of errors occurs in the pre-analytical phase, which includes activities such as patient preparation, blood sampling, labelling of samples and transport of samples for laboratory analysis [7].

Whole blood, plasma or serum is examined through blood tests. Blood is divided into two types: venous and capillary. It is important to differentiate between these two types of blood because each type has its own specific composition and is intended for different analyses and tests [8].

The venous blood sampling is usually carried out by a nurse. This blood is intended for haematological, biochemical and serological examination. Blood is most often sampled from the peripheral vein in the elbow hole in the lying position of a patient. The capillary blood sampling can be performed by a nurse or trained technician. Such blood is usually used to determine blood glucose, oral-glucose tolerance test and blood count [9]. Afterwards, the biological sample is collected and placed in a predetermined sampling vessel (Fig. 1).

An important part of the sampling is the identification, transport and storage of biological samples, which in this case are blood tubes. One of the most important steps is to identify samples with avoidance of substitution or loss. A nurse should label all tubes or other sampling vessels with a label that contains basic identification information before sampling itself. Basic information refers to a name and surname of a patient, a year of birth, a date of collection or a name of a department where the sampling takes place. This data must be legible, accurate and congruent with the prescription data. This prescription must be appropriately selected according to the type of sampling and sample. Labels on tubes should be spiral bonded for the centrifugation process. This is done by analyzing samples in a laboratory where you need to see the serum and blood cake interface. The prescription, which is sent

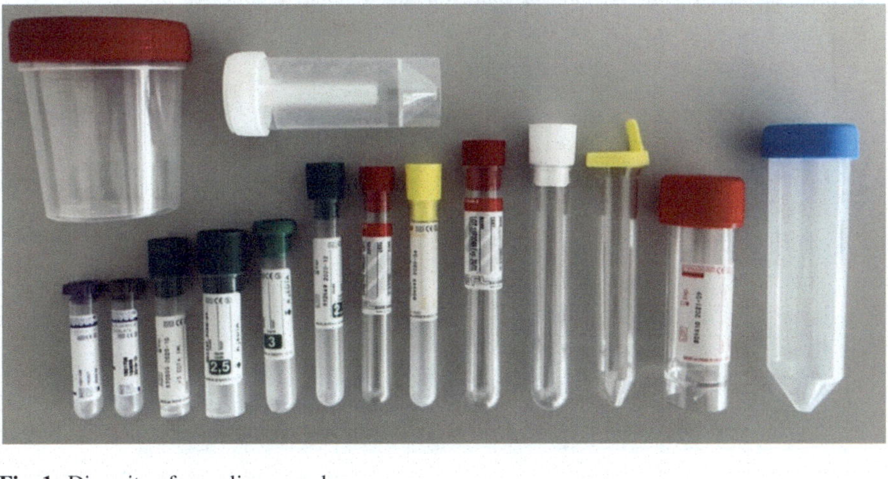

Fig. 1 Diversity of sampling vessels

together with the sample for analysis, must also be filled in legibly and contain the required information by the laboratory, such as a title, a first name, a surname, an insurance company number, a birth number, a department number, a department name, a diagnosis of the patient according to the International Classification of Diseases, a department stamp, a code of the prescribing or treating medical and their signature, a type of material, examinations required, a date, time and a method of sampling, and a department contact information [10].

2 Methodology

As mentioned earlier, it is important to pay attention to the compatibility and functionality of the system. Therefore, the required hardware must be specified at the beginning. In this case, it is a commonly available Lenovo Tab3 10 Business Model TB3-X70F tablet with NFC technology. Its choice is not a random one, since it is a cheap and affordable tablet with 10.1″ diagonal with good readability and manageability (Fig. 2).

Emphasis was also placed on quality, availability and tag price when selecting a transponder for NFC technology. NFC offers a choice of five categories: NTAG®Series, MIFARE Classic®, MIFARE Ultralight®, MIFARE®, DESFire® and ICODE® Series. All of these categories operate at 13.56 MHz; however, they do

Fig. 2 Placing NFC hardware technology into the tablet

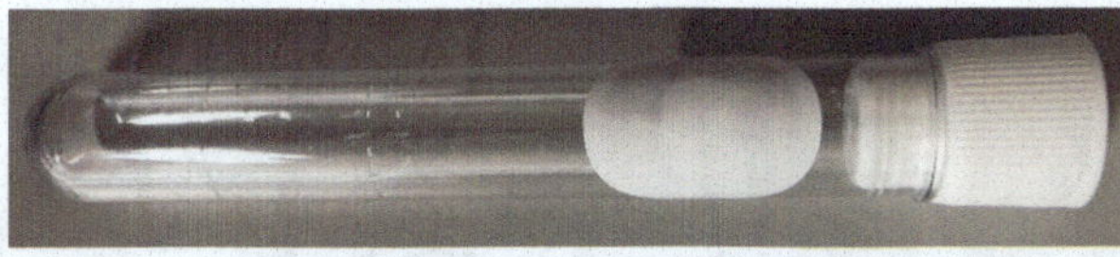

Fig. 3 Placing of MIFARE Ultralight® NTAG 213 NFC Transponder on D1038 tube

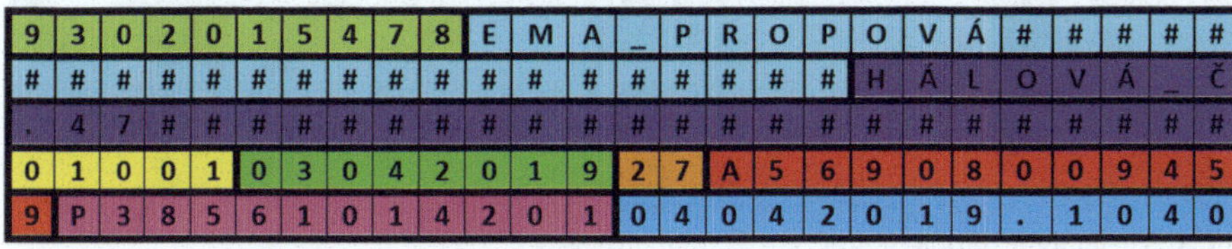

Fig. 4 The patient information sentence filled

not work in the vicinity of metals. Special NFC On-Metal tags are designed for metal materials that will not be necessary for our proposal, as tubes are normally made of glass or plastic. NTAG® Series is a tag category that is compatible and works with all NFC-enabled devices. However, MIFARE Ultralight® category transponders that are cheaper will suffice for our needs. Specifically, we will use the NTAG 213 model. It is necessary to mention a pair of parameters in this case. The first is a memory capacity of 144 bytes and a recording length of 130 characters. This information is essential to us due to the creation of the information sentence described later. We chose a D1038 tube of 16×75 mm with a volume of 8 mL from the large number of sampling tubes (Fig. 3).

We created a patient's 9-block information sentence for our needs that is based on the person's input information requirements that would be written to the NFC tag (Fig. 4). Written data is illustrative and structured as follows:

- Block 1: birth number information, which is in the form of 9302015478
- Block 2: name and surname that is Ema Propová
- Block 3: street name and number of a patient Hálová č.47
- Block 4: is intended for postal code 01001, and we can find out that it is the city of Žilina
- Block 5: the day on which the blood analysis request was made
- Block 6: information about the health insurance company
- Block 7: treating physician information (the personal number is A56908009459)
- Block 8: is about a healthcare provider (the branch identification number)
- Block 9: information about the date and time of sampling

The patient information sentence created this way will be transferred to the NFC Tools application, where several modifications are necessary to make before writing to the transponder.

A text file which is entirely sufficient for the experiment was selected for the information sentence. After copying the sentence and creating the necessary text file for writing to the transponder, the application displays its size. The experiment

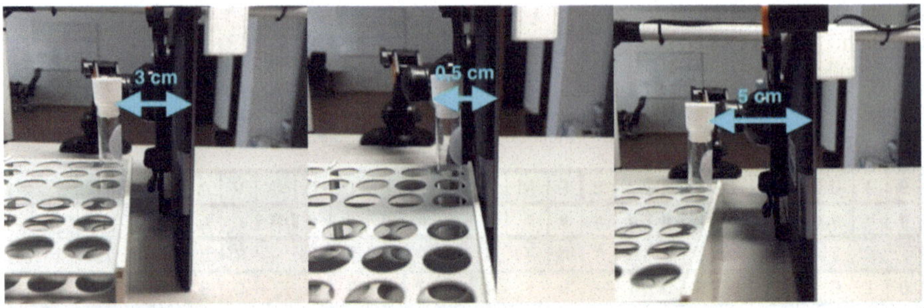

Fig. 5 The demonstration of NFC tag reading distances

itself began after the file for writing the patient's information sentence was created. It consisted in verifying the reading distance between the tablet and the tube. A gap width of 0–5 cm was selected with a graduation of 0.5 cm for the experiment. This means that 11 attempts for reading to an NFC transponder have been made. The observed factor in this measurement was transponder detection and writing distance capability. A sample of three distances between the tablet and the tube NFC transponder can be seen in Fig. 5.

3 The Results

The first measurement was made at a distance of 5 cm when NFC tools did not indicate any transponder. Measurements without transponder reading capability continued up to a distance of 3 cm when a change occurred. The first detection of the NFC tag was registered at this width. Subsequently, the actual writing of the patient's information sentence was successfully performed without any problems, which can be observed on the left side of Fig. 6. The demonstration of the patient's information sentence tag can be observed on the right side of Fig. 6.

Shifting the tube 0.5 cm closer to the tablet confirmed the ability to identify and read the transponder using NFC technology. In this experiment, the information was written on the tag that was withdrawn and then returned to the laboratory stand for reading verification. All other measurements were made in this way, where the changing factor was the shortening distance between the tube and the tablet with NFC technology. The results of all 11 measurements of the experiment can be seen in Table 1, where the distance between the elements is given as well as the information whether or not reading was carried out.

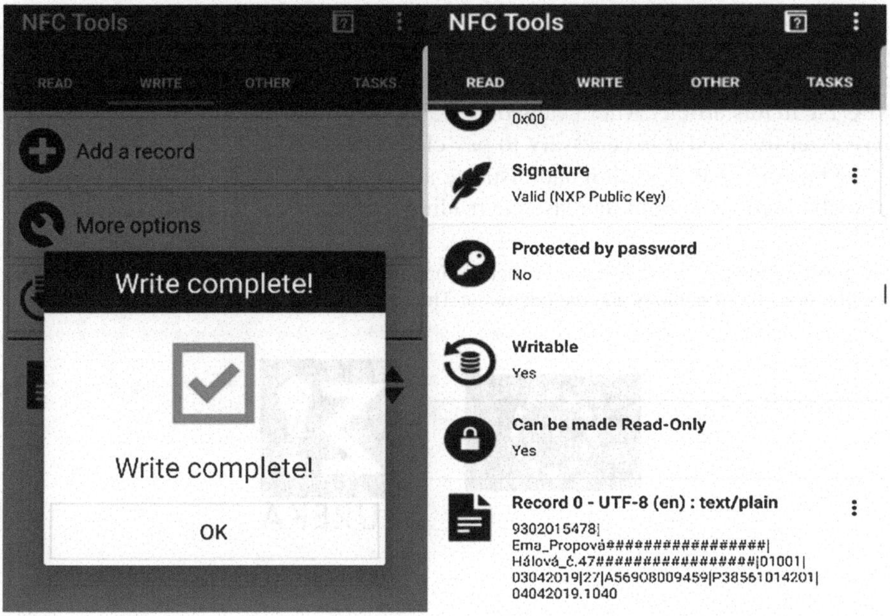

Fig. 6 Reading data to a transponder and the demonstration of read information

Table 1 Results of reading and writing of NFC transponders

Tag and tablet distance	Reading	Writing
5.0 cm	No	No
4.5 cm	No	No
4.0 cm	No	No
3.5 cm	No	No
3.0 cm	Yes	Yes
2.5 cm	Yes	Yes
2.0 cm	Yes	Yes
1.5 cm	Yes	Yes
1.0 cm	Yes	Yes
0.5 cm	Yes	Yes
0.0 cm	Yes	Yes

4 Conclusion

We can state a few facts related to NFC technology after concluding all the measurements and summarizing them. The range of information transfer within 5 cm of a standard mobile device is mentioned in the various definitions associated with NFC technology. This distance was not confirmed by the experiment because the first signs of reading were registered up to 3 cm distance. We can state the absolutely smooth function of NFC technology after overcoming the reading distance of 3 cm. These statements have also been validated for liquid-filled tubes where the findings described earlier have been confirmed. We tried to present a common variant of a simple freely available tablet and a commercially available NFC tag in this article. After evaluating the experiment, we can say that the issue is very extensive, and it is necessary to pay close attention to it. As has been verified in the practical part, location and distance affect the reading capability. Also, more powerful hardware could increase the reading distance.

Acknowledgement This article was created to support the project named EUREKA-E!11158 U-health – Auto-ID technology and the Internet of Things to enhance the quality of health services.

References

1. Husár, J., & Knapčíková, L. (2017). *Automatic identification systems* (1st ed.). Košice: TUKE.
2. About NFC, http://www.nearfieldcommunication.org/about-nfc.html, last accessed 2019/05/10.
3. Al-Ofeishat, H. A., & Al Rababah, A. A. (2012). Near field communication (NFC). *International Journal of Computer Science and Network Security, 12*(2), 93–99.
4. History of Near Field Communication, http://www.nearfieldcommunication.org/history-nfc.html, last accessed 2019/05/07.
5. Host-based Card Emulation, http://developer.android.com/guide/topics/connectivity/nfc/hce.html, last accessed 2019/05/10.
6. Zidek, K., & Hošovsky, A. (2014). Wireless device based on MEMS sensors and bluetooth low energy (LE/smart) technology for diagnostics of mechatronic systems, design. *Testing and Characteristics of Mechatronic Devices, 460*, 13–21.
7. National Transfusion Service of the Slovak Republic, http://www.ntssr.sk/o-darovani-krvi/preco-darovat-krv, last accessed 2019/05/08.
8. Blood collection for laboratory examination, http://www.nsplm.sk/dokumenty/odber%20krvi.pdf, last accessed 2019/05/08.
9. Blood collection, http://www.schneider-laboratoria.sk/70.html, last accessed 2019/05/09.
10. Principles of biological material sampling, transport and storage, http://www.synlab.sk/fileadmin/standortseiten/synlab_sk/pdf/ostatne/zasady-odberu-vzoriek.pdf, last accessed 2019/05/10.

Possibility of Identification of Transport Units Through RFID Technology at Own Level of the Company

Jiri Tengler and Jana Pekna

1 Introduction

Radio frequency identification (RFID) has been used in various forms and areas for decades. This technology has gradually been used in various industries, mainly in areas where it was necessary to load a large number of identifiers very quickly. Thus, the application was found in areas where the barcode itself was no longer sufficient and it was necessary to extend it with an RFID identifier. The biggest application of RFID technology was recorded especially in the field of logistics. The RFID technology is used to make logistics and customer-supply chain processes visible. It enables real-time tracking of relevant data in these processes. These data can be used for practical measures after correct interpretation (increasing efficiency, reducing errors, and improving quality).

One of the aforementioned areas of application is the monitoring of transport units and logistics units within the customer-supply chain. In general, a clear case is the application of RFID technology for both parties involved in supply chain. The task of the application is then to provide information about where the transport logistics unit is located, where it is directed, or where it came from. This issue can occur unless one of the parties (usually the purchaser) agrees with the application of RFID technology. Subsequently, customers have no choice but to apply RFID at their own level of a company.

There are plenty of ideas for applications of RFID technology within the supply chain. This can include streamlining logistics, managing transport units, or identifying possible defects in shipping or handling at the supplier (goods or shipping units). However, such an application can solve at least some of the

J. Tengler (✉) · J. Pekna
University of Žilina, Žilina, Slovakia
e-mail: Jiri.tengler@fpeda.suniza.sk; pekna4@stud.uniza.sk

L. Knapcikova et al. (eds.), *4th EAI International Conference on Management of Manufacturing Systems*, EAI/Springer Innovations in Communication and Computing, https://doi.org/10.1007/978-3-030-34272-2_7

above problems. There are many RFID applications operating at own level of a company. These applications are used primarily for monitoring transport units between individual logistics centers and organizational units of their own company. However, there are no problems with supply chains. However, in some specific nonstandard cases and issues, RFID can still be used. And this issue is the subject of this article.

1.1 About RFID Technology

The RFID technology is complex, combining a number of different computing and communications technologies to achieve desired objectives. Each object that has to be identified is fitted with a small object called a RFID tag stuck to it. The application at only one side of the supply chain can never bring full value information.

Each RFID tag has a unique identifier that enables additional information about each object to be stored. Devices known as RFID readers wirelessly communicate with RFID tags, with a view to identifying the attached RFID tags, as well as enabling information stored in the RFID label to be read and updated [1].

Every RFID system contains an RF subsystem (Fig. 1), and most RFID systems also contain an enterprise subsystem. An RFID system supporting a supply chain is a common example of an RFID system with an inter-enterprise. In a supply chain application, a tagged product is tracked throughout its life cycle, from manufacture

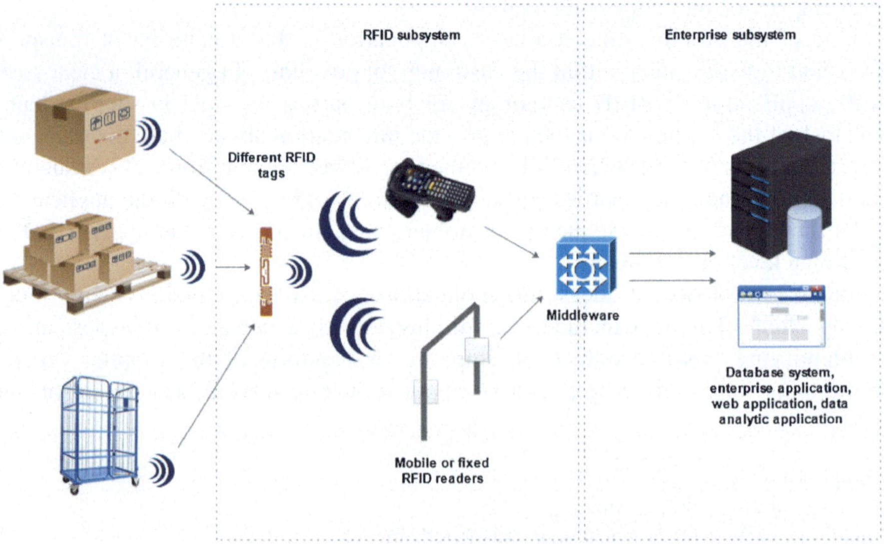

Fig. 1 The RFID system. (Source: The picture was create by author Jiří Tengler)

to final purchase, and sometimes even afterward (e.g., to support service agreements or specialized user applications). RFID is a wireless data collection technology that uses electronic tags, which store data, and tag readers, which remotely retrieve data. It is a method of identifying objects and transferring information about the object's status via radio frequency waves to a host database. RFID represents a significant technological advancement in AIDC because it offers advantages that are not available in other automatic identification and data capture (AIDC) systems such as barcodes. RFID offers these advantages because it relies on radio frequencies to transmit information rather than light, which is required for optical AIDC technologies [2].

There are three basic components of the RFID system, RFID tags, RFID readers, and middleware, which is responsible for all data transaction in the system [3].

The RFID tag is a small device that can be attached to an item, case, roll cage, or pallet, so it can be identified and tracked. It is also called a transponder. The tag is composed of microchip and antenna. These elements are attached to a material called a substrate in order to create an inlay.

Tags are divided into several categories based on their characteristics. Depending on the power source, the RFID tags can be divided into the following types:

- Active RFID tag
- Passive RFID tag
- Semipassive RFID tag
- Semiactive RFID tag

The second component in a basic RFID system is the interrogator or reader. The reader can have an integrated antenna, or the antenna can be separate. The antenna can be an integral part of the reader, or it can be a separate device. Handheld units are a combination reader/antenna, while larger systems usually separate the antennae from the readers. The reader retrieves the information from the RFID tag [2].

There is also Middleware, software that controls the reader and the data coming from the tags and moves them to other enterprise systems such as database system. It carries out basic functions, such as filtering, integration, and control of the reader. RFID systems work if the reader antenna transmits radio signals. These signals are captured tag, which corresponds to the corresponding radio signal [4, 5].

2 Problem and Solution Design Specification

For the purposes of this article, the subject of research may be closer to the example of a fictitious company. Let there be a company manufacturing bearings for rotary engines. This company supplies its products to two customers. For delivery, it uses its means of transport, that is, containers. The shipping containers are provided with a transport document before the expedition. The company lacks a system that would allow the container movement to be automatically identified. Thus, it lacks the exact information where the particular container was dispatched or

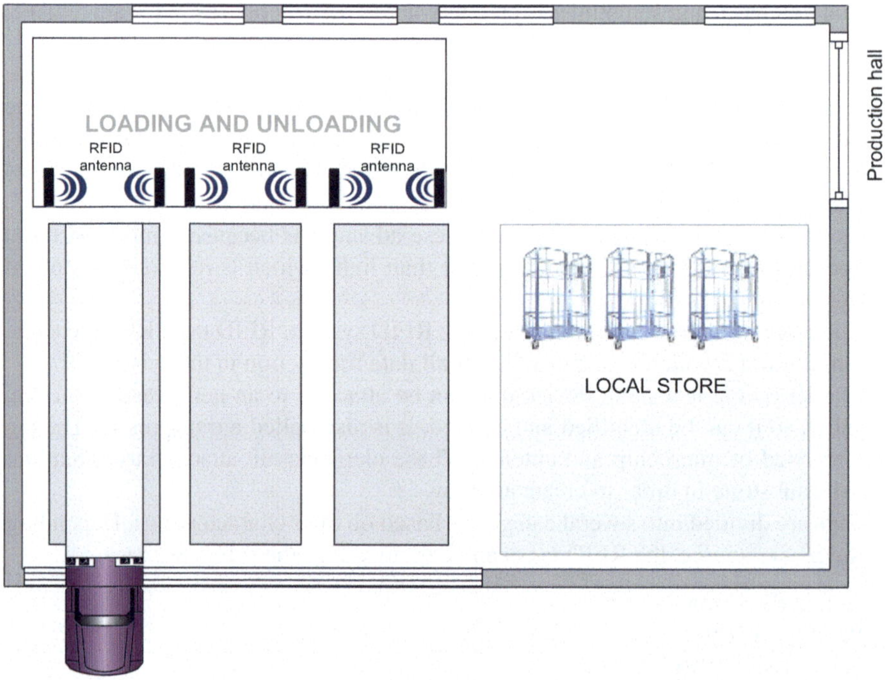

Fig. 2 Loading and unloading area. (Source: The picture was create by author Jiří Tengler)

where it came from. The transport documents are provided with a bar code and are scanned before the expedition. However, the shipping document is withdrawn upon delivery. Consequently, it is not very clear where the container came from. Application of RFID technology in both the supplier and the customer resolves the issue. Unfortunately, customers do not want to implement RFID technology [6]. What options the supplier company has is the following study. Suppose a company wants to know where the products come from and where they go. The most likely area of implementation of RFID technology seems to be an area for loading and unloading. For a better understanding, a fictitious company can be illustrated in Fig. 2. Thus, the focal point of retrieving data from RFID technology is the loading and unloading process. Unfortunately, the sample itself does not yet address the provision of the requirement.

2.1 Selected Realized Solutions

There are a large number of solutions to the chosen issue within the scope of RFID technology implementation. Only three solutions have been selected for the purpose of presenting the results of this article.

Use of RFID Printer

The first solution is to use an RFID printer. The essence of this solution consists in repeated pushing and programming of the RFID tag and their subsequent application to the container before the expedition. The RFID tag is encoded by not only identifying the container but also the direction to which it is dispatched (customer). The application of the RFID tag could also be possible in the manufacturing area where shipping documents are printed. The solution assumes to use and print and program a new RFID tag for each trip.

Use of a Permanent Container RFID Tag

Another possible solution is the application of industrial permanent RFID tag. This solution except by permanent RFID tag also envisages the use of touch in the terminal to area loading and unloading. The function of the terminal would be to supplement the retrieved data with the planned direction of the dispatched containers. Before loading the container on a transport vehicle, the employee would select a particular customer. Consequently, within the loading and retrieved container RFID tags, this information would be paired with the customer.

Use of Especially Developer RFID Box

The last solution is RFID box. An RFID box is a switchable RFID tag that allows you to switch between the active status of only one of the contained RFID tags. Each tag should contain a unique identification number of the container and also the direction where it will be dispatched (customer).

2.2 Advantage and Disadvantage of the Proposed Solutions

Based on the description of individual solutions, it is possible to specify the advantages and disadvantages of individual solutions (Table 1).

3 RFID Box/Multitag

All three solutions were elaborated in the research of the above-mentioned issues. The last solution was the most intensive. It is important to explain in more detail the principle of the RFID box and the logic of its reading through the RFID reader.

Table 1 Advantage and disadvantage of the proposed solutions

	Solution 1	Solution 2	Solution 3
Description	RFID printer	Permanent RFID tag + touch terminal	Manually switchable RFID box
Advantages	The simplicity of the solution, lower probability of errors	Disposable application of industrial RFID tags for containers	The simplicity of the solution. Lower probability of employee error when switching RFID box
Disadvantages	Large consumption of RFID tags. Possibly increased difficulty in removing an old RFID tag	Necessary to set the direction of the trip in the touch terminal by the unloader. Possible probability of employee error in the terminal	The need to switch the RFID box to the employee in production

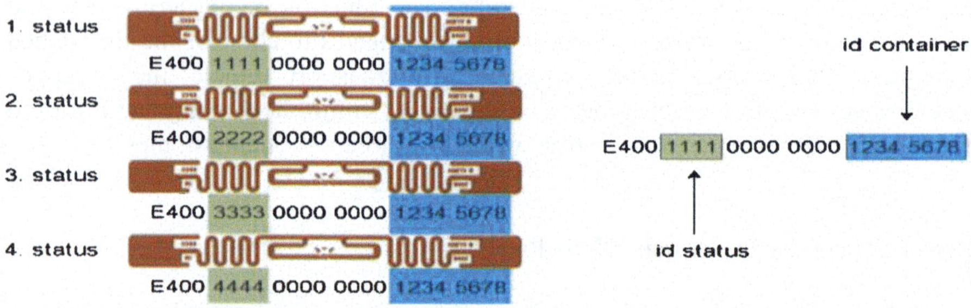

Fig. 3 Principle of stored data in RFID tag

3.1 Principle of Identification of RFID Box

The RFID box actually contains four identical RFID tags. RFID tags vary slightly with each other's information that is stored in their memory. The issue for the needs of this study can be explained without reference to the EPC standard. The developed RFID box uses 96-bit RFID tags. When you load them, we get 24 characters in the hexadecimal system. This string of characters (hexadecimal numbers) can be arbitrarily divided into two or more parts. One of these may serve as an identifier for a particular container (the unique container identifier), and the other may identify the direction the container is shipped to. In the proposed solution, four basic tags were used in the basic version, even with a larger cylinder diameter and a suitably selected RFID tag, and more RFID tags can be achieved. Thus, the container identifying portion for the tag is the same, but the status is different for each of the four RFID tags. A clearer understanding of data storage is shown in Fig. 3.

3.2 *Principle of Operation and Description of RFID Box*

The design of the RFID box since the beginning of the development was based on the idea that RFID tags would be switched through a rotating cylinder. And only that RFID tag that is rotated to the front of the RFID box allows you to reflect the electromagnetic signal. As part of this research, two versions of RFID box were designed and assembled. The boxes have been designed with respect to the use of RFID tags. The boxes were modeled using freely available modeling tools and subsequently printed through a 3D printer. Both versions of the RFID box differ in size and RFID tag used.

Small RFID Box (Fig. 4)

- Box size 110 × 27 × 27 mm
- Four pieces of AVERY DENNISON AD-226IM UHF RFID WET INLAY (NXP G2IM)

Large RFID Box (Fig. 5)

- Box size 90 × 45 × 45 mm
- Four pieces of AVERY DENNISON AD-236 U7 UHF RFID WET INLAY (NXP UCODE 7)

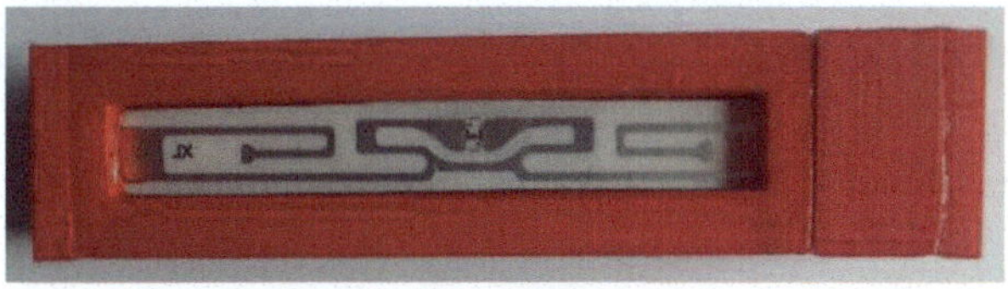

Fig. 4 Small RFID BOX/MULTITAG

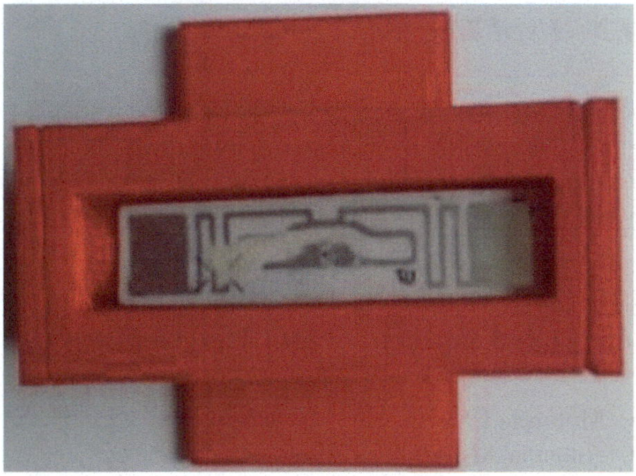

Fig. 5 Large RFID BOX/MULTITAG

Fig. 6 Antireflective treatment of both RFID boxes

Important spectra for the proper operation of the RFID box are antireflective finishes. Thus, avoid the reading of those tags that are not rotated in the direction of the RFID box opening. Antireflective aluminum foils were used for this purpose. These films were applied both inside and outside the outer cylinder section (Fig. 6).

4 Testing of RFID Boxes

The proposed solution was verified through a set of tests. The aim of the testing was to find out if the solution is usable for real traffic and to what extent the quality of reading RFID tags has changed. The tests were carried out on a real container called "Kontel." The RFID box was placed in different container positions. Fig. 7 shows the view of top and side of the container.

4.1 Monitored Parameters When Reading a Tag

- Dimensionless value RSSI (the strength of the reflected signal)

4.2 Used Tags

- Small RFID box with tags AVERY DENNISON AD-226IM
- RFID tag itself AVERY DENNISON AD-226IM
- Large RFID box with tags AVERY DENNISON AD-236 U7
- RFID tag itself AVERY DENNISON AD-236 U7

4.3 Used Equipment and Material

- Motorola FX7400
- Antenna AN480

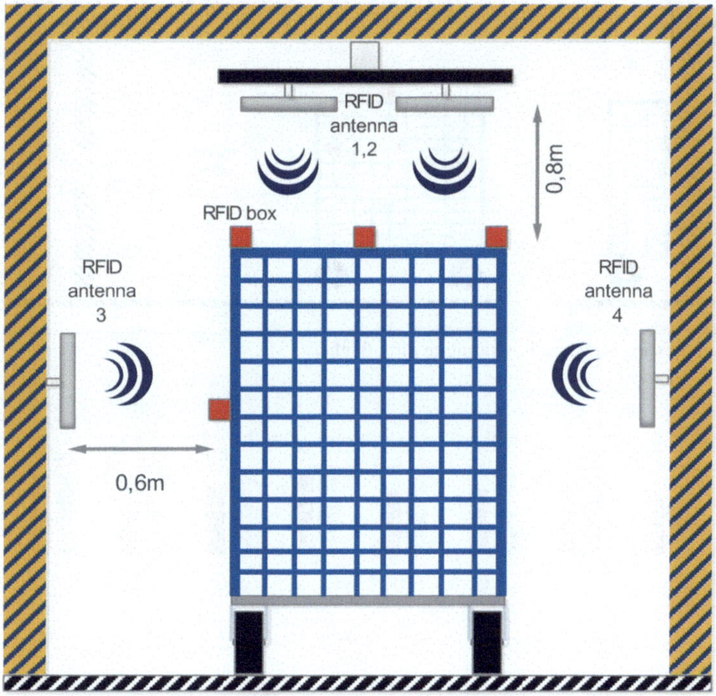

Fig. 7 View on optionality of measurement

- Middleware Aton OnID
- Database MariaDB
- Container KONTEL

4.4 Types of Tests

Two types of tests were used for testing: static and dynamic. The goal of static testing was to find out the differences between the RFID tag itself and the RFID box. Usability for real operation was verified within dynamic tests. Even within the dynamic testing, the RFID tag itself was compared with the RFID box. The principle of testing is shown in Figs. 7 and 8. The principle of static testing consisted of reading a fixed number of tags. In particular, the RSSI value was the dominant value. The principle of static testing was to move the container with the tag over the RFID antenna gate. Relevant values here were both the RSSI and the number of tags reads. In the static measurement, the so-called measurement sets were measured. Each new tag position meant a new set. Each set within a given tag position was repeatedly measured 10 times. Each set was a load of a hundred piece tag value. In

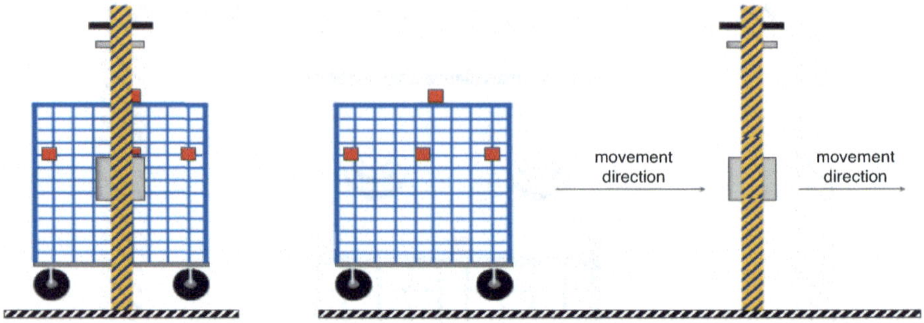

Fig. 8 Static testing (left) and dynamic testing (right)

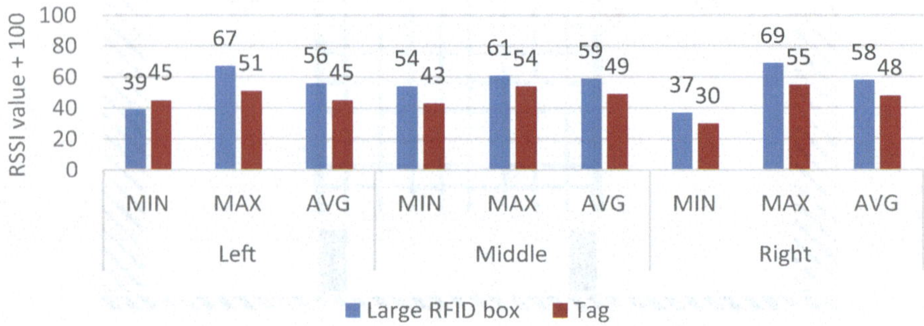

Fig. 9 Result of static testing, large RFID box, two antennas top, all position

the static measurement, readings were made at the top three positions (Figs. 7 and 8) and another three positions from the side of the container. The tags at the top of the container were read only by RFID antennas from the top, and the container side tags were read only by RFID antennas from the side. The same load settings were also made for dynamic testing. Dynamic testing measured RFID tags in motion. Thus, the container passed through a gate with RFID antennas (Figs. 7 and 8). A 10-piece transition was made for each tag location.

4.5 Result of Testing

Within the evaluated graphs, you can see the minimum, maximum, and average values of the tag itself in comparison with the RFID box. In the results, you can compare an individual tag location, either from the top or side of the container.

Sets 1–3 – large RFID box, antennas top, tags at the top of the container (Fig. 9)

Sets 4–6 – small RFID box, antennas top, tags at the top of the container (Fig. 10)

Sets 7–9 – large RFID box, antennas on side, tags on the side of the container (Fig. 11)

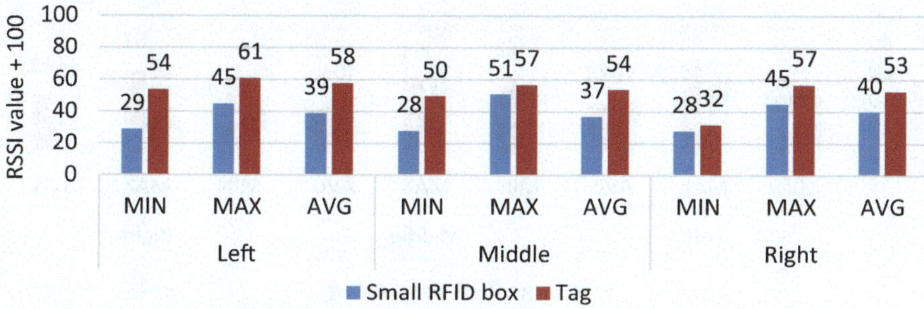

Fig. 10 Result of static testing, small RFID box, two antennas top, all position

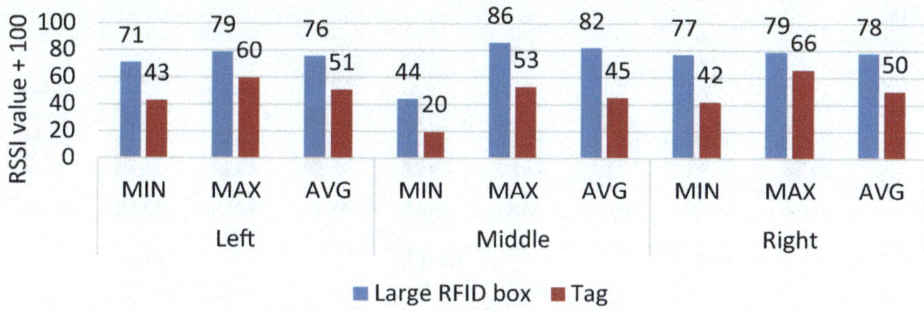

Fig. 11 Result of static testing, large RFID box, two antennas on side, all position

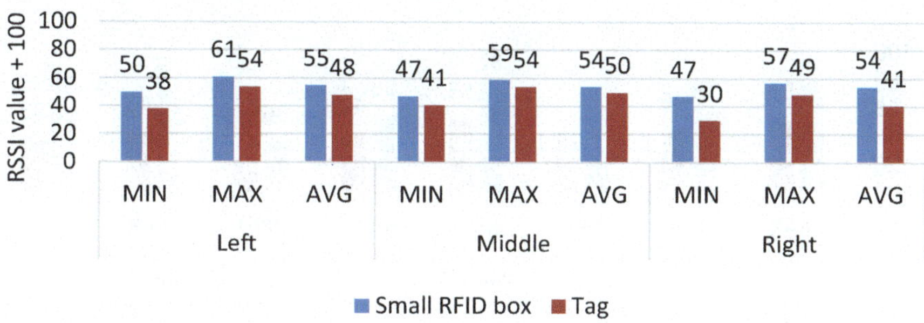

Fig. 12 Result of static testing, large RFID box, two antennas on side, all position

Sets 10–12 – small RFID box, antennas on side, tags on the side of the container (Fig. 12)

It is clear from the results of the static tests that the large RFID box is significantly better than the standard RFID tag. The reason for this phenomenon can be mainly due to sufficient space around the box tag and sufficient distance from the bovine parts of the containers. While the values of a large RFID box are better compared to a classic tag in a small RFID box, the scenario is the opposite. The small dimensions of the box do not allow the ideal electromagnetic signal propagation. That is why

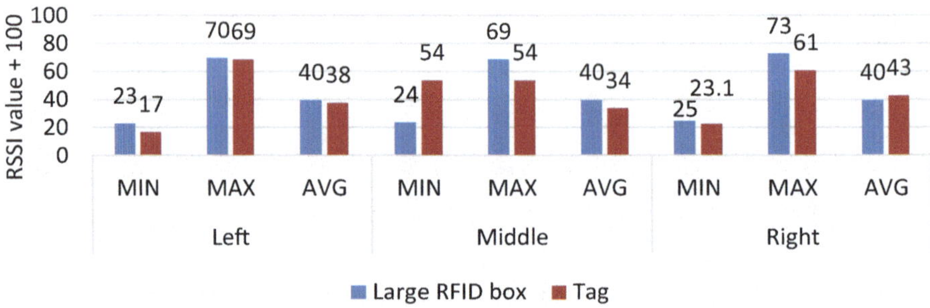

Fig. 13 Result of dynamic testing, large RFID box, two antennas top, all position

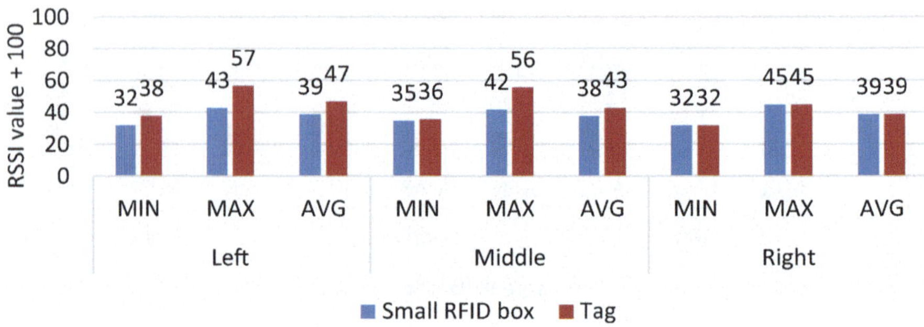

Fig. 14 Result of dynamic testing, small RFID box, two antennas top, all position

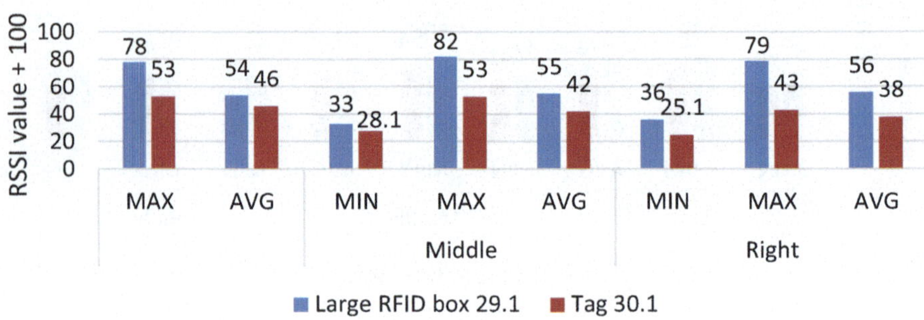

Fig. 15 Result of dynamic testing, large RFID box, two antennas on side, all position

the standard tag in some sets is significantly better. When it is loaded by antennas from the top or bottom, a large RFID box gives the best results.

Transition 1–3 – large RFID box, antennas top, tags at the top of the container (Fig. 13)

Transition 4–6 – small RFID box, antennas top, tags at the top of the container (Fig. 14)

Transition 7–9 – large RFID box, antennas top, tags on the side of the container (Fig. 15)

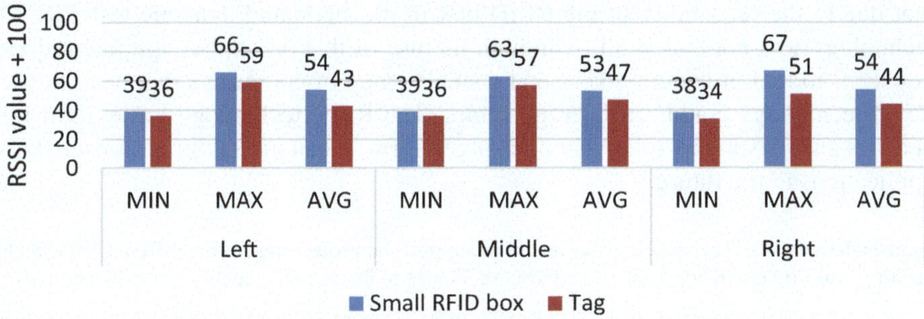

Fig. 16 Result of dynamic testing, small RFID box, two antennas on side, all position

Transition 10–12 – small RFID box, antennas top, tags on the side of the container (Fig. 16)

It is clear from the results of the static tests that the large RFID box is significantly better than the standard RFID tag. The reason for this phenomenon can be mainly due to sufficient space around the box tag and sufficient distance from the bovine parts of the containers. While the values of a large RFID box are better compared to a classic tag in a small RFID box, the scenario is the opposite. The small dimensions of the box do not allow the ideal electromagnetic signal propagation. That is why the standard tag in some sets is significantly better. Although the antennas are read from the top or bottom, the big RFID box gives the best results.

The dynamic test results show that the difference in the size of a large RFID box is called when reading in motion. The same is true when loading a small RFID box, where the difference with the classic tag becomes even stronger. Another situation is when reading tags from the side. Here at both RFID boxes, we notice a significant improvement over the classic RFID tag. Although the quality of reading of both boxes varies considerably, in both cases, the RFID box was not unread. From these laboratory tests, it is difficult to conclude that the same results could be obtained by RFID boxes even when deployed in real operation.

5 Conclusion

The subject of our research was the possibility of using RFID technology for monitoring transport units at the level of own company. More precisely, the possibilities of using RFID technology within the supply chain are analyzed, where only the supplier is interested in applying RFID technology. In the context of the issues identified, several potential solutions have been proposed, with the most significant being mentioned at the core of the article. Multitag device or RFID box is one of the solution proposed, which is the pivotal part of this study. A part of the proposal was also the verification of the solution by testing and comparison of standard with proposed classic RFID tags. From the test results, it was apparent that due to the readability of the RFID box in the static and dynamic tests, RFID technology was not bad at all. Although the use of this or similar equipment does not seem so real, it is quite likely that a small group of companies that want to use RFID technology could reach in the future. The RFID technology has existed for decades and, despite its relatively unchanged form, it can offer a lot of nonstandard applications in the future.

Acknowledgments This article was created to support the project named EUREKA-E!11158 U-health – Auto-ID technology and the Internet of Things to enhance the quality of health services.

EUREKA

References

1. Vojtěch, L., & Neruda, M. (2010). Application of shielding textiles for increasing safety airborne systems – limitation of GSM interference. In *The 9th international conference on networks* (pp. 157–161). Los Alamitos: IEEE Computer Society.
2. Beneš, F., Kubáč, L., Staša, P., & Kebo, V. (2013). RFID and augmented reality. In *Proceedings of the 14th international Carpathian control conference* (pp. 186–191). Rytro: IEEE.
3. Madleňák, R., Madleňáková, L., & Štefunko, J. (2015). The variant approach to the optimization of the postal transportation network in the conditions of the Slovak Republic. In *Transport and telecommunication* (pp. 237–245). Riga: Transport and Telecommunication Institute.
4. Hunt, V., Puglia, A., & Puglia, M. (2007). *RFID: A guide to radio frequency identification.* Hoboken: Wiley.
5. Thornton, F., & Lanthem, C. (2006). *RFID security*. Rockland: Syngress Publishing.
6. Dávid, A. (2019). Automation of handling systems in the container terminals of maritime ports. In *Transport and communications* (pp. 6–9). Žilina: EDIS – University of Žilina.

Automatic Identification Applications for Smart Railway Concept

Lucia Knapcikova, Jozef Husar, and Patrik Kascak

1 Introduction

Industry 4.0 or the Fourth Industrial Revolution has a number of definitions, and it depends on how a particular firm or initiative decides to do so. According to a relatively widespread definition, this is a term denoting the current trend of automation and data movement in production technologies using cyber-physical systems (Fig. 1) [1].

Other possible definitions may also be:

- The next stage in the digitization of the manufacturing sector, driven by four ways – a staggering increase in data volumes, computing power and connectivity (especially in new low-energy networks), new analysis and business intelligence alternatives, new possibilities for people and machine interaction such as touch interfaces and augmented reality and new forms of digital transfer capabilities such as advanced robotics or 3D printing [1, 2].
- Linking systems where smart products are able to communicate with their users, new digital business models are able to use scanned data and offer new services, and specific products are capable of managing their production process on assembly lines.

The Industry 4.0 principle was taken from the German Industrie 4.0 initiative, which was the first of its kind used at the Hannover Messe in 2011 as a future project by the German Federal Government. In 2013, 2 years later, he was introduced to the same place, the German National Initiative for the Use of New Technologies in the

L. Knapcikova (✉) · J. Husar · P. Kascak
Technical University of Košice, Faculty of Manufacturing Technologies with a Seat in Prešov, Department of Industrial Engineering and Informatics, Presov, Slovakia
e-mail: lucia.knapcikova@tuke.sk; jozef.husar@tuke.sk

L. Knapcikova et al. (eds.), *4th EAI International Conference on Management of Manufacturing Systems*, EAI/Springer Innovations in Communication and Computing, https://doi.org/10.1007/978-3-030-34272-2_8

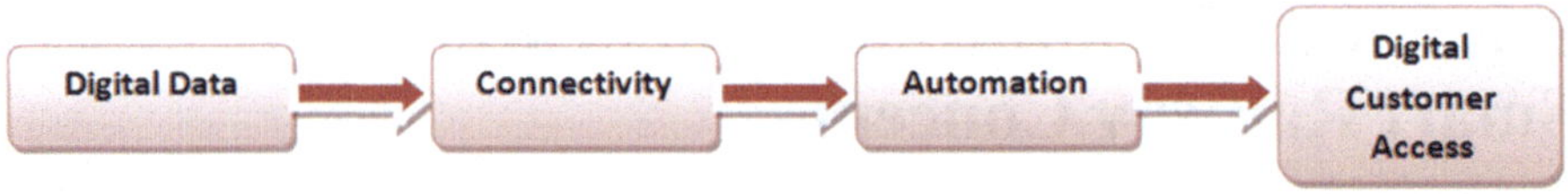

Fig. 1 Most important steps for railway digital transformation

German Industry [3]. There are two ways to look at the goals in Industry 4.0. From the point of view of the national economy, such as the coordination of technologies, the development of standards and the efficient use of resources, with the aim of strengthening the generated added value of national industries. The second is the benefit that the introduction of new technologies will bring to a particular company – the example of the automotive industry, the Roland Berger consultancy company, used in the use-time to increase by 25% points to approximately 40% [4].

As it was heard at the Hannover Fair in 2013, in the interconnected world of Industry 4.0, smart factories will be part of the so-called smart infrastructure and the Internet of Things. The horizontal, vertical and system integration of processes will achieve a new level with smart products. In Smart factories, people, machines and materials will be able to communicate with each other in a similar way to social networks. Smart products will contribute to the running and improvement of the workforce, making it more efficient, smoother and with much less errors [4, 5].

2 Current State Proposal Solutions

The railway industry is a dynamic sector with a large amount of assets, maintenance and repair equipment, depots and stations spread across a wide territory. Good internal communication, fast reactions based on equipment geo-location data, high-quality maintenance planning and regular interventions are required to keep this massive machine working [5, 6].

This article is focused on proposal solution for Industry 4.0 in a smart specialized logistics cluster for the chemical industry in the Port of Antwerp (Fig. 2):

- Concept of smart specialized logistics cluster with a new added value for rail freight services

In Fig. 3 is presented a current and future state of specialized smart logistics cluster. The current state of less than truck load trucking solution is as follows:

- Pallets are transported from site A to site E by using a truck.
- Pallets are transported from site B to site F by using a truck [7].

Project's consortium solution for future Less than Wagon Load multi-modal is presented as follows:

- First mile

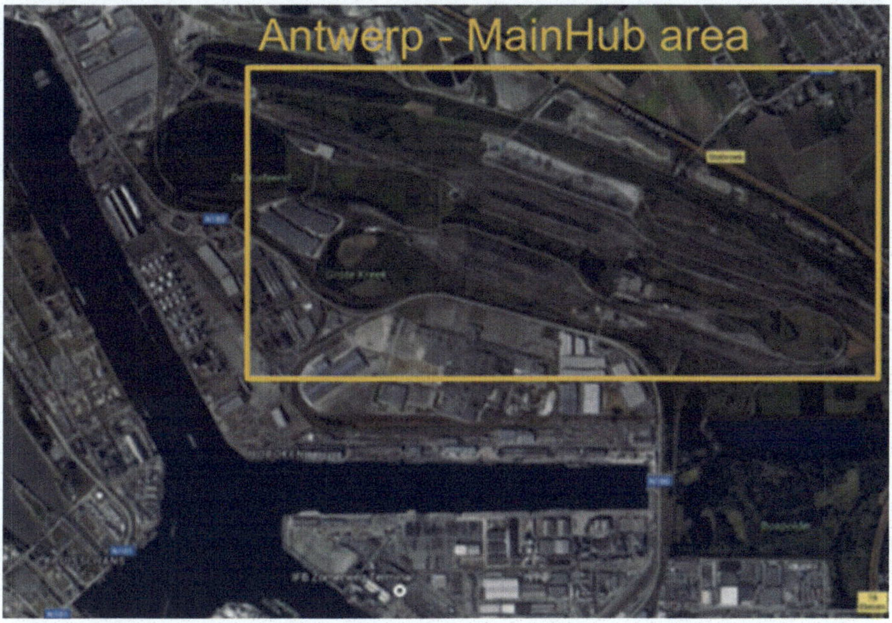

Fig. 2 Antwerp – main hub area in a SMART specialized logistics cluster [7]

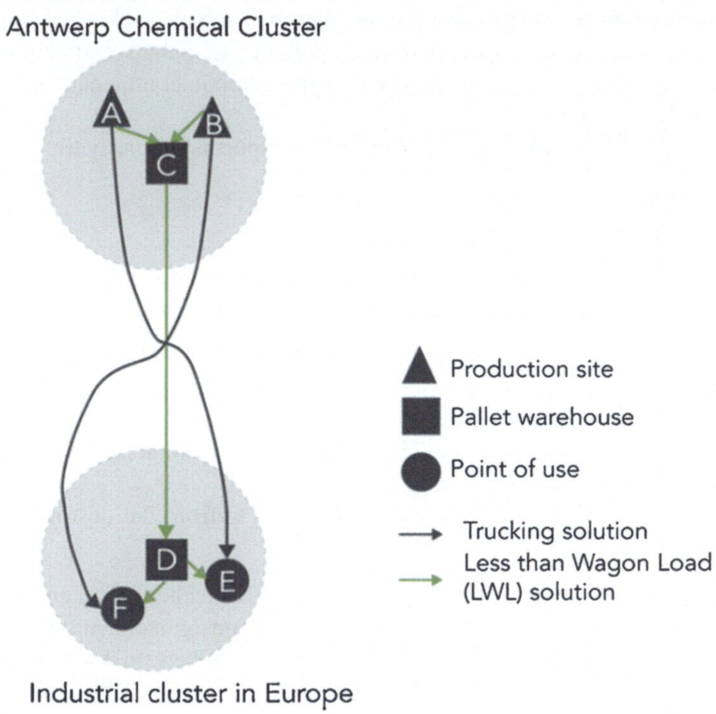

Fig. 3 Current and future state in the Port of Antwerp [7]

  - Pallets from site A are transported to cross-docking pallet warehouse C by using a truck, pallets from site B are transported to cross-docking pallet warehouse C by using a truck.

- Long haul

  - The pallets from A and B are grouped and transported in a pallet wagon from warehouse C to warehouse D by rail.

- Last mile

  - Pallets from site A are delivered by truck from cross-docking pallet warehouse D to site E and pallets from site B are delivered by truck from cross-docking pallet warehouse D to site F [7].

A system that fits into the concept of transport telematics and is very well developed is the automatic identification based on radio frequency technology. Railways are expected to be the industry drivers and a major economic leader in the transport of raw materials in the heavy and energy industries. At present, this position is weakened by the growing car traffic. Sensors are perhaps the smallest but one of the most important details when dealing with sensible topics as traffic and trains control [8]. The good news is that sensors are getting constantly cheaper and their implementation as a leading real-time data provider will generously increase. This will push condition-monitoring analysis to prevail over reactive-based approaches and irreversibly change the way assets are managed and maintained. These potential new services consist of parking, repair, picking and cleaning for chemical wagons, rail-connected cross-docking of pallets and improved rail connections by setting up mixed trains with conventional, continental and maritime container volumes [5, 9].

The use of RFID technology in rail freight transport is primarily to:

- Determine the exact location of the shipment
- Collect shipment information
- Protect the shipment against theft or loss.

By applying an RFID system, rail transport will gain additional benefits such as:

- Tracking shipments
- Evidence of the train set and individual wagons
- RFID application with sensors – information, for example, about temperature, pressure, slope and acceleration of moisture
- Checking the arrival and departure of the train to/from the destination station with shipments [10]

Tracking and collecting information using RFID technology can be used in rail freight transport for a complete train, wagon consignment, transport vehicle (container) and also for tracking a specific product. When tracking shipments in rail freight, it is also possible to use a GPS satellite tracking system that accurately locates the installed RFID tag [10].

The tracking of the railway car park is done in many countries through barcodes. Problems in detecting and verifying codes in a polluted environment and in adverse weather conditions, on greasy surfaces or non-metallic objects, along with the ability to read barcodes in direct sunlight, have moved to the act of American Railroad. Implementation of RFID was a solution to these problems and also brought benefits in the form of long-range shooting [11].

Since RFID technology is complex and multifaceted, it is important to develop common standards and standards for the application of the European Rail RFID system. The implementation of such standards must provide full interoperability and interoperability with all rail transport systems or wagon owners [12].

3 Smart Railway – Wagon Concept Using RFID Technology

Industry 4.0 in the railway context can be definite as Railway 4.0 concept and introduces integration across the supply chain through production to the end customer. Smart logistics systems are interconnected and coordinated with each other. Manufacturing processes include advanced materials, 3D printing technologies or adaptive manufacturing that minimizes waste and maximizes product customization. Production lines include new generations of robotic devices that enable collaboration as well as real-time data transmission.

Such a system allows full automation, interconnection of individual systems and M2M communications (Machine to Machine). With proper filtering, Big Data analysis and Cloud computing, real-time data can be gathered from a large volume of relevant information, promoted creativity and collaboration. At the same time, however, security will need to be strengthened when working in the cloud [7, 9]. The direct link and communication throughout the product life cycle means greater efficiency, optimizing the warehouse and waste generated by the manufacturer, while also increasing customer individualization. All this can be done by using the energy and material resources of the future (alternative raw materials, new materials, eco-energy and others). A Smart City uses intelligent solutions to optimize mobility and communication, creating safer societies with smoother traffic flows and more sustainable urban environments [3, 10].

Automating processes associated with data collection, transmission, and information recording can help improve various aspects of logistics. Developing information processes and increasing the level of automation are key to significantly improving operational efficiency. Implementing an RFID system offers greater reliability and more detailed chain information, optimizes the use of infrastructure and rolling stock, as well as enhancing product safety [11]. The information obtained helps railway undertakings to better plan the services offered in the event of a supply disruption and also to help ensure an efficient maintenance programme.

3.1 Proposal of Using RFID Technology on Railway Wagon

The implementation of RFID technology in rail transport is clarified by a directive and a standard on rules and how it can be applied in compliance with certain standards. 861/2006 entitled Technical Specification for Interoperability [3, 10]. The RFID tag on the wagon must comply with the ISO regulations to ensure that the information stored on the carrier is not read. It is necessary to pay attention to the positioning of the reading device, which must be able to read the information stored on the tag even at higher speed. The desired speed is adapted to the frequency band, optionally; there is an appropriate speed reduction mark on the track so that the reader is able to read the RFID tag on the wagon or on the load unit.

When placing an RFID tag on a wagon (Fig. 4) in rail, the location of passive RFID tags (Fig. 5) on the sides of the wagons is such that the reader can read the wagon identification number:

- A1 minimum centre height of RFID tag is 500 mm.
- A2 the maximum centre height of the RFID tag is 1100 mm.

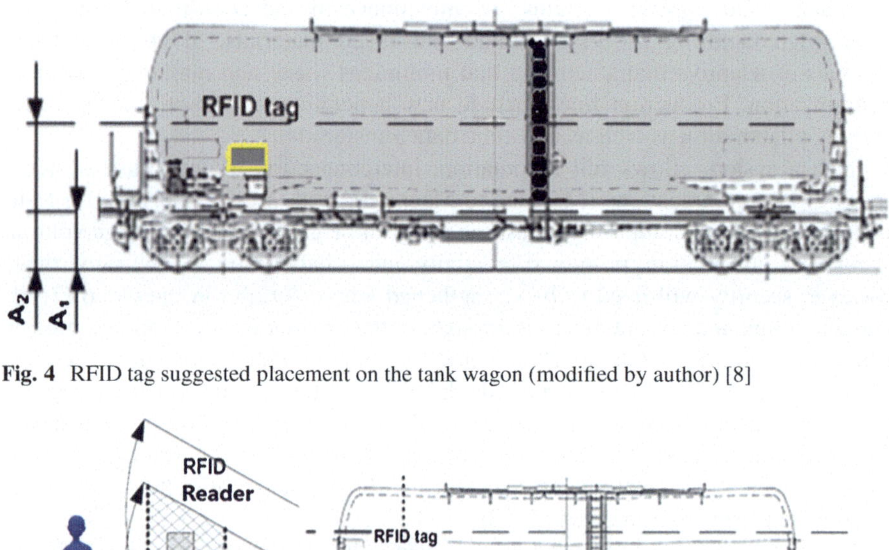

Fig. 4 RFID tag suggested placement on the tank wagon (modified by author) [8]

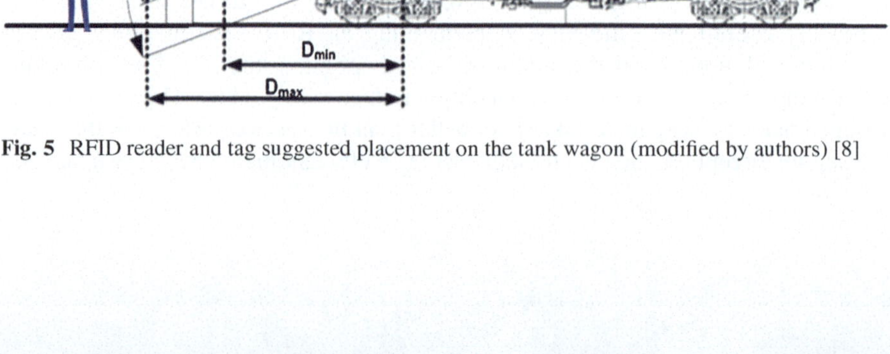

Fig. 5 RFID reader and tag suggested placement on the tank wagon (modified by authors) [8]

The location of the RFID tag meets all the requirements of Directive No. 861/2006. When placing the RFID reader next to the track, the reader location (Fig. 5) must be at a distance such that the reader can read the wagon identification number located inside the RFID tag [7]:

- D1 minimum RFID reader length is 1000 mm.
- D2 maximum length of the RFID reader is 10,000 mm.
- The maximum speed of the locomotive with wagons is 30 km/h.

The reliability of the on-board train positioning system is vital for the trains to operate safely while still allowing a much higher throughput of trains than with traditional interlocking.

For condition monitoring, automatic vehicle identification (AVI) is often installed in combination with different sensor systems (e.g. wheel flats, brake failure, hot boxes, noise levels, tyre pressure, etc.) providing combined data from the measurement and the AVI function. The system provides accurate and reliable information about a train's location and the condition of the specific equipment detected at the location. This information is forwarded to track-side IT systems, where it is used to automatically update passenger information, train control systems and input for maintenance planning [12]. For technical conditions monitoring of wagons arriving in Main hub in Antwerp, it is necessary to design a communication network between the locomotive, wagons and dispatcher (Fig. 6). It is important to implement database process management.

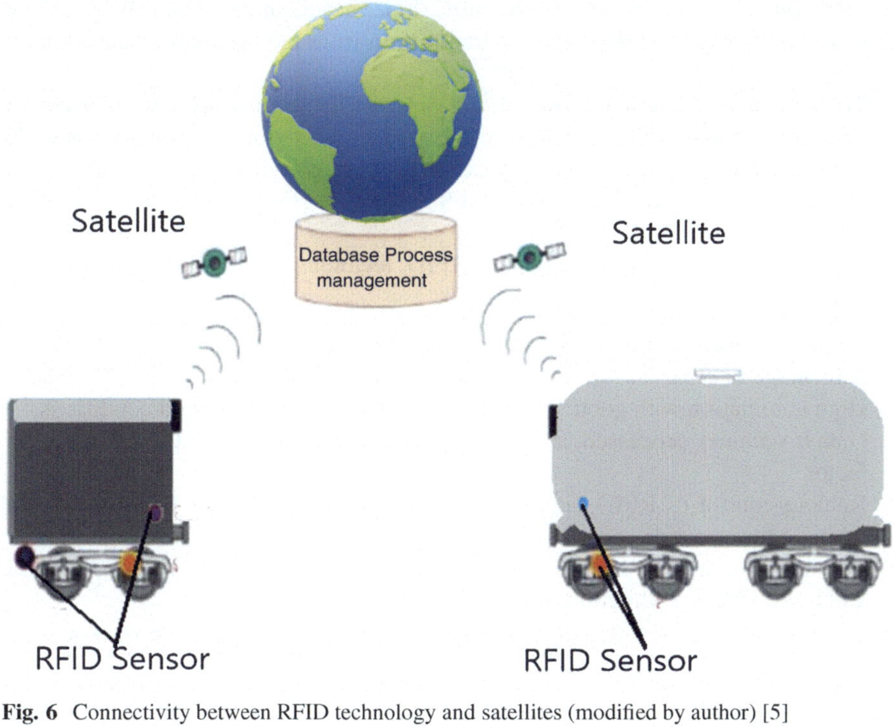

Fig. 6 Connectivity between RFID technology and satellites (modified by author) [5]

The designed RFID reader is resistant to weather and climatic conditions. Changes in climatic conditions (wind, humidity, sunlight, snow) can damage the reading device, so all aspects that may impair the functionality of the system should be considered when installing [13]. The design of the stationary reader is mounted on a solid concrete base to be extremely resistant to climatic conditions and also to theft. Interconnections (communication) between the RFID tag, the reader and the computer unit that contains the appropriate programme for processing and evaluating the received information must communicate seamlessly with each other. The transmission can be carried out using an optical cable line, which is already used by the railways, or by means of wireless transmission, which is also commonly used [14, 15]. The fibre optic cables are grounded and have a higher digital signal transmission capability, which is several times faster and better than the analogy signal. Vibration or weather conditions affect wireless transmission, but eliminating these adverse effects can be achieved by using an antenna and an amplifier that is capable of improving signal transmission several times [16].

4 Conclusion

By applying the RFID system in the smart logistics cluster in Main hub, rail transport will gain additional benefits such as tracking shipments, evidence of the train set and individual wagons, RFID application with sensors – information, for example, about temperature, pressure, slope, acceleration of moisture and by checking the arrival and departure of the train to/from the destination station with shipments.

Tracking and collecting information using RFID technology can be used in rail freight transport for a complete train, wagon consignment, transport vehicle (container) and also for tracking a specific product. When tracking shipments in rail freight, it is also possible to use a GPS satellite tracking system that accurately locates the installed RFID chip.

The main benefits of introducing RFID technology in rail freight are the following:

- Reduced workload, improved accuracy and timeliness, and incomparably higher percentage of error.
- High automation with good set-up, almost perfect reading in good conditions.
- Data is scanned, processed, archived, updated without delay, and without human factor.
- Remote control of RFID tags where a large number of items can be scanned within a few seconds.
- Possibility of writing a large amount of information on a tag of a given object and transcribing it while moving.
- The real-time stock situation is accurate, allowing all stocks to be tracked with minimal cost.

The main bottlenecks in introducing RFID technology into the enterprise are the following:

- High input costs and carrier costs, and savings will be realized in the overall usage.
- It is a relatively new technology, and innovations are emerging.
- Need for education training for staff.
- Unable to read with eyes without the scanner.
- Great time to prepare for production technology, training for software and hardware.

References

1. Balog, M., et al. (2015). Automation monitoring of railway transit by using RFID technology. *Acta Technológia, 1*(1), 9–12.
2. Balog, M., & Straka, M. (2004). Logistický informačný systém. *Transport and Logistics*. no. 6, pp. 5–10. – ISSN 1451-107X. Original in Slovak.
3. Balog, M., Minďaš, M., Szilágyi, E. (2015). Využitie RFID technológie na sledovanie technického stavu vybraných prvkov nákladných železničných vozňov. *ATP Journal*. Č. 4, pp. 12–14. – ISSN 1335–2237. Original in Slovak.
4. Balog, M., Szilágyi, E., & Minďaš, M. (2015). Utilization of RFID technology for monitoring technical state of chosen elements of cargo railway wagon. *International Journal of Engineering Research in Africa, 18*, 167–174. – ISSN 1663-3571.
5. Railway goes Smart with predictive maintenance and Industry 4.0. Cited [18 April 2019]. Available on https://www.maintworld.com/Applications/Railway-Goes-Smart-with-Predictive-Maintenance-and-Industry-4.0-CMMS.
6. Balog, M., & Minďaš, M. (2016). Informatization of rail freight wagon by implementation of the RFID technology, In: *International summit smart city 360°* Bratislava: Springer Verlag, Switzerland. pp. 592–597. ISBN 978-3-319-33680-0.
7. Less Than Wagon Load, official project web page. Cited [10 March 2019]. Available on http://lessthanwagonload.eu/addedvalue.html.
8. Balog, M., Šimeková, Ž., & Semančo, P. (2015). Smart vehicle railroad. *Applied Mechanics and Materials: Logistics Development: CLC 2014, 708*, 148–152.
9. Jones, E. C., & Chung, C. A. (2008). *RFID in logistics*. Boca Raton: CRC Press Taylor & Francis Group.
10. Behunova, A., Soltysova, Z., & Behun, M. (2018). Complexity management and its impact on economy. *TEM Journal, 7*(2), 324.
11. Miles, S. B., et al. (2008). *RFID Technology and Applications*. Published in the United States of America by Cambridge University Press, New York.
12. Sensor-enabled RFID tag handbook. Cited [10 April 2019]. Available on: http://www.bridgeproject.eu/data/File/BRIDGE_WP01_RFID_tag_handbook.pdf.
13. Banks, J. (2007). *RFID applied* (p. 528). Hoboken: John Wiley and Sons.
14. Behúnová, A., Mandičák, T., Krajníková, K., Mesároš, P., & Behún, M. (2018). Impact of enterprise information systems on selected business performance indicators in construction industry. In: *MMS 2018: 3rd EAI International Conference on Management of Manufacturing Systems*. – Gent (Belgicko): European Alliance for Innovation (p. 1–10).
15. Hunt, D., Puglia, A., & Puglia, M. (2007). *RFID a guide to radio frequency identification*. Hoboken: John Wiley.
16. Hanny, D., et al. (2007). *RFID applied*. Hoboken: John Wiley.

Simulation and Modelling of Transport Processes for the Needs of Mineral Resources Delivery Support

Martin Straka, Janka Saderova, and Peter Bindzar

1 Introduction

Dynamic changes in business environment require a well-organized supply chain, and this requires proper organization of logistics processes within the enterprise [1].

The transport of minerals is carried out by various types of vehicles. Most often, for on- and off-site transport of bulk raw materials, road trucks, wagons, ships, long-distance belt transport and others are used [2]. Choosing appropriate means of transport depends on several factors. Besides the raw materials property, the factors include route length and the amount of transported material [3, 4]. For the transport of smaller amount of raw materials within company's premises (quarries) and short distance transport (e.g. construction, mining, agriculture, food industry), trucks and various types of conveyors (belt, bulk-conveyor) are mostly used. For the horizontal transport in the mining conditions, mining rail transport and belt conveyors are used. Vertical mining transport uses transport cage towing equipment and conveyors [5]. Wagons are used for transporting huge amount of materials over long distances, and ships transport bulk material. Before the actual transportation takes place, it is necessary to load and unload the suitable means of transport at the intended destination. The method of loading and unloading is again dependent on the type of minerals, means of transport and unloading facilities available [6].

Several authors deal with the loading and unloading process in the literature, and it is not only for the bulk materials. That issue is most elaborated for port terminals; articles are dedicated to modelling and simulation of bulk material or containers

M. Straka (✉) · J. Saderova · P. Bindzar
Institute of Logistics and Transport, BERG Faculty, Technical University of Kosice, Kosice, Slovakia
e-mail: martin.straka@tuke.sk; janka.saderova@tuke.sk; peter.bindzar@tuke.sk

L. Knapcikova et al. (eds.), *4th EAI International Conference on Management of Manufacturing Systems*, EAI/Springer Innovations in Communication and Computing, https://doi.org/10.1007/978-3-030-34272-2_9

[7–13]. Wagon unloading is wide because this type of transportation uses various types of wagons for which different unloading equipment.

The modelling process can serve as a basis for selecting unloading process or rationalization of the existing system, which has a significant impact on the evaluation of the performance of the transport system or the entire enterprise [14–20].

Article is focused on the transport processes as a means of the mineral resources' delivery support. The problem is related to the amount of downtime with unloading of concrete mineral resources. The system of implementation of transport processes is limited by a lack of suitable means of transport, creates a lot of downtime and restriction of planning for the needs of the efficient use of means of transport and of logistics ensuring. The aim is to streamline and configure the transport system to a level that would be effective as economically, technically and logistics.

For the transport of mineral resources by rail, the types of wagons marked by capital letters are as follows:

E – Open Top (High wall) wagon of ordinary construction with a flat floor and the possibility of frontal or side tipping (wagons are designed to carry the bulk and general cargo goods that do not require the carriage of covered space and protection from the weather).

F – Open wagon of special construction (wagons are designed to carry bulk of powdered bulk goods like coal, limestone, gravel, etc.). Wagon construction enables double-sided gravity unloading of goods.

G – Covered wagon of normal construction (wagons are designed to carry the palletized goods, general cargo, bulk grain or other bulk substrate like industrial salt that must be protected from the weather. The wagons allow the transport of live animals).

T – Wagon with openable roof (Tds) – wagon of special construction with a convertible roof (wagon is designed to transport bulk goods requiring weather protection). The wagon construction enables double-sided gravity unloading of goods.

Unloading of minerals from railway wagons under real conditions can be carried out in various ways depending on the type of rail wagon, type of minerals, loading mechanisms to be available and forms of transloading.

2 Research Methodology

Modelling and model creation is one of the basic cybernetic approaches to study, analysis, design and design of systems. The modelling is the process of replacing a dynamic system by its model. The model represents a simplified object or process and is created on an artificial or real object [21, 22]. Several authors deal with the modelling and model creation in various sectors across process technology, transport, handling, services and so on [23–27].

Fig. 1 A formalized scheme of unloading process

The unloading process consists of several activities. Figure 1 shows the simple formalized scheme of the process of unloading (see Fig. 1).

The type of minerals in the railway wagon and their physical and chemical properties should be considered as an input data for individual sub-processes.

The first activity in the process of unloading is the 'in-feed motion' – supplying the wagons at the place of unloading (unloading ramp). Before this activity, there should be done operations such as announcement about the arrival of wagons to be unloaded. When doing transhipment between railway wagons (form 3 – direct transhipment), it is also necessary to make an order for rail wagons that will be loaded.

After wagons are ported, wagon unloading takes place. During the unloading process, wagons are either still or move in certain times or continuously depending on the type of unloading equipment device.

When unloading with a gripper crane, a wagon is still, and the raw material is unloaded directly to another means of transport (e.g. right on the truck, through the hopper onto a conveyor belt, etc.). When unloading with a wagon tippler, wagons are shifted to the rotary tilter where they are disconnected from the set, tilted, put out of the tilter and subsequently joined to the wagon set again. Material falls from the rotary tilter into the containers that are then moved by conveyors. When using bucket unloaders, the wagons are still or continuously move depending on unloaders' construction.

Unloading of minerals is followed by its removal (displacement) to the destination such as operational stock, to customer, stock for input materials, to the manufacturing process, etc. Figure 2 shows an example of simplified flowchart of unloading of minerals from the railway wagons.

3 Computer Simulation of Transport System of Mineral Resources Support

Before the model itself was created, it had to be determined based on Table 1, a number of trucks that are able to manage the performance of unloader NT. Based on capacitive conversion of truck transportation, the required number of trucks on distance of 450 m is NT = 2.1 [5]. Figure 3 shows the model of unloading where transport of unloaded mineral resources will be done by two trucks. The model was made based on the assumption that the loading process starts at the moment the unloader has arrived to the first wagon as indicated earlier. It can be seen on graphic model a process of wagon unloading RW1 to RW10. The unloading of each railway

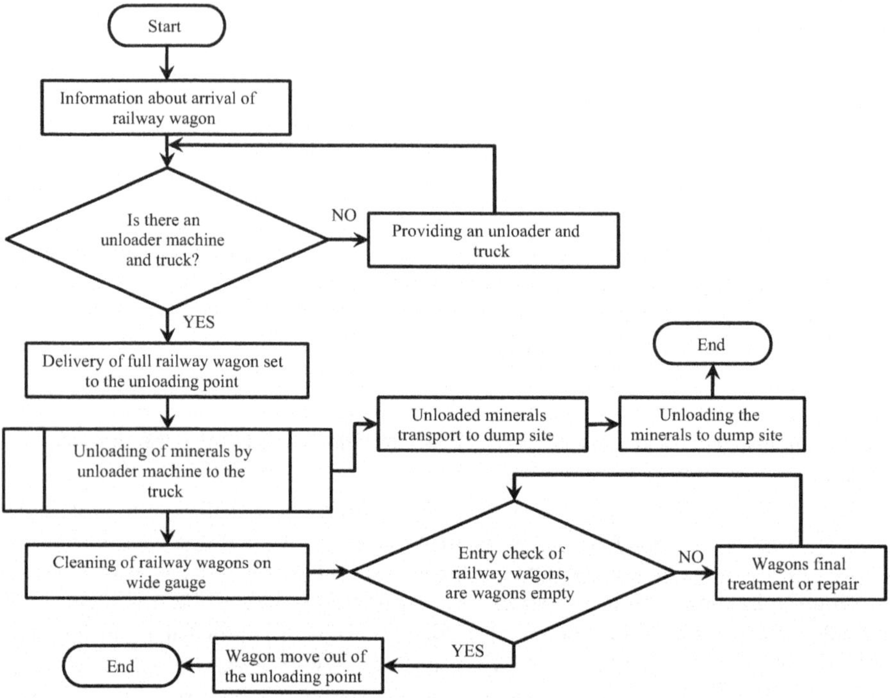

Fig. 2 A formalized scheme of unloading process

Table 1 Input data for modelling of unloading process

Parameter	Value
Truck capacity	K = 7.77 t (capacity of 9 claws)
Time for loading a one truck	$t_n = 243$ s
Route length for transport of mineral resource	L = 450 m
Truck cycle time (transport, unloading, return journey to loading point)	$t_o = 270$ s

wagon consists of two lines that are represented by trucks T1 and T2 providing removal of the mineral resource to the destination. Black stripes represent unloading activities that alternate depending on the truck to be loaded. Red stripes represent down-time 't_i' of an unloader that is caused by waiting for the trucks in operation (T1 or T2). Green stripes represent moving of an unloader to the next wagon. Grey stripes represent truck circulation (driving time out and back, time of unloading and its bringing again to rail wagon). Blue colour stripes are a truck waiting for loading.

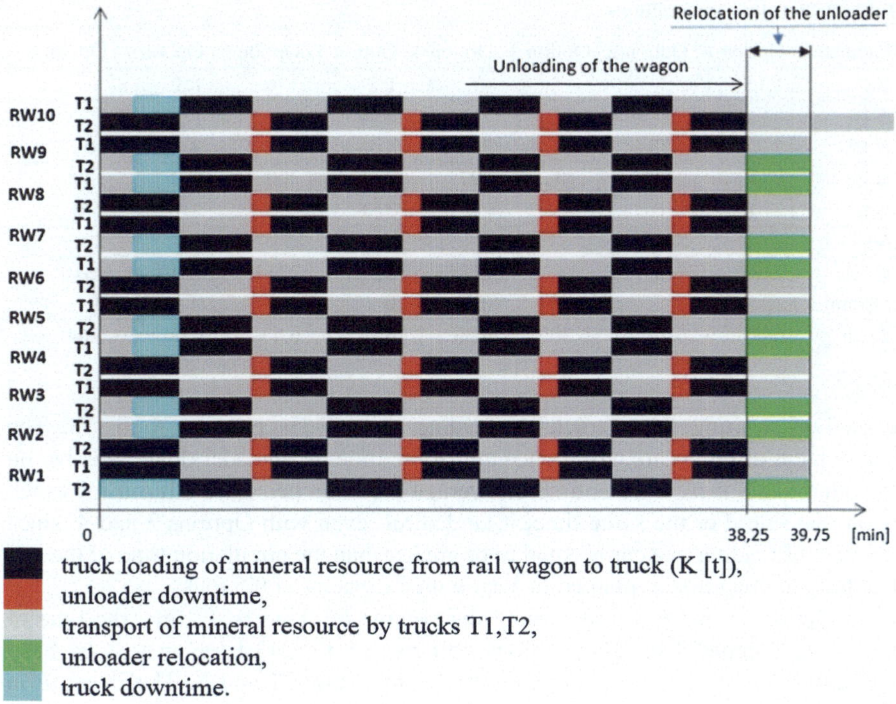

Fig. 3 The graphic model

4 Results and Discussion

The graphic model in Fig. 3 results in the following: 9 trucks are needed to unload 1 rail wagon, every wagon can be unloaded within 2295 s = 38.25 min; this time includes 4 unloader downtimes (totally 108 s) caused by waiting of trucks T1 (resp. T2); 90 trucks are needed to unload mineral resource from 10 rail wagons; 6.075 h netto is needed to unload 10 rail wagons; totally 40 downtimes is formed while unloading 10 wagons; it is 0.3 h (18 min), the unloader is moved – times during the unloading; it is 0.25 h (15 min), time of unloading T_T is 6.60 h.

Time for unloading a train set can be reduced by deploying multiple loaders and the corresponding number of means of transport. Losses caused by waiting for a truck can be eliminated in several ways:

1. By deploying a multiple truck (Option 2)
2. By increasing a truck capacity (Options 3 and 4)
3. By increasing a loader capacity (Options 5–8)

Similar models have also been developed for these options and the results are found in Table 2. In option 2 (deploying a 3 trucks), each wagon is unloaded in t_u = 2187 s = 36.15 min. It is similar to Option 1, but there are not downtimes

Table 2 Results of modelling

Parameter	Option 1	Option 2	Option 3	Option 4	Option 5	Option 6	Option 7	Option 8
V_D	1.1				1.2		1.3	
N [t]	0.86				0.94		1	
K [t]	7.77	7.77	8.64	9.50	8.46	9.40	9	10
t_n [s]	243	243	270	297	270	300	297	330
t_o [t]	270	270	270	270	270	270	270	270
N_T [−]	2	**3**	2	**2**	2	2	2	2
t_u [s]	2187	2187	2187	2187	2234	2234	2310	2310
t_i [min]	18	0	0	0	0	0	0	0
T_T [h]	6.60	**6.30**	6,30	**6.30**	6.43	6.43	6.64	6.64

caused by waiting for a truck. Unloading time T_T is reduced to 6.3 h. The deployment of three cars will reduce unloading time by unloader downtime, but on the other hand, there is downtime on average 3.60 min per vehicle turnover, which is on one side. For the same time, it landed raft even with Options 3 and 4, since the time of loading one car is equal to or greater than the circulation time of the car, resulting in downtime remained in a car with a capacity of 9.5 tons.

In Options 5 and 6, by increasing the volume of claws to 1.2 m^3, the time of unloading compared to other options will rise up to 6.43 h or up to 6.64 h by increasing the volume of claws to 1.30 m^3 in options 7 and 8. The increase in this case is due to the time increase in the unloading cycle and not by unloader's downtime. Deploying grabs with a larger volume negatively impacts the time utilization of trucks. However, such downtime provides enough time in case the route needed for unloading the material takes longer than expected. Table 2 shows recommended colour-coded options which should be applied when unloading at given input data.

A graphical representation of simulation is shown in Fig. 4. It is an unloading of 10 wagons by one unloader using 3 trucks (Lorries). The blue line represents the unloading of a wagon to 9 trucks while the red line represents the completion of unloading the wagon from train set.

The simulation results show that the lowest unloading time is reached when unloading by one unloader working with three or four lorries. Extra downtimes are present when using four lorries.

5 Conclusion

This article was aimed to use a model approach – create an algorithm and creation of graphic and simulation model for the needs of modelling the unload process from railway wagons using the grab unloader followed by transfer of minerals by trucks. In order to create an algorithm, there was a short overview of the process of

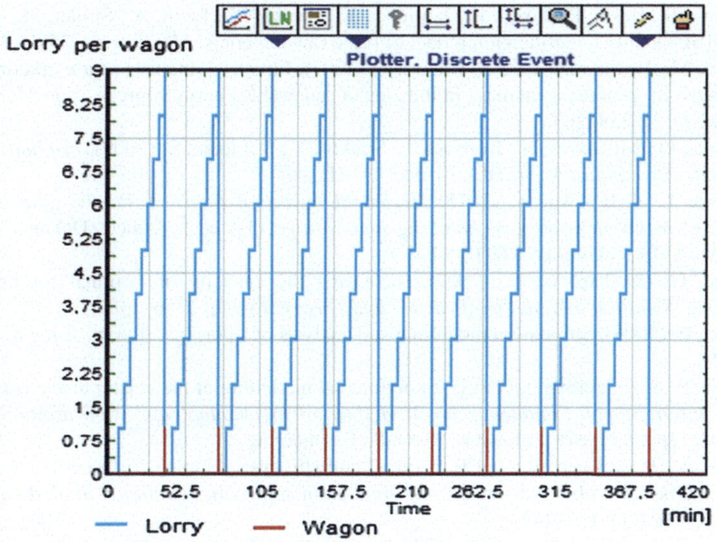

Fig. 4 Graph of the simulation of the transport processes

unloading described in the article's introduction: forms, methods, types of unloading facilities and types of wagons used to transport raw materials.

This algorithm forms a basis for creation the specific model in graphic form as well as a simulation model. The algorithm can be modified easily for any way and form of unloading. This is only one of many procedures which can be used to address the problem. The others methods such as simulation, software applications can be used to solve the problem and the decision-making process with use of multi-criteria analysis. That job does not end with the creation of the algorithm and forming specific actions that are necessary. Next, it is necessary to monitor and evaluate the process using the tools of controlling as well as to ensure the efficiency of the operation.

Acknowledgement The submitted work is a part of the project VEGA 1/0317/19, '*Research and development of new smart solutions based on principles of the Industry 4.0, logistics, 3D modelling and simulation for production streamline in the mining and building industry*', funded by the Scientific Grant Agency of the Ministry of Education, science, research and sport of the Slovak Republic and the Slovak Academy of Sciences.

References

1. Malindzak, D., & Takala, J. (2005). *Design of logistic systems* (1st ed.). Košice: Express Publications.
2. Marasova, D., Petruf, M., Taraba, V., Saderova, J., Fedorko, G., Bindzar, P., & Husakova, N. (2007). *Logistics of transport* (1st ed.). Kosice: TUKE.

3. Vilamova, S., Besta, P., Kozel, R., Janovska, K., Piecha, M., Levit, A., Straka, M., & Sanda, M. (2016). Quality quantification model of basic raw materials. *Metalurgija, 55*(3), 375–378.
4. Tausova, M., Horodnikova, J., & Khouri, S. (2007). Financial analysis as a marketing tool in the process of awareness increase in the area of renewable energy sources. *Acta Montanistica Slovaca, 12*(2), 258–263.
5. Marasova, D., Mikusova, N., Boroska, J., Strakos, V., & Lukac, S. (2009). *In-plant transport in the extractive industry* (1st ed.). Kosice: TUKE.
6. Saderova, J., & Rosova, A. (2015). *Technical means of logistics. Passive and active the logistic elements in transport, handling and storage* (1st ed.). Kosice: Dekanat – Edicne stredisko/AMS, Fakulta BERG TUKE.
7. Bugaric, U., & Petrovic, D. (2007). Increasing the capacity of terminal for bulk cargo unloading. *Simulation Modelling Practice and Theory, 15*(10), 1366–1381.
8. Demirci, E. (2003). Simulation modelling and analysis of a port investment. *Simulation, 79*(2), 94–105.
9. Trebuna, P., & Halcinova, J. (2012). Experimental modelling of the cluster analysis processes. In *5th International Conference on Modelling of Mechanical and Mechatronics Systems (MMaMS)* (pp. 673–678). Slovakia: Procedia Engineering.
10. Cartenì, A., & de Luca, S. (2012). Tactical and strategic planning for a container terminal: Modelling issues within a discrete event simulation approach. *Simulation Modelling Practice and Theory, 21*(1), 123–145.
11. Kia, M., Shayan, E., & Ghotb, F. (2002). Investigation of port capacity under a new approach by computer simulation. *Computers & Industrial Engineering, 42*(2–4), 553–540.
12. Pavolova, H., Khouri, S., Cehlar, M., Domaracka, L., & Puzder, M. (2016). Modelling of copper and zinc adsorption onto zeolite. *Metalurgija, 55*(4), 712–714.
13. Trebuna, P., Kliment, M., Edl, M., & Petrik, M. (2014). Creation of simulation model of expansion of production in manufacturing companies. *Procedia Engineering, Modelling of Mechanical and Mechatronic Systems, 96*, 477–482.
14. Rosova, A., & Balog, M. (2012). Traditional and modern methods and approaches to the evaluation of company performance. In *Carpathian Logistics Congress 2012* (pp. 1–7). Jesenik: Tanger.
15. Haverland, J., & Besta, P. (2018). Determination of importance of ore raw materials evaluation criteria. *Acta Logistica, 5*(2), 39–43.
16. Ho, W., Xu, X. W., & Dey, P. K. (2010). Multi-criteria decision making approaches for supplier evaluation and selection: A literature review. *European Journal of Operational Research, 202*(1), 16–24.
17. Pan, R., Zhang, W., Yang, S., & Xiao, Y. (2014). A state entropy model integrated with BSC and ANP for supplier evaluation and selection. *International Journal of Simulation Modelling, 13*(3), 348–363.
18. Straka, M. (2005). *Simulation of discrete systems and simulation languages* (1st ed.). Kosice: FBERG TUKE.
19. Stas, D., Lenort, R., Wicher, P., & Holman, D. (2015). Green transport balanced scorecard model with analytic network process support. *Sustainability, 7*(11), 15243–15261.
20. Knapcikova, L., Duplakova, D., Radchenko, S., & Hatala, M. (2017). Rheological behavior modelling of composite materials used in engineering industry. *TEM Journal-Technology Education Management Informatics, 6*(2), 242–245.
21. Malindzak, D., Strakos, V., Kostial, I., Lenort, R., Straka, M., Macurova, P., Rosova, A., & Cibulka, V. (2009). *Modelling and simulation in logistics, theory of modelling and simulation* (1st ed.). Kosice: Karnat.
22. Pekarcíkova, M., Trebuna, P., & Markovic, J. (2015). Simulation as part of industrial practice. *Acta Logistica, 2*(2), 5–8.
23. Straka, M., Malindzakova, M., Rosova, A., & Trebuna, P. (2016). The simulation model of the material flow of municipal waste recovery. *Przemysl Chemiczny, 95*(4), 773–777.
24. Saderova, J., & Bindzar, P. (2014). Using a model to approach the process of loading and unloading of mining output at a quarry. *Gospodarka Surowcami Mineralnymi, 30*(4), 97–112.

25. Loucanova, E., & Olsiakova, M. (2019). Distribution flow identification in cooperation and support for ecological innovation introduction in Slovak enterprises. *Acta Logistica, 6*(1), 5–8.
26. Malindzak, D., Zimon, D., Bednarova, L., & Pitonak, M. (2017). Homogeneous production processes and approaches to their management. *Acta Montanistica Slovaca, 22*(2), 153–160.
27. Saniuk, S., Samolejova, A., Saniuk, A., & Lenort, R. (2015). Benefits and barriers of participation in production networks in a metallurgical cluster – research results. *Metalurgija, 54*(3), 567–570.

Testing the Replenishment Model Strategy Using Software Tecnomatix Plant Simulation

Peter Trebuna, Miriam Pekarcikova, and Marek Kliment

1 Introduction

The process of inventory replenishment in dynamic models is usually deterministic (inputs are always delivered in ordering quantity and time), but it can also be random in time as well as in quantity. In these models, it is not a determination of time and the size of one order and it also cannot be planned a series of orders (an order cycle) [1, 2].

The aim is to find a way of managing the stock size by orders. The random dynamic models include P-system and Q-system, whose specification is given below. On the other hand, static models are characteristics with the determination of one unique delivery that do not repeat. It is also possible to include here models determining the amount of inventory of spare parts, inventory to cover the one-time sale [3, 4].

2 Demand-Driven Inventory Replenishment Models

The models showing the dynamic inventory management problems with determined movement probabilistically incorporate all types of costs: costs associated with the acquisition of supplies, maintaining inventory costs, costs of lack of safety stock and cost of excess inventory. Some of them have new features, for example, the cost of excess inventory does not have such a costly impact as in the case of static

P. Trebuna · M. Pekarcikova (✉) · M. Kliment
Technical University in Kosice, Faculty of Mechanical Engineering, Institute of Management, Industrial and Digital Engineering, Košice, Slovakia
e-mail: peter.trebuna@tuke.sk; miriam.pekarcikova@tuke.sk; marek.kliment@tuke.sk

L. Knapcikova et al. (eds.), *4th EAI International Conference on Management of Manufacturing Systems*, EAI/Springer Innovations in Communication and Computing, https://doi.org/10.1007/978-3-030-34272-2_10

models. This is because of the long unlimited time period of stocks maintenance and so inventory excess generated in one period can be used in the next period by a reduction in orders. This means that the cost of inventory excess is shown in the form of the high cost of maintaining inventory [5–7]. The probabilistic nature of demand and consumption of inventory accept fluctuation in the consumption of supplies. In the event that causes a lack of prompt supply and significant losses in the supply system, the company secures itself by increasing inventory corresponding to the expected consumption of so-called safety stock to ensure a supply system to random fluctuation in consumption. This safety stock reduces the shortage risk of prompt supplies, but it binds capital and also requires the storage, keeping, and thus causes additional costs of maintaining inventory. The maintaining costs of safety stock stand against the costs from the shortage of prompt supplies in the dynamic probabilistic models [8–10]. Therefore, a new agent is created that acts in these models which is the stock replenishment interval. Managing of inventory state in the dynamic models can be influenced by two types of interventions [11–13]:

- Orders' frequency – Q-system
- Orders' sizes – P-system

The number of orders 'N' and their size 'x' are bound together by precise quantitative relation: x = Q/N. This relation is valid next only when it comes to middle values of these variables. Particular sub-periods lead to random deviations from the true consumption of its middle value and thus inventory fluctuations around its expected size. These effects need to balance which is possible, for example, by changing the frequency of orders at a constant order size – it is 'Q' system, or by changing the amount of orders at a fixed interval between particular orders – then it is 'P' system. These are two basic approaches in inventory management used for the absorption of probabilistic fluctuations in consumption [14–16].

2.1 Q-System of Inventory Replenishment

This system operates with a fixed order size, which is indicated by the symbol 'Q' (from the word 'quantity'). Fluctuation in the stock consumption is balanced by changes in the frequency of orders. Q-system works on the basis of determining the amount of inventory that is used to cover consumption during the particular period so-called signal level of stocks. If the actual stock reaches that level, then the new order is executed. Random fluctuations in consumption and cost of the lack of emergency stock are taken into account when determining the signal level. Q-system is also known as '*a system of two stores*'. The name comes from the use of the mentioned system in practice, where each item is stored in two separate stores. The major part of inventory is kept in the first one, and in the second store inventory is maintained, and it also serves to cover consumption for the time period of replenishing. The new order is realized when the stock in the first store is used. By the time a new delivery arrives to the warehouse, the consumption is covered from the second warehouse. After the arrival of new delivery, the second warehouse

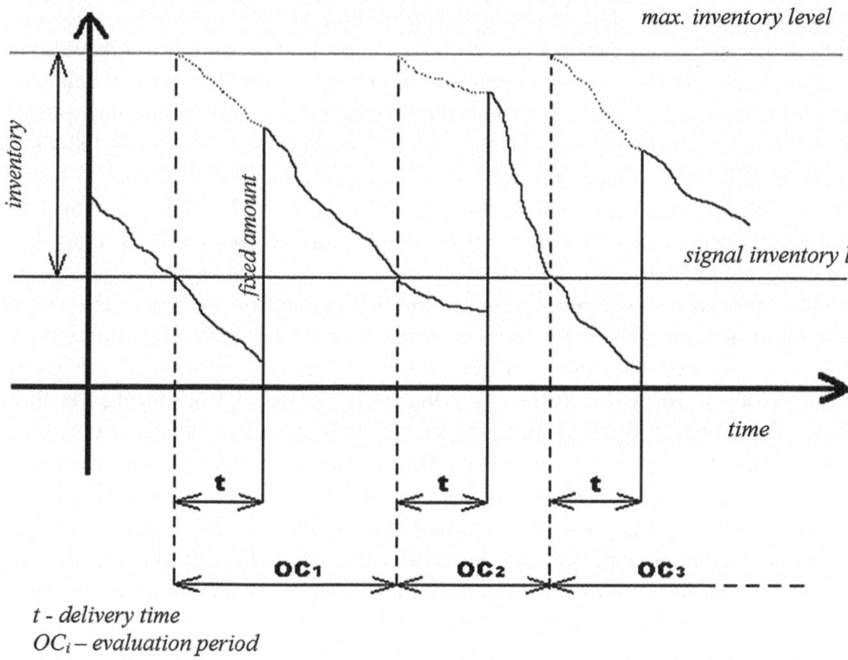

Fig. 1 Graphic's schema of the Q-system of inventory replenishment [20]

is first supplied for the signal level and the rest of supply is placed in the first store. It is essential to monitor whether the remaining stock does not drop under the signal level which is, in case of several stock items, quite difficult [17, 18].

Application of the Q-system in practice means determining the constant size of the order and inventory signal level for each item. Since this is the system where consumption fluctuation is expressed by changes in the ordering cycle, it is not necessary to maintain safety stocks for the random increased consumption during the ordering cycle – in this case, the actual supply drops to the signal level, and, thus, there is a realization of a new order. This, however, cannot be applied during the time of order realization, so to cover possible consumption fluctuation exactly during this time, the safety stock is required (dependency exists between the size of the safety stock and the length of ordering cycle); see Fig. 1 [19, 20].

2.2 P-System of Inventory Replenishment

Stock replenishment is realized by orders in fixed delivery times, usually with constant intervals between orders. The system is referred as 'P' (from the word 'period') inventory management system. It is the second basic inventory management system with probabilistically determined movement. It is based on the fact that orders of variable size are carried in advance fixed ordering times. The size of particular orders is set so that the sum of the actual stock size at the time of the order was equal to a predetermined variable of which size takes into account the consumption fluctuation. Fluctuation in actual consumption around its middle value is balanced by the size of particular orders. It is possible by using this model to find the optimal length of ordering cycle and the optimal ordering level at which the actual stock is replenished. The size of a particular order is determined as the difference between the stock ordering level and actual stock at the time of order. This system does not require a permanent inventory control which is an advantage; it only needs a periodic control [17–19].

It is possible to find out by comparing the mechanism of the Q and P systems that the main difference between them is in the way they receive sales fluctuation. While in the Q-system shortening of the ordering cycle induces higher demand and the safety stock is formed only to cover the higher consumption during the time of stock acquisition, in the P-system, the safety stock must compensate fluctuation in sales during the entire ordering cycle. During this time, there is an interaction between the size of the order of one period and the shortage risk of prompt stock in next periods. The P-system uses a simplified assumption which is to determine the safety stock only with respect to one ordering cycle plus the next interval of stock acquisition. Q-and P-systems form the main inventory management systems where the practical use is relatively simple; see Fig. 2 [19, 20].

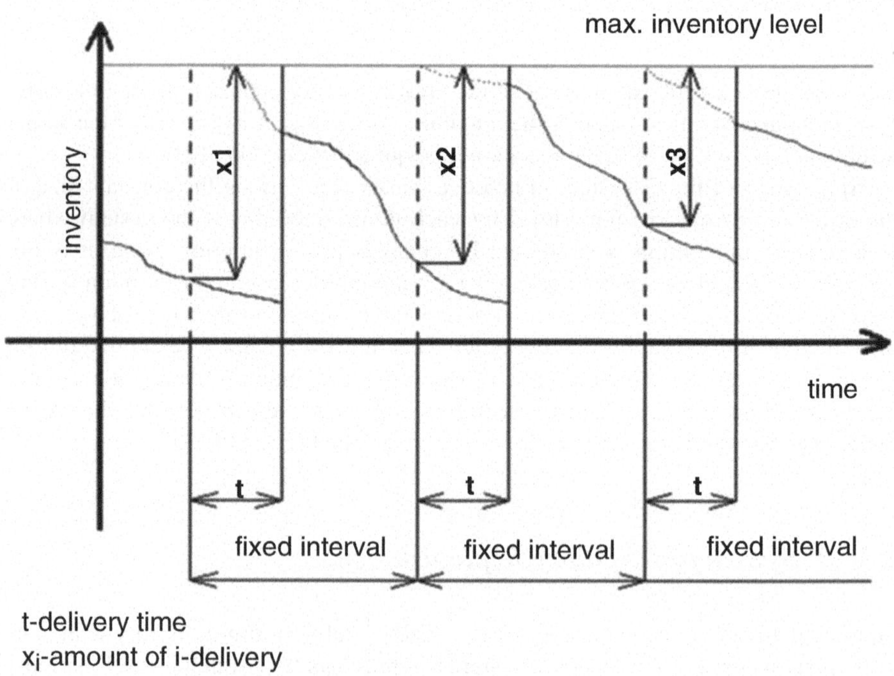

Fig. 2 Graphic's schema of the P-system of inventory replenishment [20]

3 Testing Replenishment Strategy by Using Tecnomatix Plant Simulation

Simulation model that is processed below is oriented on testing the replenishment strategy. Case study was made in the area of simulation software Tecnomatix Plant Simulation (Siemens). Specifically, it is the simulation model of procurement for demand-driven consumption, that is, customer order simulation and consumption, which consume the inventory and initiate the supply process from the supplier when reaching buffer minimum of inventory level. The way of representation of inventory consumption is viewed as a silo.

Input data for case study: maximum inventory level = 5000 pcs, delivery size = 2000 pcs, delivery cycle = 2 days.

The storage strategy is part of the process of deciding on the quantity and time of the inputs acquisition, that is, the process of initializing the purchase and supply of material, depending on the length of the delivery time. The supply method can be implemented in two variants in terms of time [20]:

- Order time will depend on the stock, i.e. ordering will be realized upon acquisition of the order point, resp. the warehouse will be filled at regular intervals (P-model – with relatively constant demand, or relatively high insurance stocks).
- As far as the quantity is concerned, the filling of the warehouse can be carried out in fixed quantities (Q-model); resp. amount will vary depending on actual consumption. The essence of the supply is to reduce the storage in the storage tank to a minimum level, which necessitates the implementation of the supply.

The way of showing the consumption of supplies is in the form of silos. It is an illustrative depiction of the process of replenishment and consumption of inventory by simultaneously displaying the maximum, average, and resp. minimum stock levels by using colours (green – maximum, yellow – average and red – minimum). The simulation ends when the silo is empty. This is due to the set replenishment method and to the amount of consumption, delivery time and order size that is constant. According to the results of the simulation, it is possible to set the optimum size of the order, which would be closer to the needs of customer requirements. Figures 3, 4 and 5 show simulation models for the above-mentioned consumption and replenishment phases. There is also an inventory sawtooth diagram showing the consumption progress in each phase of the simulation run [21–23].

However, efforts are being made to keep stocks to a minimum, that is, it is necessary to know when the level of inventory is approaching the minimum level and to enter a higher order requirement in order to cope with future consumption and avoid stock shortages. In the case study, there is a supply option whose output is as follows: Decrease in the level of inventories in a way of supply and consumption, and decrease to zero level in about 37 days. Therefore, it can be stated that in the case of demand fluctuations considered and with respect to the delivery period of 2 days, this order will have to be increased.

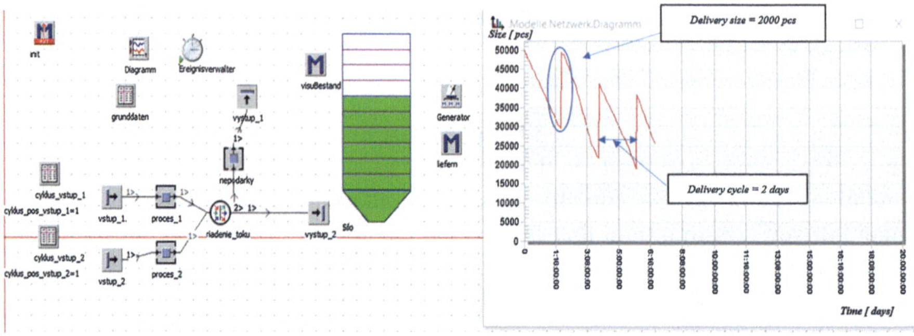

Fig. 3 Simulation model – maximum inventory level

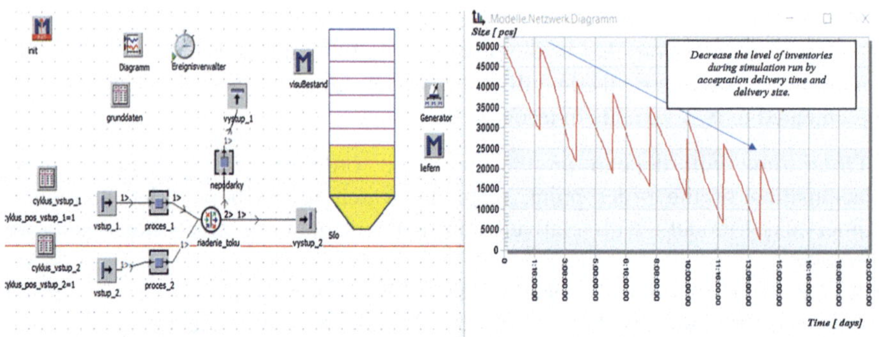

Fig. 4 Simulation model – average inventory level

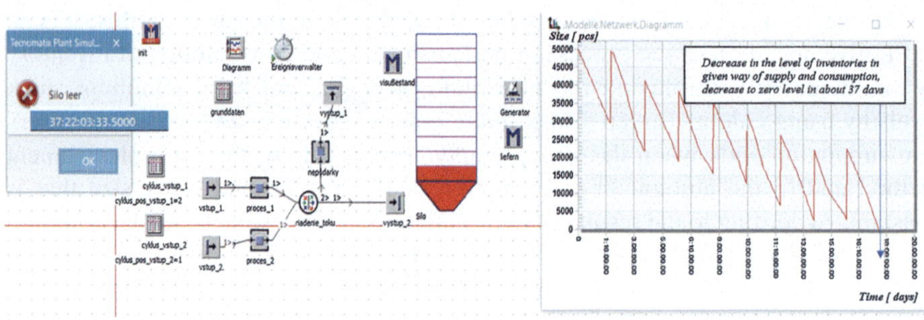

Fig. 5 Simulation model – empty buffer

4 Conclusion

Computer simulation is an attempt to model a real, resp. hypothetical situation, process or thing using a computer with the goal to study it and see how it works by different and changing conditions. The model has to represent key characteristics and own behaviour of system to find the answers on tasked questions or verify a result that is otherwise costlier or impossible to achieve. Tecnomatix Plant simulation performs a specific type of simulation, defined by NASA as 'a tool to model organizational activities and system responses to discrete events in the operational flow, supports Monte Carlo analysis and can be used for manufacturing flows, operational processing flows, supply chain flows, and flows of information through an organization' [24]. Implementation and use the simulation tools greatly strength the potential in connection with decision-making competencies of managers in logistics.

Acknowledgement This article was created by implementation of the grant project VEGA 1/0708/16 Development of a new research methods for simulation, assessment, evaluation and quantification of advanced methods of production, KEGA 030TUKE-4/2017 Implementation of innovative instruments for increasing the quality of higher education in the 5.2.52 Industrial engineering field of study and APVV-17-0258 Digital engineering elements application in innovation and optimization of production flows.

References

1. Edl, M., & Kudrna, J. (2013). *Methods of industrial engineering* (1st ed.). Plzen, Czech Republic: Smart Motion.
2. Lenort, R., Stas, D., & Samolejova, A. (2012). Heuristic algorithm for planning and scheduling of forged pieces heat treatment. *Metalurgija, 51*, 225–228.
3. Dulina, L., Rakyta, M., Sulirova, I., & Šeligova, M. (2017). Improvement of the production system. In *Smart City 360°. 2nd EAI international Summit: Revised selected papers*. Ghent: EAI.
4. Gregor, M., Hodon, R., Biňasová, V., Dulina, Ľ., & Gašo, M. (2018). Design of simulation-emulation logistics system. *MM Science Journal, 2018*, 2498–2502.
5. Seebacher, G., Winkler, H., & Oberegger, B. (2015). In-plant logistics efficiency valuation using discrete event simulation. *International Journal of Simulation Modelling, 14*, 60–70.
6. Poór, P., Šimon, M., & Karková, M. (2015). CMMS as an effective solution for company maintenance costs reduction. In *Production Management and Engineering Sciences* (pp. 241–246).
7. Straka, M., Malindzakova, M., Trebuna, P., Rosova, A., Pekarcikova, M., & Fill, M. (2017). Application of EXTENDSIM for improvement of production logistics' efficiency. *International Journal of Simulation Modelling, 16*, 422–434.
8. Laciak, M., & Sofranko, M. (2013). Designing of the technological line in the SCADA system PROMOTIC. In *Proceedings of the 2013 14th International Carpathian Control Conference, Rytro, Poland.*
9. Malindzakova, M., Rosova, A., Baranova, V., & Futo, J. (2015). Modelling of outbursts and ejections occurrences during steel production. *Metalurgija, 54*, 247–250.
10. Saniuk, S., Saniuk, A., Lenort, R., & Samolejova, A. (2014). Formation and planning of virtual production networks in metallurgical clusters. *Metalurgija, 53*, 725–727.
11. Straka, M., Bindzar, P., & Kadukova, A. (2014). Utilization of the multicriteria decision-making methods for the needs of mining industry. *Acta Montanistica Slovaca, 19*, 199–206.
12. Straka, M. (2005). *Simulation of discrete systems and simulation languages* (1st ed.). Kosice: TU Kosice.
13. Badida, M., Gombar, M., Maslejova, A., Sobotova, L., Kmec, J., & Vagaska, A. (2015). Evaluation of zinc coating quality by statistical methods. *Przemysl Chemiczny, 94*, 2146–2149.

14. Rosova, A., & Malindzakova, M. (2014). Material flow – starting point for recovery of inputs in the production company. In *International Multidisciplinary Scientific GeoConference Surveying Geology and Mining Ecology Management, 14th International Multidisciplinary Scientific Geoconference and EXPO, SGEM 2014; Albena; Bulgaria SGEM Vol. 3, no. 5.*
15. Lerher, T., Edl, M., & Rosi, B. (2014). Energy efficiency model for the mini-load automated storage and retrieval systems. *International Journal of Advanced Manufacturing Technology, 70*(1–4), 97–115.
16. Wicher, P., Staš, D., Karkula, M., Lenort, R., & Besta, P. (2015). A computer simulation-based analysis of supply chains resilience in industrial environment. *Metalurgija, 54*(4), 703–706.
17. Pekarčíková, M., Trebuňa, P., & Markovič, J. (2014). Case study of modelling the logistics chain in production, conference: 6th conference on Modelling of Mechanical and Mechatronic Systems (MMaMS) location: Vysoke Tatry, Slovakia, modelling of mechanical and mechatronic systems book series. *Procedia Engineering, 96*, 355–361.
18. Klos, S., & Ptalas-Maliszewska, J. (2013). The impact of ERP on maintenance management. *Management and Production Engineering Review, 4*, 15–25.
19. Gregor, T., Krajčovič, M., & Wiecek, D. (2017). Smart connected logistics. In *Procedia Engineering. vol. 192. Transcom 2017 12th International Scientific Conference of Young Scientists on Sustainable, Modern and Safe Transport. High Tatras, Grand Hotel Bellevue, Slovakia* (pp. 265–270).
20. Trebuňa, P., Fiľo, M., & Pekarčíková, M. (2013). *Supply and distribution logistics* (1st ed.). Ostrava: Amos.
21. Bangsow, S. (2016). *Tecnomatix plant simulation: Modeling and programming by means of examples*. Switzerland: Springer-Verlag Berlin Heidelberg.
22. Siemens. Process Simulate: Manufacturing process verification in powerful 3D environment. Retrieved 11 April, 2019, from https://www.plm.automation.siemens.com/en/products/tecnomatix/manufacturing-simulation/robotics/process-simulate.shtml#lightview%26url=/en_us/Images/7457_tcm1023-80351.pdf%26title=ProcessSimulate%26description=Process Simulate Fact Sheet%26docType=pdf. (2018).
23. Siemens. Tecnomatix. Retrieved 16 April, 2019, from http://www.sova.sk/sk/produkty/tecnomatix. (2017).
24. SE Research Consortium, Retrieved 20 April, 2019, from https://www.nasa.gov/consortium/DiscreteEventSimulation

Identification of the Relevant Parameters for Modeling the Ecosystem Elements in Industry 4.0

Dragan Perakovic, Marko Perisa, Ivan Cvitic, and Petra Zoric

1 Introduction

With the development of new innovative information and communication services (IICS), it is possible to contribute more efficiently to all processes of system operation in the fields of networking and providing all relevant information to the stakeholders of the business process. Today's concept used in manufacturing processes and the development of logistics processes based on modern information and communication (IC) technologies is called Industry 4.0. It focuses on the process of transition from traditional business to digital business transformation [1]. To establish an ecosystem and to conduct all business processes and models, it is important to choose the optimal information and communication service (ICS), where it is necessary to define and link infrastructure elements, people, and other stakeholders. Unlike previous research, this research is focused on the identification of important elements of the ecosystem with the aim of selecting and forming optimal elements. For this purpose, it is necessary to satisfy all the requirements (technical and technological, organizational and legislative) in the function of selecting the appropriate elements of the ecosystem. This research is focused on the following elements: users, device (sensors and user device), communication network and technology, environment and content. Necessity for integration and connection of the all the above-mentioned ecosystem elements into a complex with the establishment of new value chains in the delivery of services is needed.

The aim of this research is to identify the relevant parameters required for modeling ecosystem elements within the Industry 4.0 concept. The identification of relevant parameters provides a starting point in the field of modeling ecosystem

D. Perakovic (✉) · M. Perisa · I. Cvitic · P. Zoric
Faculty of Transport & Traffic Sciences, University of Zagreb, Zagreb, Croatia
e-mail: dperakovic@fpz.hr; mperisa@fpz.hr; icvitic@fpz.hr; pzoric@fpz.hr

L. Knapcikova et al. (eds.), *4th EAI International Conference on Management of Manufacturing Systems*, EAI/Springer Innovations in Communication and Computing, https://doi.org/10.1007/978-3-030-34272-2_11

elements for the purpose of creating a unique integrated system. In the process of designing a unique integrated system, it is important to create new business models with the aim of more efficient business within the concept of Industry 4.0. A scientific literature review provides a comparative analysis of the business transition from traditional business to digital business through the presentation of current business models.

2 Literature Review

The term Industry 4.0 first appeared in 2011 at the technology fair in Hanover, Germany, where it was presented as an idea of strengthening the competitiveness of the German industry. The initiative became an integral part of the High-Tech Strategy 2020 for Germany, which aimed to promote Germany as a global leader in the field of technological innovations [2]. In the field of modeling and development of innovative services, numerous researches are focused exclusively on partial results, which refer to individual elements of the ecosystem.

Therefore, ecosystem users need to be segmented appropriately (user requirements for business processes and end users) and meet certain requirements needed for delivering content [3–5]. Sensors and user devices play a key role in creating an integrated ecosystem based on the Internet of Things (IoT) concept where Device-to-Device (D2D) communication is important. Researching the possibilities of the integration and interoperability of D2D communication, authors have put the focus on network layer functionality such as addressing, routing, mobility, security and resource optimization, resulting in the identification of the elements of the network layer limitations for the purpose of D2D communication in IoT concept [6]. The sensors used in the IoT concept can be classified into three basic categories: motion sensors, ambient sensors, and position sensors [7]. Devices in Industry 4.0 represent cybernetic-physical systems, that is, integrated systems with information and communication components. These are autonomous systems that can make their own decisions based on machine learning algorithms and real-time data collection, analysis of results, and historical data on system behavior [8]. Communication networks and technologies that can connect these elements are divided into three categories. The first category includes widely used short- or medium-range radio technology. This category includes technology and networks: Automatic Identification and Data Capture (AIDC) technology, IEEE 802.15.14 (e.g., Zigbee), IEEE 802.11 (Wi-Fi), LoRa, and others [9]. Radio-frequency identification (RFID) technology, which belongs to a group AIDC technology, represents the most common technology within the concept of Industry 4.0. Numerous authors conducted various experimental measurements where the regression models of the external factors influencing the operation of RFID technology were created [10, 11]. The second category represents technology and networks based on mobile communication systems (e.g., 2G/GSM, 3G/UMTS, 4G/LTE, 5G) that can provide greater coverage, far greater bandwidth (>100 Mbps), but also consume significant energy for the device itself. The third category represents networks with wide area coverage (IoT networks) and with low power emitting, such as LPWAN (Low Power Wide Area Network), NB-IoT, Sigfox, and others. Cloud computing as data storage and data processing technology and the IoT network are the basic support networks and technologies of the Industry 4.0. Typical resources such as employees, devices, materials, and manufacturing activities in the Industry 4.0 turn into smart manufacturing facilities, so they can feel and react to each other in an intelligent environment [12]. A major impact on the application of innovative services through Industry 4.0 concept relates on business models and processes as well as on the future company's business development plans [13–15]. The field of artificial intelligence also has a significant contribution in supporting the work of a highly automated system which allows them greater autonomy in decision-making and defining digital services [16]. The research on sustainable systems as a support in designing services considers three key segments: the system's sustainability and complexity, the dynamics of the communication network development, and the integration of these segments through the ecosystem [17]. The concept of digital innovation and Big data in the development of IC services has great significance in the application of contemporary IC concepts such as IoT; hence, the conceptual methodological framework has been developed for this purpose [18].

3 Business Process and Information Framework Tool

For the purpose of modeling innovative services, it is necessary to include a certain degree of generic models to assist in the understanding of the work process.

For this purpose, it is possible to use the defined TM forum frameworks: Business Process Framework (eTOM), Information Framework (SID), and The Application Framework (TAM). Visualization of business processes and scenarios takes place using the 3D tool (Transformation Accelerator) available through TM forum web pages and corresponding licenses [19]. eTOM represents a set of categorized business process specifications of a typical Internet and Computer Service Providers (ICSP). SID defines comprehensive standardized and categorized information definitions (aligned with eTOM process categories). TAM is a reference map that contains a list of categorized New Generation Operations Systems and Software applications (aligned with eTOM process categories).

An example of using this tool is shown in Fig. 1. In the example of the Smart City ecosystem platform, all the components of the ecosystem and their connections are shown depending on the tasks and processes. With a detailed view of a particular stakeholder, it is possible to define specific tasks and connections depending on the user requirements (the service provider or the end user).

The possibility of modeling the business processes is shown in Fig. 2 (image shows a small part of whole processes); all elements that affect the business model development are: key partners, key activities, value propositions, customer relationships, customer segments, key partners, key resources, channels, cost structure, and

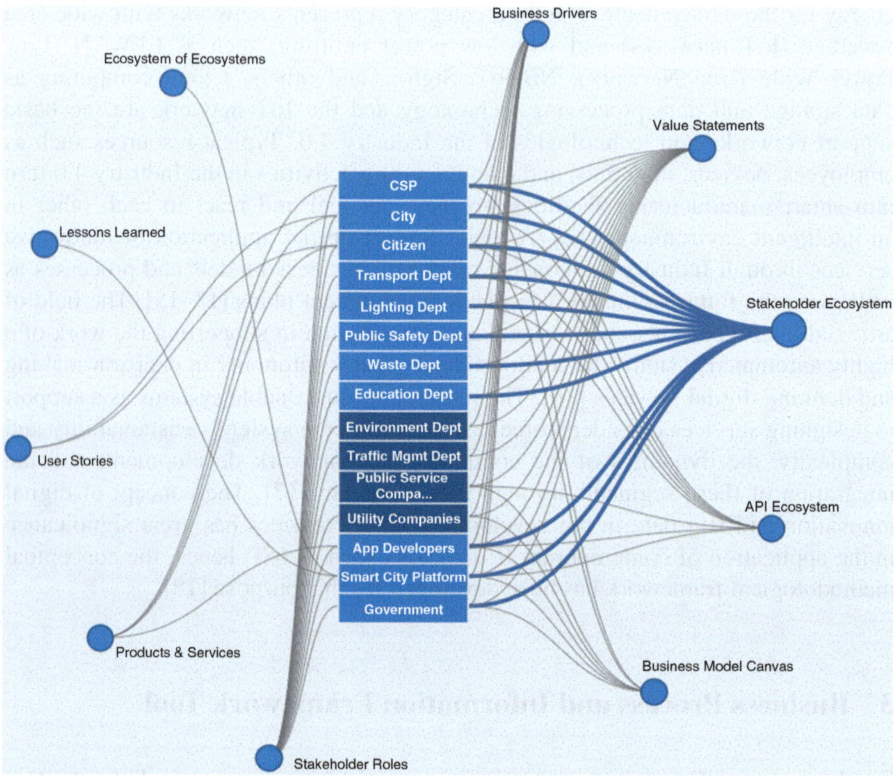

Fig. 1 The use of 3D modeling process tool according to the TM forum guidelines

revenue streams. Each of the above-mentioned elements has the ability to configure to accurately defined business models (Business 2 Business – B2B, Business 2 Consumer – B2C, or some of the B2B2X derivatives).

The entire ecosystem with predefined processes and usage scenarios is shown in Fig. 3. All defined stakeholders (Transport department, Lighting department, Public safety department, Waste department, and Education department, Communication Service Provider – CSP, City, Government, Citizen, and App Developers) and their connections to Smart City platform are visually displayed. The Smart City platform is the area for integrating all relevant data into the Smart City environment network. If there is a defined contract between certain stakeholders, it can also be displayed.

Citizens are shown as stakeholders and contain precisely defined processes for the use of Smart City services. It is also possible to display the technology of using (Cloud Computing) defined service. When defining tasks of individual services (Data and Cloud Services) and applications used by the citizens, it is possible to immediately define the necessary framework (eTOM, SID, and TAM).

This way of visualization provides a clear insight into all relevant processes and business models for a precisely defined service or information system.

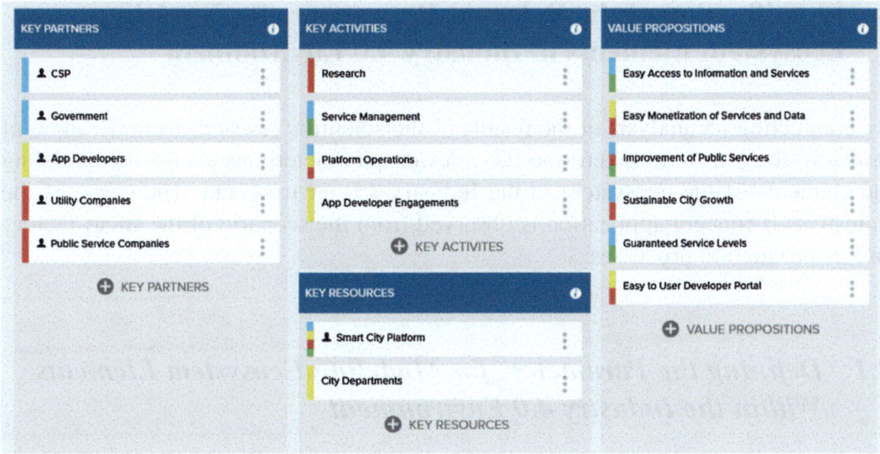

Fig. 2 Defining a business model with the Transformation Accelerator tool

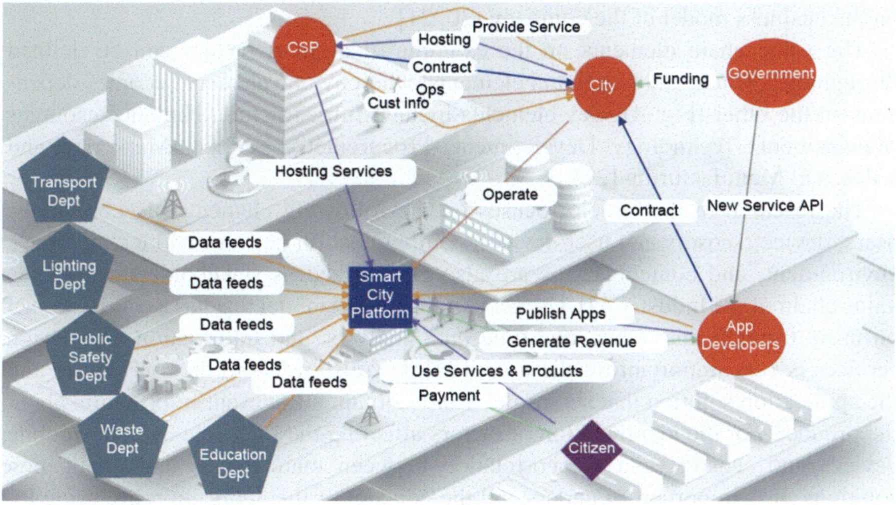

Fig. 3 An overview of the entire ecosystem on the Smart City platform example

Using the above-mentioned solutions in the development of innovative services for Industry 4.0 environment processes, it is possible to create more efficient process management system, especially in the business transition from traditional to digital business.

4 Identification of the Relevant Parameters for Modeling Ecosystem Elements in Industry 4.0 Environment

By conducting an analysis of the results of the scientific research from the topic of research, the aim is to determine the relevant parameters necessary for modeling the elements of the ecosystem in the Industry 4.0 environment. The scope of the Industry 4.0 concept application is observed from the scenario of the smart factory and its production processes.

4.1 Defining the Parameters for Modeling Ecosystem Elements Within the Industry 4.0 Environment

By defining the scenario (smart factory) of the modeling ecosystem elements, a starting point for modeling of the relevant ecosystem elements is created. For this purpose, it is necessary to define the stakeholders and their roles that have an impact on the business model of the company [20, 21].

The value chain elements on the example of a smart factory can be defined through the vertical and horizontal elements, whereby one element becomes dependent on the other (Fig. 4). Key elements include Infrastructure, Human Resources Management, Technology Development, Procurement, Service, Marketing and Sales, and Manufacturing [22].

The research in this article focuses on the following elements of ecosystems: users, device (sensors and user device), communication network and technologies, environment, and content. Users are also categorized as stakeholders within the value chain of the Industry 4.0. User experience (UX) model that has a role to impact on users through marketing, real-time data analysis, and information and design services is very important for the users [23]. UX also refers to the user device and the application solution that serves the user to obtain all relevant information about the product. For the identification of factors affecting the life cycle of smart mobile devices and sensors, and the correlation between maintenance and application solutions, it is important to understand the concept of the application life cycle [5, 24]. Sensors are used to enable data collection from the ambient or the environment, and their role in creating a smart factory is extremely important [25].

The communication network has the role of connecting all elements of the ecosystem. The scientific research in this field points to IoT networks: LoRaWan, Lora, NB-IoT, and Sigfox, as reference networks for delivering defined services in the field of Industry 4.0 [26–31]. Also, research on the impact of the 5G network and the ability of Blockchain technology to develop Industry 4.0 has been presented [32–34]. Interoperability in the concept of a smart factory, Cloud Computing, and IoT networks is important from the point of view of access to data regardless of location and coverage network [35].

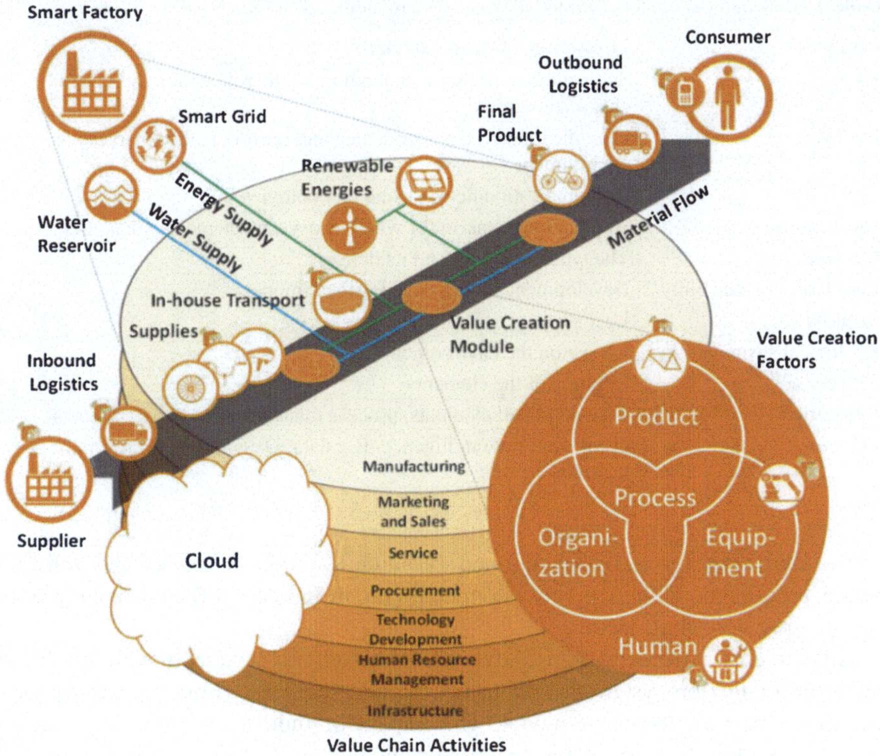

Fig. 4 Micro perspective of Industry 4.0 [22]

For the modeling of ecosystem elements, the user safety area in the smart factory environment is also important, as well as the reliability of the service. Services based on concepts of IoT and Cloud Computing improve the automation processes in the fields of logistics and Industry 4.0. Absence or unavailability of such services may result in negative user experience in using certain types of services and may also represent a potentially high security risk and endangered user safety.

The level of availability of the service in such cases is influenced by various factors identified and classified by the authors in the analyzed research through guidelines for achieving the required level of availability of information and communication services [36].

Detection of DDoS (Distributed Denial of Service) traffic is of great importance for the availability of services and other information and communication resources [37]. The research presented in the analyzed work shows the application of artificial neural networks in developing models of detection and classification of three types of DDoS attacks and legitimate network traffic [38].

Table 1 Relevant parameters for modeling ecosystem elements in Industry 4.0

Parameter	Description of the parameter
User	Segmentation of users, depending on the role within the system: customer, worker
Devices	Classification on the user device and sensors in the smart factory environment
Technology	Communication network and technologies
Environment	Application scenario for which ecosystem elements are modeling
Content	The type of content and its delivery
Economic sustainability	Development strategy and business models
Security	User security, reliability of service
Ecological sustainability	Impact on the environment
Life cycle	Lifetime of the elements
Integration	Connecting all elements, process management, security aspects
Forecast	Using artificial intelligence, Big data, Data mining, Machine learning

By analyzing the above-mentioned researches, Table 1 shows the relevant parameters for modeling the ecosystem elements in Industry 4.0 environment with the example of a smart factory.

All elements of the ecosystem and their connectivity to the integrated information and communication system are important for all processes within a smart factory, and therefore for a business model. The impact of Industry 4.0 development on business models has an increasing importance, mostly due to automation and business process optimization [39–41].

4.2 Overview of the Parameters with the 3D Modeling Process Tool

Relevant parameters are through a simulation tool used for overview of the eTOM, SID, and TAM processes. The stakeholders defined in Fig. 5 are a part of the smart factory environment: worker, suppliers, users, product, organization, equipment, development engineers, I 4.0 platform, and CSP. I 4.0 platform is the integrator of all the data needed for delivering the required services in the smart factory environment.

When defining the stakeholders, it is necessary to state all their tasks and their interrelationships. The key values in defining a business model are shown in Fig. 6. It is important to note that Fig. 6 shows only part of the value as well as the roles of some of the defined stakeholders.

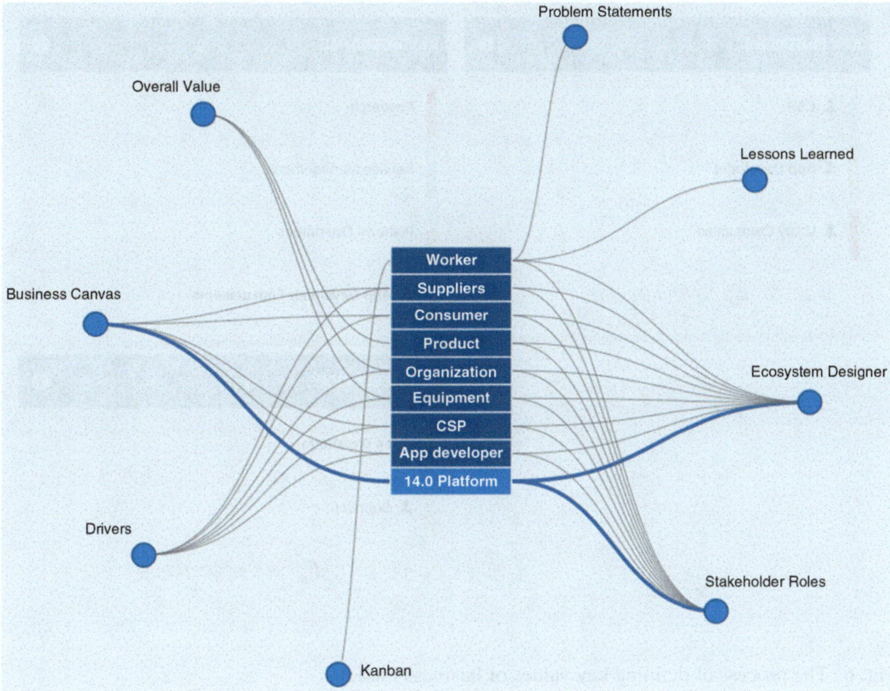

Fig. 5 An overview of the stakeholders and their mutual connections

The values to be entered are divided by key partners, activities, resources, propositions, customer relationships, advertising channels, structure of the costs, and revenues. The tool also has the ability to use the Unified Modeling Language (UML) use case diagram to show all functionality within the ecosystem and its elements. Using the visual elements of the UML diagram, an ecosystem model was created, as shown in Fig. 7. The next step allows using the reference model Digital Services Reference Architecture (DSRA), which provides creation of:

- Illustrating how some example business scenarios implement the DSRA concepts
- Mapping TM Forum APIs to Support Services
- Introducing the Digital Service component and Digital Service Management API
- Extending some of the Support Service Descriptions

An overview of all eTOM, SID, and TAM processes is generated at the end of the modeling, while the last phase defines a document describing all the tasks and roles of each process.

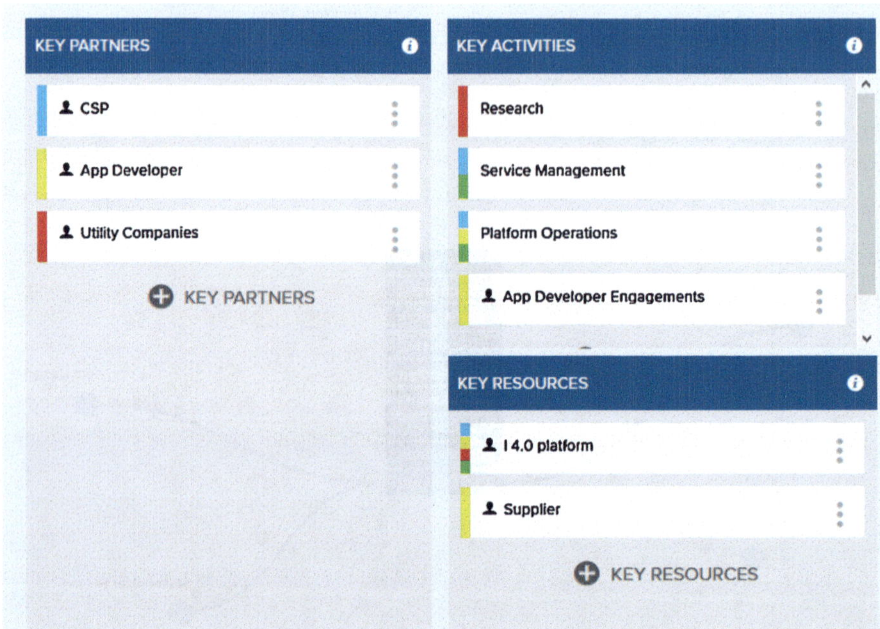

Fig. 6 The process of defining key values of business model

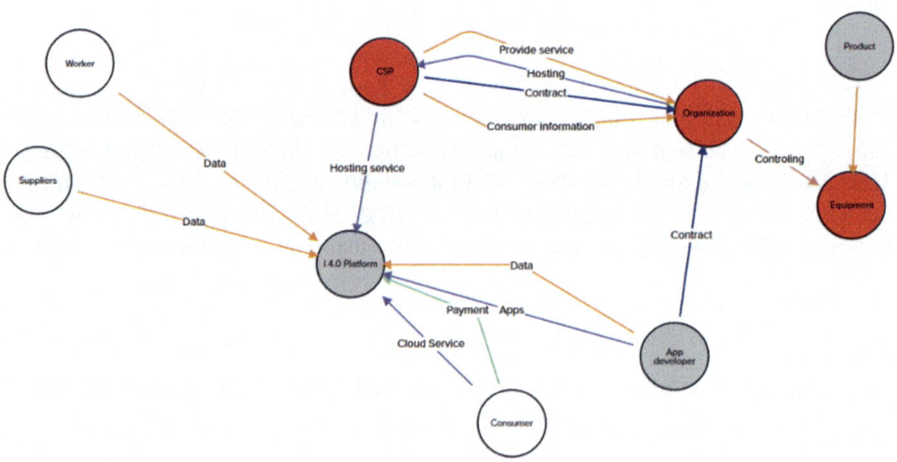

Fig. 7 An overview of the ecosystem model

5 Conclusion

This article presents the identification of relevant parameters for the modeling of ecosystem elements in Industry 4.0 environment. By this, a starting point for the modeling ecosystem elements in order to create a unified integrated system is enabled. The great opportunity to develop innovative services within Industry 4.0 has enabled the transition from traditional to digital business. The use of relevant tools for modeling the example of smart factory in operation, whose business is based on the Industry 4.0 concept, allows possibility to propose an ecosystem model that will be optimal for business processes that are performed. It is important to mention that the relevant parameters are key not only to designing innovative services but also to modeling all relevant elements of the ecosystem. Moreover, during the process of creating an effective ecosystem within the Industry 4.0 environment, it is necessary to take care of selecting the optimal communication network and technology as well as other elements. Also, business process represents an important factor in the process of implementing the concept of Industry 4.0. Included information and findings presented in this article point to the necessity of further research on these issues where the processes will be further modeled and presented in a scientific journal article.

References

1. Oztemel, E., & Gursev, S. (2018). Literature review of Industry 4.0 and related technologies. *Journal of Intelligent Manufacturing*, 1–56. https://doi.org/10.1007/s10845-018-1433-8.
2. Klitou, D., Conrads, J., & Rasmussen, M. (2017). *Germany: Industrie 4.0 fact box for Germany's Industrie 4.0 policy initiative*. Retrieved from https://ec.europa.eu/growth/tools-databases/dem/monitor/sites/default/files/DTM_Industrie%204.0.pdf
3. Takeda, A., & Hatakeyama, Y. (2016). Conversion method for user experience design information and software requirement specification. In A. Markus (Ed.), *Design, user experience, and usability: Design thinking and methods* (pp. 356–364). Cham: Springer. https://doi.org/10.1007/978-3-319-40409-7_34.
4. Guo, Y., Wu, J., Yang, K., & Yu, L. (2017). Research on requirement elicitation model of high-end equipment based on requirement classification under Internet and big data environment. In *Advances in computer science research* (Vol. 71, pp. 685–692). doi:https://doi.org/10.2991/icmmita-16.2016.127.
5. Häikiö, J., & Koivumäki, T. (2016). Exploring digital service innovation process through value creation. *Journal of Innovation Management, 4*(2), 96–124.
6. Bello, O., Zeadally, S., & Badra, M. (2017). Network layer inter-operation of Device-to-Device communication technologies in Internet of Things (IoT). *Ad Hoc Networks, 57*, 52–62. https://doi.org/10.1016/j.adhoc.2016.06.010.
7. Sikder, A. K., Petracca, G., Aksu, H., Jaeger, T., & Uluagac, A. S. (2018). *A survey on sensor-based threats to Internet-of-Things (IoT) devices and applications*. Retrieved from http://arxiv.org/abs/1802.02041
8. Rojko, A. (2017). Industry 4.0 concept: Background and overview. *International Journal of Interactive Mobile Technologies, 11*(5), 77. https://doi.org/10.3991/ijim.v11i5.7072.

9. Sethi, P., & Sarangi, S. R. (2017). Internet of things: Architectures, protocols, and applications. *Journal of Electrical and Computer Engineering, 2017*, 1–25. https://doi.org/10.1155/2017/9324035.
10. Balog, M., Szilágyi, E., Dupláková, D., & Minďaš, M. (2016). Effect verification of external factor to readability of RFID transponder using least square method. *Measurement, 94*, 233–238. https://doi.org/10.1016/j.measurement.2016.07.088.
11. Kolarovszki, P. (2014). Research of readability and identification of the items in the postal and logistics environment. *Transport and Telecommunication Journal, 15*(3), 196. https://doi.org/10.2478/ttj-2014-0017.
12. Ren, L., Zhang, L., Tao, F., Zhao, C., Chai, X., & Zhao, X. (2015). Cloud manufacturing: From concept to practice. *Enterprise Information Systems, 9*(2), 186–209. https://doi.org/10.1080/17517575.2013.839055.
13. Immonen, A., Ovaska, E., Kalaoja, J., & Pakkala, D. (2016). A service requirements engineering method for a digital service ecosystem. *Service Oriented Computing and Applications, 10*(2), 151–172. https://doi.org/10.1007/s11761-015-0175-0.
14. Abeywickrama, D. B., & Ovaska, E. (2017). A survey of autonomic computing methods in digital service ecosystems. *Service Oriented Computing and Applications, 11*(1), 1–31. https://doi.org/10.1007/s11761-016-0203-8.
15. Sklyar, A., Kowalkowski, C., Tronvoll, B., & Sörhammar, D. (2019). Organizing for digital servitization: A service ecosystem perspective. *Journal of Business Research, 104*, 450–460. https://doi.org/10.1016/j.jbusres.2019.02.012.
16. Pakkala, D., & Spohrer, J. (2019). Digital service: Technological agency in service systems. In *Proceedings of the 52nd Hawaii International Conference on System Sciences* (Vol. 6, pp. 1886–1895).
17. Barile, S., Lusch, R., Reynoso, J., Saviano, M., & Spohrer, J. (2016). Systems, networks, and ecosystems in service research. *Journal of Service Management, 27*(4), 652–674. https://doi.org/10.1108/JOSM-09-2015-0268.
18. Chae, B. K. (2019). A General framework for studying the evolution of the digital innovation ecosystem: The case of big data. *International Journal of Information Management, 45*, 83–94. https://doi.org/10.1016/j.ijinfomgt.2018.10.023.
19. Tr3Dent. (n.d.). *Digital transformation Accelerator*. Retrieved from https://www.tr3dent.com/
20. Flatscher, M., & Riel, A. (2016). Stakeholder integration for the successful product–process co-design for next-generation manufacturing technologies. *CIRP Annals – Manufacturing Technology, 65*(1), 181–184. https://doi.org/10.1016/j.cirp.2016.04.055.
21. Thoben, K.-D., Wiesner, S., & Wuest, T. (2017). "Industrie 4.0" and smart manufacturing – A review of research issues and application examples. *International Journal of Automation Technology, 11*(1), 4–16. https://doi.org/10.20965/ijat.2017.p0004.
22. Stock, T., & Seliger, G. (2016). Opportunities of sustainable manufacturing in Industry 4.0. *Procedia CIRP, 40*, 536–541. https://doi.org/10.1016/j.procir.2016.01.129.
23. Zheng, P., Wang, H., Sang, Z., Zhong, R. Y., Liu, Y., Liu, C., & Xu, X. (2018). Smart manufacturing systems for Industry 4.0: Conceptual framework, scenarios, and future perspectives. *Frontiers of Mechanical Engineering, 13*(2), 137–150. https://doi.org/10.1007/s11465-018-0499-5.
24. Raihanian Mashhadi, A., & Behdad, S. (2018). Ubiquitous Life Cycle Assessment (U-LCA): A proposed concept for environmental and social impact assessment of industry 4.0. *Manufacturing Letters, 15*, 93–96. https://doi.org/10.1016/j.mfglet.2017.12.012.
25. Nihtianov, S., & Luque, A. (2018). *Smart sensors and MEMS intelligent sensing devices and microsystems for industrial applications* (2nd ed.). Duxford: Woodhead Publishing.
26. Mekki, K., Bajic, E., Chaxel, F., & Meyer, F. (2018). Overview of cellular LPWAN technologies for IoT deployment: Sigfox, LoRaWAN, and NB-IoT. In *2018 IEEE International Conference on Pervasive Computing and Communications Workshops, PerCom Workshops 2018, (March)* (pp. 197–202). doi:https://doi.org/10.1109/PERCOMW.2018.8480255.
27. Sinha, R. S., Wei, Y., & Hwang, S. H. (2017). A survey on LPWA technology: LoRa and NB-IoT. *ICT Express, 3*(1), 14–21. https://doi.org/10.1016/j.icte.2017.03.004.

28. Aernouts, M., Berkvens, R., Van Vlaenderen, K., & Weyn, M. (2018). Sigfox and LoRaWAN datasets for fingerprint localization in large urban and rural areas. *Data, 3*(2), 13. https://doi.org/10.3390/data3020013.
29. Periša, M., Sente, R. E., Cvitić, I., & Kolarovszki, P. (2018). Application of innovative smart wearable device in Industry 4.0. In *Proceedings of the 3rd EAI International Conference on Management of Manufacturing Systems* (pp. 1–10). EAI. doi:https://doi.org/10.4108/eai.6-11-2018.2279105.
30. Peraković, D., Periša, M., & Sente, R. E. (2019). Information and communication technologies within industry 4.0 concept. In V. Ivanov, Y. Rong, J. Trojanowska, J. Venus, A. Liaposhchenko, J. Zajac, & D. Perakovic (Eds.), *Advances in design, simulation and manufacturing* (pp. 127–134). Cham: Springer International Publishing. https://doi.org/10.1007/978-3-319-93587-4_14.
31. Peraković, D., Periša, M., & Zorić, P. (2020). Challenges and issues of ICT in Industry 4.0. In *Advances in design, simulation and manufacturing II* (pp. 259–269). Cham: Springer. https://doi.org/10.1007/978-3-030-22365-6_26.
32. Jovović, I., Husnjak, S., Forenbacher, I., & Maček, S. (2019). Innovative application of 5G and blockchain technology in Industry 4.0. *EAI Endorsed Transactions on Industrial Networks and Intelligent Systems, 6*(18), 157122. https://doi.org/10.4108/eai.28-3-2019.157122.
33. Cheng, J., Chen, W., Tao, F., & Lin, C. L. (2018). Industrial IoT in 5G environment towards smart manufacturing. *Journal of Industrial Information Integration, 10*(March), 10–19. https://doi.org/10.1016/j.jii.2018.04.001.
34. Boyes, H., Hallaq, B., Cunningham, J., & Watson, T. (2018). The industrial internet of things (IIoT): An analysis framework. *Computers in Industry, 101*(June), 1–12. https://doi.org/10.1016/j.compind.2018.04.015.
35. Mittal, S., Khan, M. A., Romero, D., & Wuest, T. (2019). Smart manufacturing: Characteristics, technologies and enabling factors. *Proceedings of the Institution of Mechanical Engineers, Part B: Journal of Engineering Manufacture, 233*(5), 1342–1361. https://doi.org/10.1177/0954405417736547.
36. Cvitić, I., Peraković, D., & Kuljanić, T. M. (2017). Availability factors in delivery of information and communication resources to traffic system users. In J. Mikulski (Ed.), *Smart solutions in today's transport* (pp. 28–41). Springer International Publishing, Cham, Switzerland.
37. Cvitić, I., Peraković, D., Periša, M., & Botica, M. (2019). Novel approach for detection of IoT generated DDoS traffic. *Wireless Networks, 1*, 1–14. https://doi.org/10.1007/s11276-019-02043-1.
38. Perakovic, D., Perisa, M., Cvitic, I., & Husnjak, S. (2017). Model for detection and classification of DDoS traffic based on artificial neural network. *Telfor Journal, 9*(1), 26–31. https://doi.org/10.5937/telfor1701026P.
39. Ibarra, D., Ganzarain, J., & Igartua, J. I. (2018). Business model innovation through Industry 4.0: A review. *Procedia Manufacturing, 22*, 4–10. https://doi.org/10.1016/j.promfg.2018.03.002.
40. Mittal, S., Khan, M. A., Romero, D., & Wuest, T. (2018). A critical review of smart manufacturing & Industry 4.0 maturity models: Implications for small and medium-sized enterprises (SMEs). *Journal of Manufacturing Systems, 49*(November), 194–214. https://doi.org/10.1016/j.jmsy.2018.10.005.
41. Sjödin, D. R., Parida, V., Leksell, M., & Petrovic, A. (2018). Smart factory implementation and process innovation: A preliminary maturity model for leveraging digitalization in manufacturing moving to smart factories presents specific challenges that can be addressed through a structured approach focused on people. *Research Technology Management, 61*(5), 22–31. https://doi.org/10.1080/08956308.2018.1471277.

Evaluation of the Workplace in Order to Reduce Waste in Material Flow

Daniela Onofrejova, Jaroslava Kadarova, Dusan Simsik, and Jaroslava Janekova

1 Introduction

1.1 Value Management for Lean Enterprise

Leanness of an enterprise is a system of methods that lead to the elimination of waste in the manufacturing process; simplified, it means performing just what is needed and performing it right the first time, with faster achievement than others and less expenditure. Leanness means increasing the performance of a company by producing more than competitors in a field, producing more added value with a given number of people and equipment than others, and handling more orders at a given time, using less time for individual business processes and activities. The leanness of the enterprise means adapting exactly to customer's demand, with the minimum number of activities that do not increase the value of a product or service [1].

Lean manufacturing facilitates self-controlled production and concentrates on cutting costs through uncompromising efforts to achieve perfectionism. Each day in production includes the principles of kaizen activities, flow analysis, and Kanban systems. This effort must be implemented by all employees at the company, and change must be accomplished from top management to production workers [1].

James P. Womack [2], an expert of lean processes and lean thinking, defines Lean production as a five-step process:

D. Onofrejova (✉) · J. Kadarova · D. Simsik · J. Janekova
Technical University of Kosice, Faculty of Mechanical Engineering, Kosice, Slovakia
e-mail: daniela.onofrejova@tuke.sk; jaroslava.kadarova@tuke.sk; dusan.simsik@tuke.sk; jaroslava.janekova@tuke.sk

L. Knapcikova et al. (eds.), *4th EAI International Conference on Management of Manufacturing Systems*, EAI/Springer Innovations in Communication and Computing, https://doi.org/10.1007/978-3-030-34272-2_12

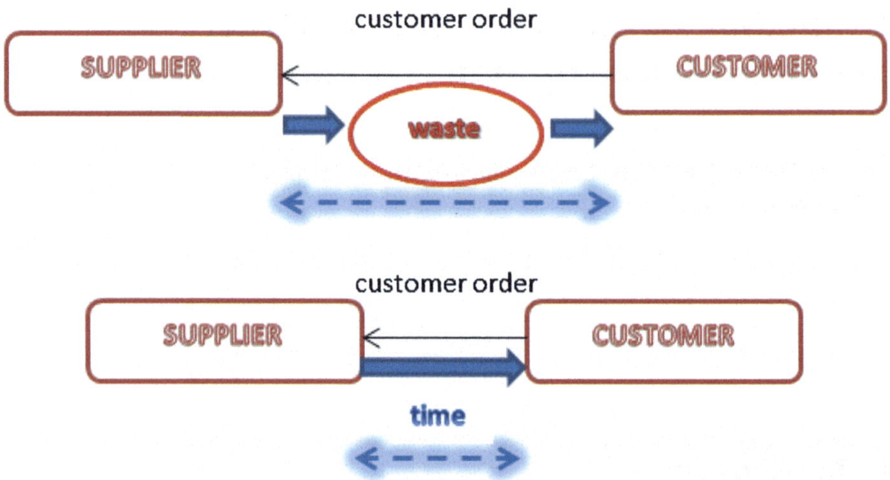

Fig. 1 Reduce time between customer and supplier and eliminate waste

- Definition of value from the customer's perspective
- Defining a value stream
- Achieving a smooth flow
- Applying a tensile principle
- Striving for excellence

To achieve high quality, low cost, short lead times, and flexibility in manufacturing process, the continuous flow of material through the entire production process, standardization of processes, and the elimination of losses have to be implemented (Fig. 1).

The concept of lean manufacturing defines the value considering the customer's requirements and separates the process of adding value from activities that do not add value. Value-adding activities are considered operations that convert a product to a customer's demand. Activities that do not add value include all activities that consume resources, but their implementation does not change anything on the product.

1.2 Smoothing the Value Flow for Lean Logistics

The basic prerequisite for introducing "leanness" is to create a continuous flow in the production process as well as in service processes. Flow means that when a customer hands over an order, the process of providing the inputs necessary to satisfy that customer's order is initiated. The entire process should take several hours or days, not weeks or months. The solution is to properly synchronize flows. Only products upon the customer's requirements in the demanded quantity, time, and quality are produced [2, 3].

Significant is the definition of customer requirements, that is, the speed of delivery agreed with customers. When converting customer requirements into existing production capacity, that is indicator for the production cycle. Production tact – the working rhythm represents the pace at which customers buy a product. The tact determines which products, in what quantities and in terms, must be manufactured to meet customer requirements.

Ensuring smooth flow is conditioned by removing the waste. Lean production can be featured as a set of methods whose use leads to waste elimination in the manufacturing process. The methods facilitate building a lean workplace with measures, standards, and action (5S, visual control, and standardized work) as well as lean equipment (TPM, SMED, Poka-Yoke, Andon, Jidoka). Each method focuses on one or more typical sources of waste [2, 3].

Behavior of batch production method results in large stocks of semifinished products. This approach was seen as a waste as many enterprises do not have sufficient production areas and storage facilities available. Solution for such a waste has been seen in continuous production, known as one-piece flow. Production smoothing is an often used process which targets mitigating the uncertainty in demand (the peaks and troughs) while minimizing waste and driving manufacturing efficiencies. Production smoothing is one of the most important tactical planning activities for the efficient operation of mixed-product just-in-time (JIT) manufacturing systems.

2 Materials and Methods

2.1 Autodesk Factory Design Tools as an Evaluation Tool for Material Flow

The issue analyzed in this article focuses on following forms of waste:

- Waiting – waiting for parts, material, information, the next step in the processing process, waiting for stock depletion, downtime and failures, and capacity problems.
- Transport or relocation that is not necessary.
- Redundant stocks of raw materials, semifinished products, or finished products cause longer running times, obsolescence, damage, transportation and storage costs, and delays [4–8].

Software tool Autodesk Factory Design (further referred to as AFD) was used to create process optimization, design of digital manufacturing process models. The entire program is based on a CAD system [9, 10]. It allows to digitally display and to optimize the manufacturing building and interior, as well. The benefits of

AFD consist of system of several modules that are interconnected and used not only for 2D display but also for the 3D display of manufacturing systems as closely as possible [8, 11–13]. It also allows creating and displaying a conveyor system, robotized workplaces, and lines. Software package offered by AFD can be used to design workplaces, buildings, and operations or simulate logistic and material flow in 2D view.

3 Problem Description

3.1 Description of a Drumsticks Production Process

AutoCAD Architecture module was used for creating digital factory design of a future production in the wood-processing industry. The product wooden sticks are specifically wooden drumsticks used in the music industry.

The initial production process begins with entering of the material into production. In this case, material remains of large pieces of potted trees (wooden logs) that are cut into smaller individual pieces to produce wooden cuttings and wooden boards suitable for further processing. The entire cutting process takes place in areas that are outside the production hall, and, therefore, the individual cut parts must be conveyed by conveyors to the plant where these parts have to be further processed and adapted to the dimensions suitable for working. During the process of cutting, the tree trunk is separated into individual boards, and it is further specified which plates and cuttings are suitable for further processing. Cutting of the trunk is realized by gradual separation into the required approximate dimension in the vertical direction.

Cut pieces of wood are stacked on pallets and transported to a special room with temperature and humidity control, so that the quality of the wood is not impaired. Stacking in this room takes about 2 weeks until the moisture content of the wood reaches 15%. After this cycle, the cut wood is put into production and processed to the required size [10].

The third step in the manufacturing process is to make from prism a profile with perfect edges. With a small curvature, the wooden prism is removed from the process. The wooden prism is machined on a special machine that cuts the prism from all four sides. After exiting the machine, the prism is moved to the lathe where a cylindrical stick is made from the prism. Using the feeder, the sticks are moved to a control where various deficiencies on the wood surface are sought.

The following operation is to create a drumstick shape using a lathe. First, the head of the stick is turned and then the body. Subsequently, the semifinished product goes to operation – printing the logo and paint coating, from where it is moved to a special conveyor on which the weight of the sticks is checked. The drumsticks are thrown out of the conveyor based on a weight testing [10].

Sound is one of the most important factors needed for quality of drumsticks. Therefore, this process is performed in a specially separated room. The drumsticks are tapped with a wooden peg in front of a microphone that captures their sound, and the computer software records the kind of that sound. Subsequently, the computer software evaluates the sound and pair the best candidates to each other. Those drumsticks that do not find the match are included in the next pairing. After the pairing process, the sticks move to the final step of the manufacturing system – packaging. The packaging starts when the sticks are mounted by the worker in special cardboard packaging and the individual pairs are superimposed on each other. The sticks are tied together and slid over the conveyor belt through a shrink wrapping machine, where are wrapped in a foil due to warm air in a hot air chamber or a tunnel. After passing the tunnel, the drumsticks continue to the carton packaging process and then subsequently are taken to the warehouse [10].

3.2 Variants of Machine Disposition Prior to Analysis

The created design consists of three manufacturing work stations, two cutting and two control stations, and one packaging station (Fig. 2). The production process in this variant starts in the drying room (position 1), from where unprocessed wooden logs are transferred using automatic guided vehicles (AGV) to station 2. An automated robot and conveyor belt are activated using the proximity sensor. The wooden logs are justified and then formed into a shape of wooden cylinder. After this treatment, the wooden cylinders are directed to the mechanical treatment (station 3), and the semifinished products are moved to station 4 and subsequently are transferred to station 5.

Station 6 is a control station where a sound test and subsequent pairing is performed. Using the AGV, semifinished products are moved to station 7, where packaging is performed. The following process after packing is the removal of loaded pallets to the warehouse. There the products are based on stacking units and then removed as needed.

In the layout of variant B (Fig. 3), we will deal in more detail with the technical elements on the packaging line and the ways of transporting the packaged products to the warehouse and then putting them in stacking units. Layout B consists of eight workstations for the manufacture of wooden sticks.

When designing variant B, a few technical elements that are useful in creating a digital factory were implemented in an operation. The workplace of the packaging is one of the last workplaces where the unpackaged material itself is in contact. As mentioned in Chap. 3, packing is done by stacking the material on top of each other and sliding it into the machine, which fits and seals the film on the material for protection, reinforcement, and better handling. In the conveyor belt, there is a worker who folds the boxes and moves them along the belt. These boxes serve for storing foil sticks. The drumsticks are placed in boxes using an automated robot. A computer-controlled robot uses sensors to detect the position of the material and

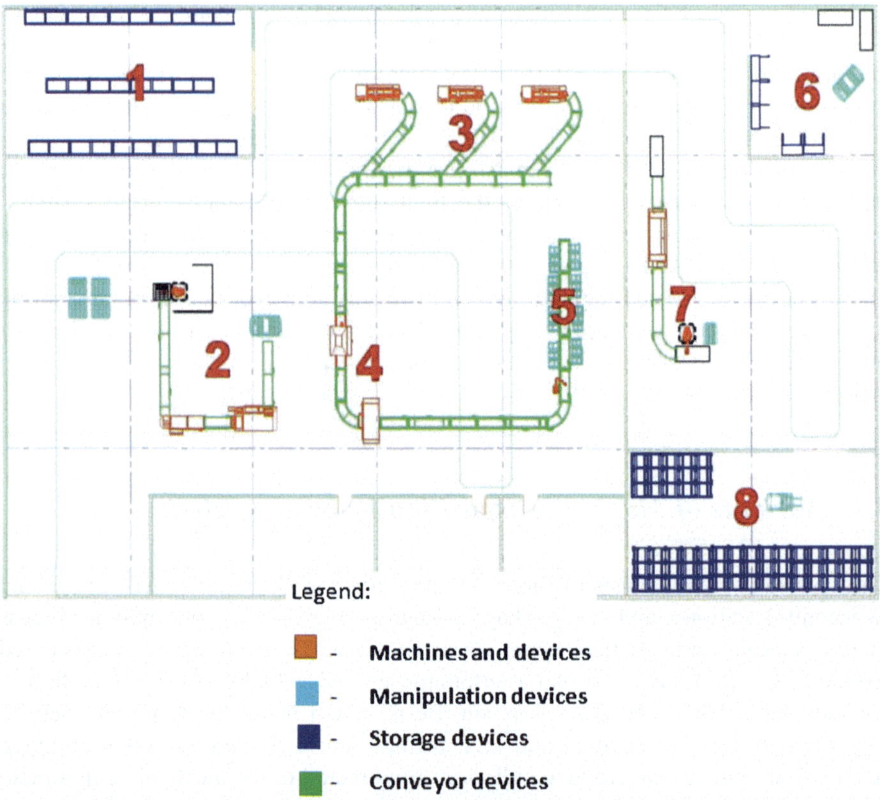

Fig. 2 Layout A of digital factory for drumstick production. Legend: (1) – Drying room, (2) – Automated robot and conveyor belt, (3) – Mechanical treatment, (4) – Paint coating and labeling station, (5) – Weight control station, (6) – Sound control station, (7) – Shrink wrapping station, (8) – Warehouse

grips it into its gripper. The attached material moves into the box if one of its sensors located at the grip face does not detect that the box is full. Wrapped boxes continue to slide along the belt, where they are transferred to the next conveyor belt leading to the output warehouse with the help of a portal robot.

4 Results

Based on the analysis of the proposed material flow, the differences in the proposals that were created were monitored. As input data, the following items were used: machine setup costs, machine work costs, workplace utilization, and consumed energy. Nearly the same conditions were created for each workplace [14]. Also,

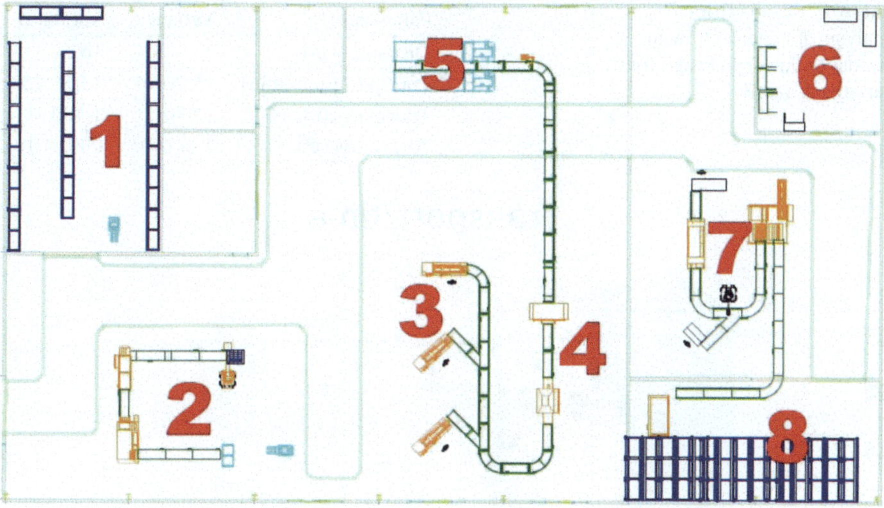

Fig. 3 Layout B of digital factory for drumstick production. Legend: (1) – Drying room, (2) – Automated robot and conveyor belt, (3) – Mechanical treatment, (4) – Paint coating and labeling station, (5) – Weight control station, (6) – Sound control station, (7) – Shrink wrapping station, (8) – Warehouse

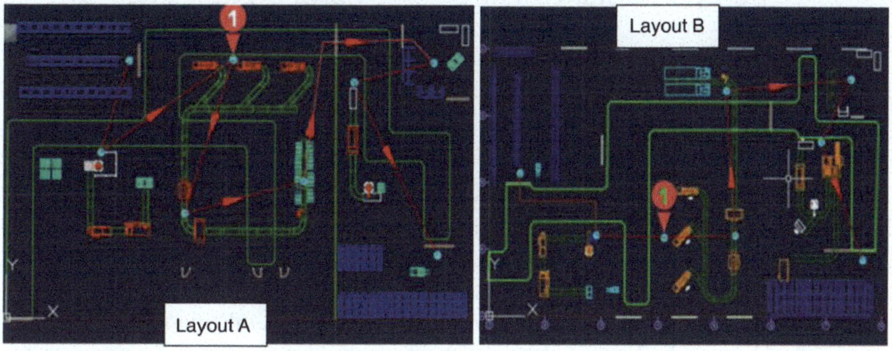

Fig. 4 Analysis of material flow in both variants: (**a**) and (**b**)

the data needed for the production system, such as the production batch and the total number of units produced, were also entered to generate material flow results.

Selected possibility of evaluation was the choice of material flow calculation [15–17]. The generated program results are shown in Fig. 4. Table 1 shows monitored parameters – distance, transport, time, transport costs, and total costs. The calculations estimate that the total cost of transport is 1530 € for two-shift operation. The total cost, together with operating costs, is 20,915 €. The transport time of the material on this proposal was calculated by the AFD software at 15:20:42 min. The total material path is 25.8 km.

Table 1 Results from analysis of material flow in Autodesk Factory Design for variants A and B

Parameters	Variant A	Variant B
Distance (km)	25.3	16.2
Transport time (min)	15:20:47	13:58:22
Transport costs (€)	1.530	1.400
Total costs (€)	290.150	289.810

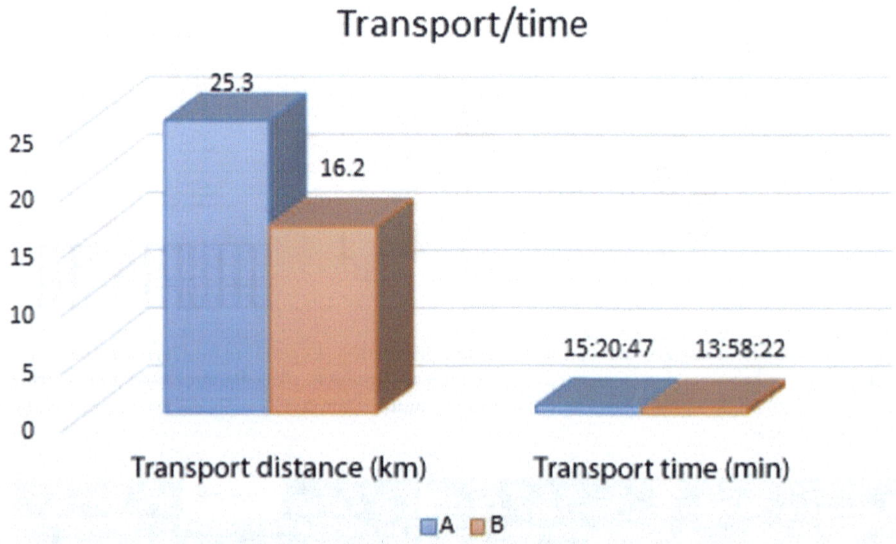

Fig. 5 Evaluation of transport distance and transport time between variants A and B

The following graph (Fig. 5) shows the difference between the total travel distance traveled and the time needed to transport the material.

The design A implies that the distance to transport the material is much larger and thus takes up more time than design B: the costs of transport consumed on the two-shift operation and the costs necessary for the overall management of the enterprise. The graph (Fig. 5) shows that due to the longer transport distance for design A, transport costs are higher than for design B, and this also affects the total cost of running the business and also shows a higher value than for design B.

The design of variant B for the given results is based on a better possibility to evaluate the implementation of this type of production. The distance that passes through the B design is 36.95% less than design A. The costs that are only needed for in-process transport as such are less by 130 €. The overall operating costs of the business and workplaces are smaller by 340 €.

5 Conclusion

The following graph shows the difference between the total travel distance traveled and the time needed to transport the material. The design A implies that the distance to transport the material is much larger and thus takes up more time than design B: the costs of transport consumed on the dual-shift operation and the costs necessary for the overall management of the business. The following graph shows that due to the longer transport distance for design A, transport costs are higher than for design B, and this also affects the total cost of running the business and also shows a higher value than for design B.

The design of variant B for the given results is based on a better possibility to evaluate the implementation of this type of production. The distance that passes through the B design is 36.95% less than design A. The costs that are only needed for in-process transport as such are less by 130 €. The overall operating costs of the business and workplaces are smaller by 340 €.

Finally, from results can be concluded that, out of the two proposals that were created for the purpose of smart environments, a B design is better and more economically advantageous on the basis of a material flow assessment. In general, the design of such a manufacturing system with elements belonging to the Industry 4.0 is economically burdensome, especially when it comes to purchasing machines and being included in IoT (Internet of Things) implementation, but on the other hand, it creates a system with which these resources have an expected return on energy savings due to intelligent building management and control, bringing comfort and convenience, and time-saving service activities [18, 19]. Therefore, changes of a similar nature will be introduced gradually in enterprises, depending on the availability of enterprise resources [20].

Acknowledgments This work has been supported by the Slovak Grant Agency KEGA 026TUKE-4/2017 Implementation of innovative educational approaches and tools to enhance the development of the core competencies graduate study program Industrial Engineering.

References

1. Kosturiak, J., & Frolik, Z. (2006). *Lean and innovative enterprise* (1st ed.). Praha: Alfa Publishing, s.r.o.. (in Czech).
2. Womack, J. P., & Jones, D. T. (2003). *Lean thinking banish waste and create wealth in your corporation* (2nd ed.). New York: Free Press, Simon & Schuster, Inc..
3. Liker, J. K. (2004). *The Toyota way. 14 management principles from the world greatest manufacturer* (1st ed.). New York: McGraw-Hill.
4. Bauerhansl, T., Hompel, M., & Vogel-Heuser, B. (2014). *Industrie 4.0 in Produktion, Automatisierung und Logistik* (1st ed.). Springer Vieweg, Springer Fachmedien Wiesbaden, Heidelberg.
5. Onofrejova, D., & Janekova, J. (2016). Design concept of digital production system. *Transfer of Innovations, 20*(33), 169–173.
6. Ottomotors.com: What is the smart factory and its impact on manufacturing, https://ottomotors.com/blog/what-is-the-smart-factory-manufacturing. Accessed 02 Mar 2018.
7. Fei, T., Jiangfeng, C. H., Qinglin, Q., Meng, Z., He, Z., & Fangyuan, S. (2018). Digital twin-driven product design, manufacturing and service with big data. *The International Journal of Advanced Manufacturing Technology, 94*(9–12), 3563–3576.

8. Simsik, D., et al. (2017). *Identification, modelling and simulation of systems* (1st ed.). Kosice: Edition of educational literature. (in Slovak).
9. Onofrejova, D., Onofrej, P., & Simsik, D. (2014). Model of production environment controlled with intelligent systems. In *Procedia engineering: Modelling of mechanical and mechatronic systems MMaMS 2014: 25th–27th November 2014, High Tatras, Slovakia. No. 96* (pp. 330–337).
10. Onofrejova, D., & Simsik, D. (2019). Change in concept from conventional to digital factory of the future. *MM Science Journal, 12*(5), 3453–3457.
11. Synergycad: Feige filling: Autodesk factory design suit increases competitive edge, https://synergiscadblog.com/2013/09/19/feige-filling-autodesk-factory-design-suite-increases-competitive-edge/. Accessed 05 Apr 2018.
12. Panda, A., Jurko, J., & Pandova, I. (2016). *Monitoring and evaluation of production processes an analysis of the automotive industry* (1st ed.). Switzerland: Springer International Publishing.
13. Straka, M., et al. (2017). Application of EXTENDSIM for improvement of production logistics' efficiency. *International Journal of Simulation Modelling, 16*(3), 422–434.
14. Onofrejova, D., & Janekova, J. (2017). Critical drivers that influence an operation performance. *Studia i Materialy, 6*(37), 48–52.
15. Fabianova, J., Janekova, J., & Onofrejova, D. (2017). Cost analysis of poor quality using a software simulation. *Amfiteatru Economic, 19*(44), 181–196.
16. Janekova, J., et al. (2018). Product mix optimization based on Monte Carlo simulation: A case study. *International Journal of Simulation Modelling, 17*(2), 295–307.
17. Trebuna, P., et al. (2017). Economic evaluation of investment project in the area of sheet metal processing. *Metalurgija, 2*(56), 245–248.
18. Vagas, M., et al. (2018). Safety as necessary aspect of automated systems. In *ICETA 2018: Proceedings: 16th IEEE International Conference on Emerging eLearning Technologies and Applications* (pp. 617–622). New Jersey: Institute of Electrical and Electronics Engineers.
19. Vagas, M., Simsik, D., & Onofrejova, D. (2019). Factors for successfully implementation of automated solutions based on industry 4.0. In *ARTEP 2019 Automatization and control in theory and praxis: 13. Annual conference of researchers from universities, educational high schools and companies* (pp. 1–8). Kosice: Technical University of Kosice.
20. Pachnikova, L., Janos, R., & Sidlovska, L. (2013). Manufacturing systems suitable for globalized market. *Applied Mechanics and Materials., 282*, 230–234.

Digital Twin of Experimental Workplace for Quality Control with Cloud Platform Support

Kamil Zidek, Jan Pitel, Ivan Pavlenko, Peter Lazorik, and Alexander Hosovsky

1 Introduction

The inspection technologies can be divided into two groups: contact and contactless. By using contactless technologies, it is not necessary to stop a fixture during quality control operation. Advanced contact technologies are coordinate measuring systems (CMS), but they work in semi-automatic mode and the reached measuring time is not suitable for full production batch check, which is the main requirement of Industry 4.0 concept. Contactless technologies used for quality control are vision systems and RFID tags communication.

Contactless data acquisition from product after production can be realized by MEMS sensors. These sensors have minimal power consumption and can be powered from battery during product operation. We can monitor product overheating, vibration, and status of environment continuously.

These digital data can be accumulated in PLC (programmable logic controllers) or SCADA (supervisory control and data acquisition) system, but they cannot be stored in long-term horizon because industrial control and monitoring systems are quite limited in storage space. The possible solution is using Cloud Platform combined with PLC system used as IoT device. Cloud Platforms are used for data analysis and visualization (data mining).

K. Zidek (✉) · J. Pitel · P. Lazorik · A. Hosovsky
Technical University of Kosice, Faculty of Manufacturing Technologies with a Seat in Presov, Department of Industrial Engineering and Informatics, Prešov, Slovakia
e-mail: kamil.zidek@tuke.sk; jan.pitel@tuke.sk; peter.lazorik@tuke.sk; alexander.hosovsky@tuke.sk

I. Pavlenko
Sumy State University, Faculty of Technical Systems and Energy Efficient Technologies, Department of General Mechanics and Machine Dynamics, Sumy, Ukraine
e-mail: i.pavlenko@omdm.sumdu.edu.ua

L. Knapcikova et al. (eds.), *4th EAI International Conference on Management of Manufacturing Systems*, EAI/Springer Innovations in Communication and Computing, https://doi.org/10.1007/978-3-030-34272-2_13

2 Related Works

Industry 4.0 (commonly referred as the fourth industrial revolution) is the current trend of automation, control, monitoring, and data exchange in manufacturing technologies [1]. It includes cyber-physical systems (CPS), the Internet of Things (IoT), cloud computing, cognitive computing, and other related disciplines [2]. The main task is digitalization, data transfer, analysis, and knowledge extraction. There are many papers with research results concerning cyber-physical systems in Industry 4.0 concept, Big Data processing [3], and combination of CPS with IoT (Internet of Things) systems in concept of Industry 4.0 [4]. Industry 4.0 can provide many methods and technologies usable in product customization [5], because customer needs are directed toward unique products. The next requirement is a low-cost product with maximized customization.

A modern approach to products identification is using RFID technology. Some research on RFID system is described in articles about security of tags [6], detection of missing tags [7], and new searching protocol [8].

Some research in monitoring devices by MEMS sensors are published in the articles about monitoring of mechatronic system by MEMS [9], vibration analyses by MEMS [10], and structural monitoring by MEMS [11].

A dynamic virtual representation of a physical object or system is a very important part of Industry 4.0 concept. This digital replica of physical assets is called digital twin, which continuously learns and updates itself from multiple sources. Case studies on digital twins are described in the articles about digital twin ergonomic optimization [12], digital twin commentary [13], learning experiences by digital twins [14], automatic generation of simulation-based digital twins [15], digital twins for legacy systems [16], and possibilities of digital twins technology [17].

Some research in cloud systems and data mining are described in the papers concerning Clouds and Big Data connection [18], cloud robotics data [19], clouds in industrial automation [20], and chaos theory combination with clouds systems [21]. The research in area of industrial process simulation by Tecnomatix software was published in [22].

3 Problem Formulation

The main problem is methodology for data transformation from quality control device to digital twin and cloud system for next analyses.

We need some experimental device to create digital 3D virtual model for next transformation to digital twin. The next task is specification of technologies for data acquisition from control process. We selected three technologies for data collection (vision system, RFID reader, and MEMS sensors). The concept of quality control is designed universally for any assembly product which can be placed on the fixture in

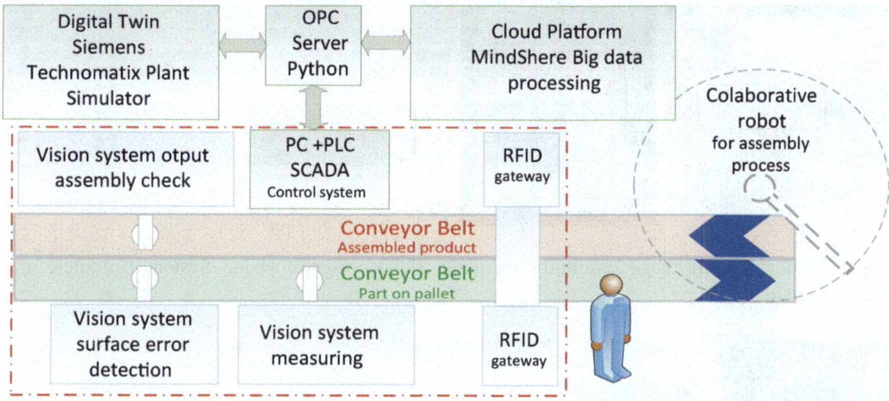

Fig. 1 Concept of experimental quality control system with data synchronization to digital twin and Cloud Platform

conveyor system. The main data transformation element between industrial control system and digital twin and Cloud Platform is OPC Server written in Python Language. For data security, we established between Internet and quality control network gateway Mind Connect IOT2040.

Scheme of components used in automated quality control system with data transformation to digital twin and Cloud Platform is shown in Fig. 1. Technologies in dash-dotted rectangle are described in next subchapters.

4 Problem Solution

The problem definition is divided into three subsections and solves the following quality control problems:

- Data digitalization case study from vision system, RFID readers, and MEMS sensors
- Methodology of transformation 3D models of product and device to digital twins
- Principle of data storage in different Cloud Platforms with analyses of critical status of product or production

4.1 Data Digitalization

Vision System

An example of data digitalization from measuring in production system is shown in Fig. 2.

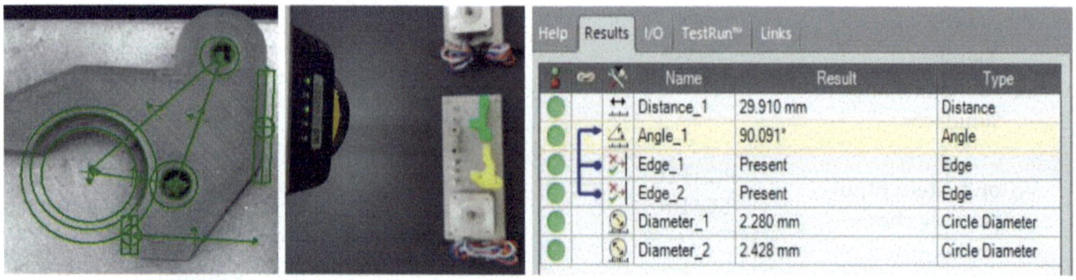

Fig. 2 Dimension check of assembly elements (left), Cognex vision system (middle), and measured data (right)

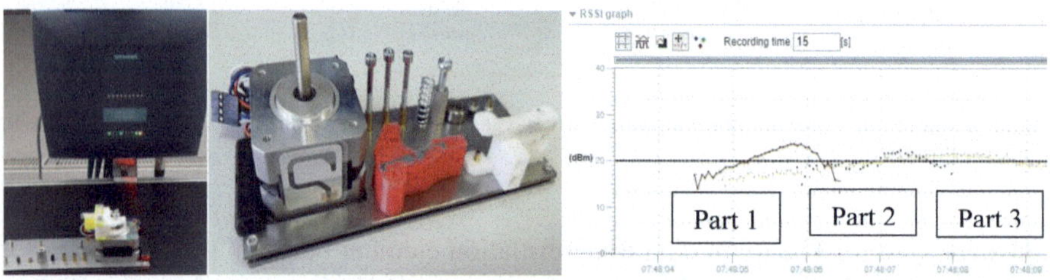

Fig. 3 RFID antenna (left) measured signal from assembly parts tags (middle), antenna, and RSSI signal acquired from motor and parts (right)

RFID

RFID tag transponder 1 is used for the stepper motor. Tags 2 and 3 are RFID label placed on two rapid prototyping parts. RFID signal intensity (RSSI) from transponder or tags can help to acquire exact position of tags according signal intensity value. Examples and principle of part position identification are shown in Fig. 3. Some research in RFID technology used for train identification was published in [23].

MEMS

MEMS sensors are suitable for integration into final product. This enables long-term monitoring and diagnostics of the product at the customer. MEMS usually contain an internal memory storage and wireless communication for data transport to the manufacturer.

Our proposal for the final product data collection (lifetime monitoring) is integration of MEMS sensors combined with MCU controller to store values locally. The connection with manufacturer can be realized in the fixed interval to send cumulated data to Cloud Platform. We have selected two variants of testing boards: first with ESP32 MCU connected to external MEMS IMU (accelerometer, gyroscope, magnetometer, temperature) with Wi-Fi connection, and second with Sigfox communication module. The first variant needs wireless connectivity (Wi-Fi)

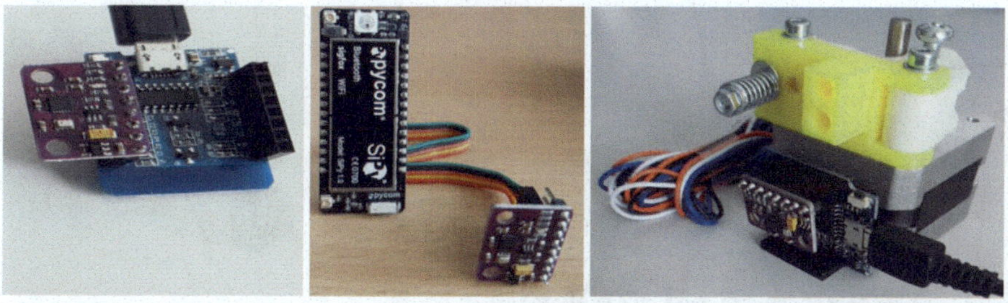

Fig. 4 ESP32 with an external MEMS sensors IMU (left), Sigfox module in the development board with sensors (middle), and ESP32 fixed to the stepper motor (right)

Table 1 Comparisons of devices for data digitalization

Technology	Device/Product presence	Solved tasks
Vision systems	–/X	measuring, object or error presence
RFID	–/X	position detection, object presence, product personalization
MEMS	X/X	product or device monitoring,

at customer and additional configuration for connection. This problem can be solved by specialized IoT devices networks (SigFox, Lora-Wan) which can be set up and managed by product manufacturer. We select the SigFox global IoT network, which can be established by part producer. The wireless IoT devices used for experiments with data collection from MEMS sensors are shown in Fig. 4.

Summary and comparison of used data digitalization devices are shown in Table 1.

4.2 *Input 3D Models and Digital Twin of Quality Control System and Inspected Product*

For the full digitalization, we need to create digital twin for quality control system and personalized digital twins for every assembled product and its parts. Digital twin 3D model of quality control system was created in Autodesk Inventor. The real experimental system is placed in the laboratory of lean production at our department and it is shown in Fig. 5.

We have selected for experiments a simple subassembly used in rapid prototyping printer (filament guide for extruder). The example of the real product and 3D models with fixture are shown in Fig. 6.

The digital twin of the system has been established in simulation software Tecnomatix plant simulator. The digital twin of the product is stored in Cloud Platform because it requires Big Data feature for data processing. The dataflow

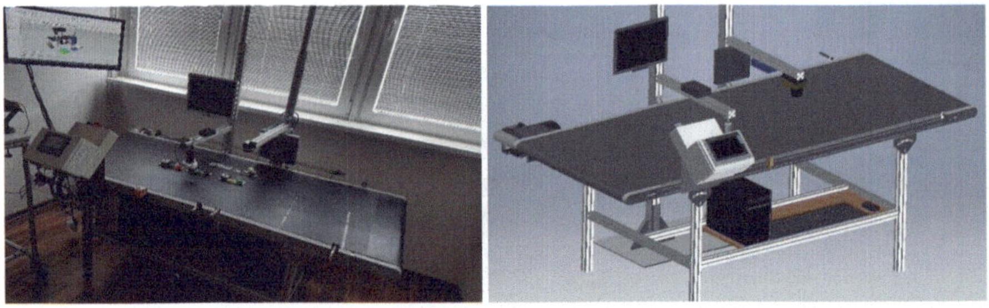

Fig. 5 Experimental quality control system (left) 3D model of device (right)

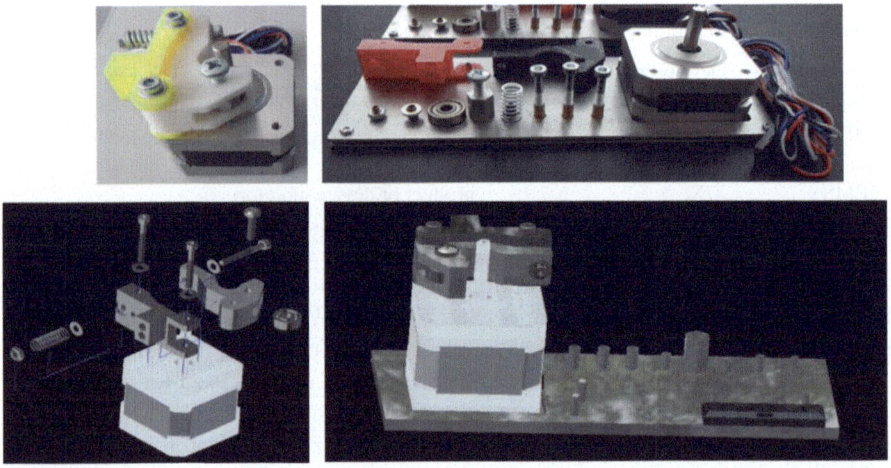

Fig. 6 Product and position in fixture (top), 3D model of the product and assembled (bottom)

principle to extend digital twin of the quality control system and the customized product is shown in Fig. 7.

The simulation of the systems for quality control and its process is usually executed offline, before the production of products starts. The main improvement in our approach to simulation is a partial optimization, which can be realized after established production. We can connect our simulation (digital twin of the system) to the control system of the real plant and to synchronize any optimization in very short time. OPC server and 3D simulation model of the active digital twin designed in Tecnomatix plant simulator are shown in Fig. 8.

4.3 *Cloud Platform Data Analyses*

The acquired data about dimension can provide information about the degradation of precision in the machining process. This degradation can predict maintenance time

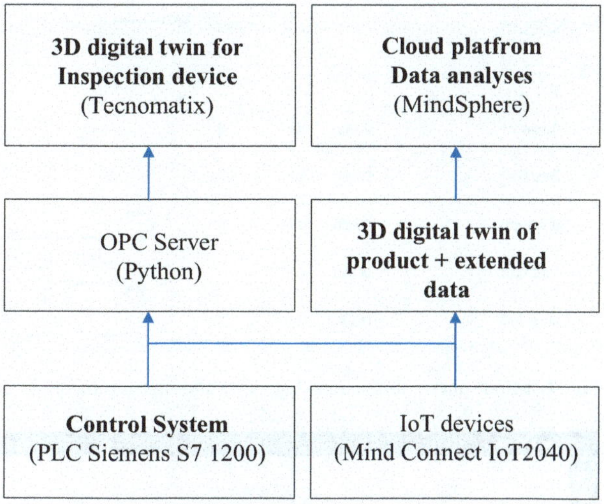

Fig. 7 Simplified dataflow from quality check device to digital twins and clouds

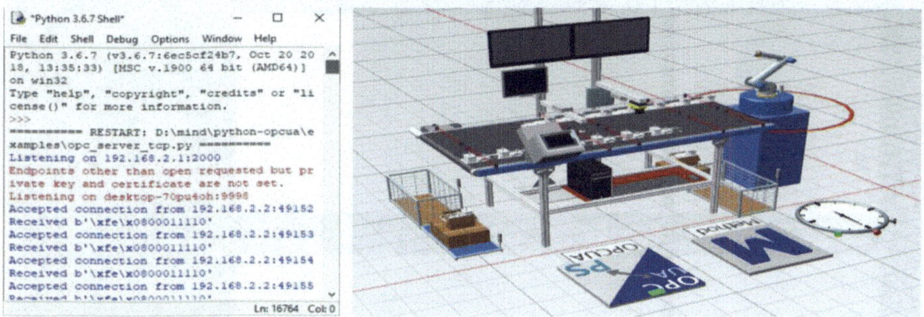

Fig. 8 OPC communication (left) and digital twin simulation in Tecnomatix (right)

for the production machine and to reduce the production of faulty parts. Measured data (dimensions from vision systems) are stored in MindSphere Cloud Platform and they create extension data in the digital twin model for all produced parts. An example of Cognex OPC server for data transfer to MindSphere Cloud Platform is shown in Fig. 9.

MEMS sensors data collection from the product established at customer can offer an automatic alarm about some problems with the product or its end-of-life signalization with minimal cost of maintenance staff. Decoded MEMS data can be visualized by IBM Watson IoT Cloud dashboard timeline as it is shown in Fig. 10.

There is also possibility to create an application for prediction of some production problems from indirect data (environmental sensors), for example, to measure the environment quantities (noise, vibration, etc.) in the production area. An example of data acquisition by Microsoft Azure IoT and ESP8266 with external temperature and humidity sensors is shown in Fig. 11.

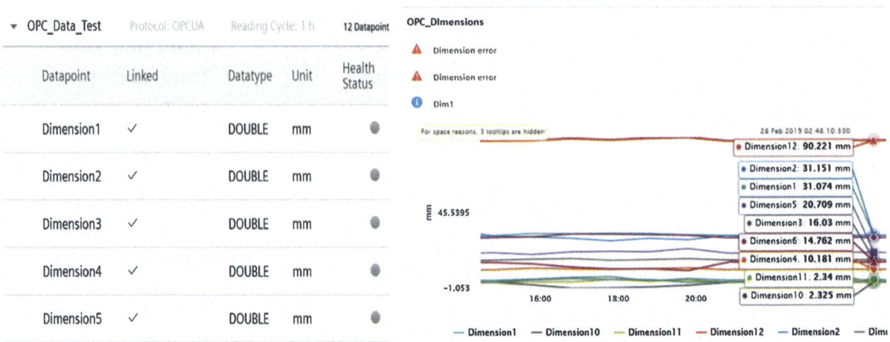

Fig. 9 MindSphere dimensions list (left) and measurements in timelines (right)

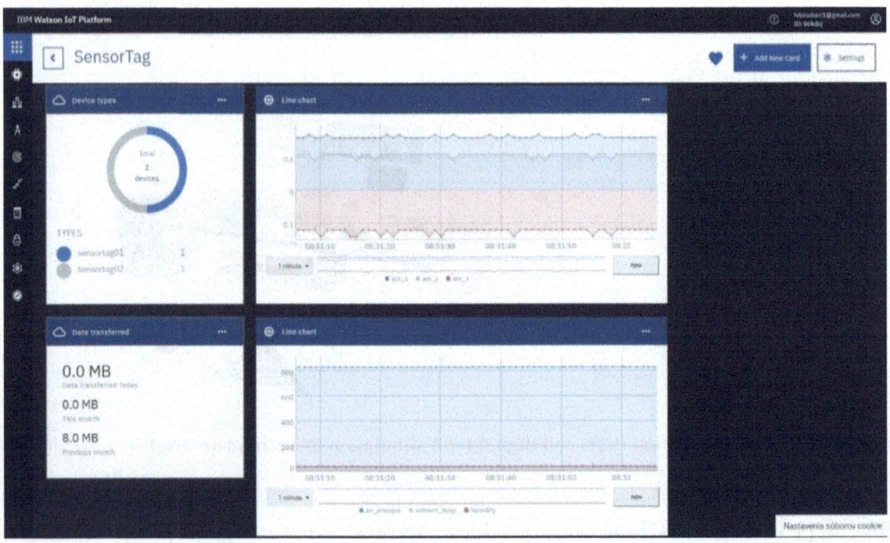

Fig. 10 IBM Watson IoT Cloud Platform for data collection from MEMS sensors

5 Conclusion and Discussion

All used devices for data digitalization have own limitation (MEMS sensors, IoT, vision systems, RFID tags). MEMS sensor data must be filtered by advanced filters (e.g., Kalman filter), because output is very noisy. MEMS outputs bandwidth is not compatible with currently used Cloud Platforms. UHF RFID system cannot be used for very small parts, because the tag size is limited by antenna length. LF or HF RFID tags are the solutions for tagging smaller parts, but they are not primarily developed for industrial parts (mainly for food industry). Industrial vision system is closed source system to algorithm modification.

Fig. 11 Microsoft Azure IoT cloud example of data representation (temperature, humidity) for the quality control system environment monitoring

This paper can provide ideas for maximizing the use of digitalized data from the product quality control process, because many companies do not store and use this data. There is described an approach to data digitalization and transport from production process to the extended digital 3D twin model, possibilities to personalize product by RFID technology and its tags, to monitor product by MEMS sensors at customer, and to store/analyze all data by Cloud Platforms. The bridge for transfer data in both directions was performed by OPC technology, mainly OPC-UA Server (OPC DCOM). OPC server was written in Python programming language customized for data collection from many sources.

The digital twin of inspection and identification system was designed for online connection to synchronize data from quality control process. Extended data from digital twin is also online synchronized with Cloud Platform.

RFID gate personalizes all products with acquired dimension data stored in the main assembly RFID tag label. RFID system is used to localize parts on the conveyor line by RSSI signal from tags.

Measured data from product were stored MindSphere Cloud Platform. MEMS sensor solution was tested with two Cloud Platforms (IBM Watson IoT and Azure IoT). The product vibration and temperature have been measured. Quality control device has been monitored by temperature and humidity.

Further experimental works on these topics will be long-term data collection and reliability verification and next implementation to some existing quality check systems in real production with automated lines.

Acknowledgments This work was supported by the Slovak Research and Development Agency under contract No. APVV-15-0602 and also by the Project of the Structural Funds of the EU, ITMS code: 26220220103.

References

1. Židek, K., et al. (2018). Data optimization for communication between wireless IoT devices and Cloud platforms in production process. In: *MMS 2018, Dubrovnik, November 06–08, 2018*, 8 p.
2. Xu, L. D., & Duan, L. (2019). Big data for cyber physical systems in industry 4.0: A survey. *Enterprise Information Systems, 13*(2), 148–169.
3. Li, G., Tan, J., & Chaudhry, S. S. (2019). Industry 4.0 and big data innovations. *Enterprise Information Systems, 13*(2), 145–147.
4. Sanin, C., et al. (2019). Experience based knowledge representation for Internet of Things and Cyber Physical Systems with case studies. *Future Generation Computer Systems, 92*, 604–616.
5. Panda, A., et al. (2011). Optimalization of heat treatment bearings rings with goal to eliminate deformation of material. *Chemické listy, 105*(S), 459–461.
6. Han, W., Liu, W., Zhang, K., Li, Z., & Liu, Z. (2019). A protocol for detecting missing target tags in RFID systems. *Journal of Network and Computer Applications, 132*, 40–48.
7. Lee, C.-C., Chen, S.-D., Li, C.-T., Cheng, C.-L., & Lai, Y.-M. (2019). Security enhancement on an RFID ownership transfer protocol based on cloud. *Future Generation Computer Systems, 93*, 266–277.
8. Liu, C.-G., Liu, I.-H., Lin, C.-D., & Li, J.-S. (2019). A novel tag searching protocol with time efficiency and searching accuracy in RFID systems. *Computer Networks, 150*, 201–216.
9. Židek, K., & Hošovský, A. (2013). Wireless device based on MEMS Sensors and Bluetooth Low Energy (LE/Smart) technology for diagnostics of mechatronic systems. *Applied Mechanics and Materials, 460*, 13–21.
10. Varanis, M., et al. (2018). MEMS accelerometers for mechanical vibrations analysis: A comprehensive review with applications. *Journal of the Brazilian Society of Mechanical Sciences and Engineering, 40*(11), 527.
11. Dumont, M., &Wolf, D. (2019). Usage of MEMS capacitive acceleration sensors for structural monitoring. In *Dynamics of Civil Structures*, Vol 2. Conference Proceedings of the Society for Experimental Mechanics Series. Cham: Springer.
12. Caputo, F., Greco, A., Fera, M., & Macchiaroli, R. (2019). Digital twins to enhance the integration of ergonomics in the workplace design. *International Journal of Industrial Ergonomics, 71*, 20–31.
13. Tomko, M., & Winter, S. (2019). Beyond digital twins – A commentary. *Environment and Planning B: Urban Analytics and City Science, 46*(2), 395–399.
14. David, J., Lobov, A., & Lanz, M. (2018). Learning experiences involving digital twins. In *IECON 2018* (pp. 3681–3686).
15. Martinez, G. S., et al. (2018). Automatic generation of a simulation-based digital twin of an industrial process plant. In *IECON 2018* (pp. 3084–3089).
16. Khan, A., Dahl, M., Falkman, P., & Fabian, M. (2018). Digital twin for legacy systems: Simulation model testing and validation. In *2018 IEEE 14th International Conference on Automation Science and Engineering (CASE)*. (pp. 421–426). IEEE.
17. Shubenkova, K., et al. (2018). Possibility of digital twins technology for improving efficiency of the branded service system. In *2018 Global Smart Industry Conference (GloSIC)* (pp. 1–7). IEEE.
18. Lu, Y., & Xu, X. (2019). Cloud-based manufacturing equipment and big data analytics to enable on-demand manufacturing services. *Robotics and Computer-Integrated Manufacturing, 57*, 92–102.
19. Aissam, M., Benbrahim, M., Kabbaj, M. N. (2019). Cloud robotic: Opening a new road to the industry 4.0. In *New Bioprocessing Strategies: Development and Manufacturing of Recombinant Antibodies and Proteins* (pp.1–20).

20. Mahmoud, M. S. (2019). Architecture for cloud-based industrial automation. In *Third International Congress on Information and Communication Technology*. Advances in Intelligent Systems and Computing, vol 797. Singapore: Springer.
21. Hu, Y., Zhu, F., Zhang, L., Lui, Y., & Wang, Z. (2019). Scheduling of manufacturers based on chaos optimization algorithm in cloud manufacturing. *Robotics and Computer-Integrated Manufacturing, 58*, 13–20.
22. Hrehova, S., & Vagaska, A. (2018). Design of study support to get skills in plant simulation tecnomatix environment. In: *INTED 2018, Valencia, 5th–7th of March* (pp. 7942–7945).
23. Balog, M., Husár, J., Knapčíková, L., & Šoltýsová, Z. (2015). Automation monitoring of railway transir by using RFID technology. *Acta Tecnologia, 1*(1), 9–12.

Hazardous Work and Destructive Elements in the Working Environment and Risk Categories in a Mining Company

Michaela Ocenasova, Lubica Kozakova, Csaba Farkas, Petra Pleskova, and Martin Paska

1 Current Status of the Occupational Health and Safety Risk Assessment in Slovakia

The global economic growth results in continuous developing of new technologies introduction of which often requires adopting changes in types and numbers of job positions regardless of the branch of industry. Moreover, new technologies produce also unanticipated health hazards, and often documented in the working environment are major occupational injuries or diseases arising out of employment. Hence, it is more than necessary that the employers would at all times pay close attention to establishing conditions ensuring occupational health and safety. An important aspect of supporting occupational health and safety is continuous assessment of the working environment condition and subsequent preventive actions that are based on both foreseeing the existing hazard factors and their elimination. Legislative, social, economic, technical and organizational measures are necessary to create and improve a safe working environment in society [1, 2].

Relating to the health and safety aspects in the Slovak Republic (SR) is a set of the European Union (EU) directives. From the directives resulting obligations are further applied within the Slovak Republic legislation. Most important for the SR occupational health and safety issue are the following acts: Act No. 311/2011 Coll.

M. Ocenasova (✉) · L. Kozakova
Technical University of Košice, Faculty of Mining, Ecology, Process Control and Geotechnologies, Institute of Earth Resources, Kosice, Slovakia
e-mail: michaela.ocenasova@tuke.sk; lubica.kozakova@tuke.sk

C. Farkas · P. Pleskova · M. Paska
Technical University of Kosice University of Košice, Faculty of Mining, Ecology, Process Control and Geotechnologies, Institute of Logistics and Transport, Kosice, Slovakia

L. Knapcikova et al. (eds.), *4th EAI International Conference on Management of Manufacturing Systems*, EAI/Springer Innovations in Communication and Computing, https://doi.org/10.1007/978-3-030-34272-2_14

(hereinafter referred to as the "Labour Code"), Slovak National Council (NR SR) acts Nos. 124/2006 Coll., 335/2007 Coll. and 309/2007 Coll.

The Act on Protection, Support and Development of Public Health defines *risk* as 'the probability of occurrence of a harmful effect on a human caused by exposure to a dangerous factor' [3].

Risk can be discussed in more detail and known as [4, 11]:

- Danger (probability) of suffering a loss, at certain level of being aware of the environment condition
- Possibility of occurring of an event that would prevent or endanger attainment of goals by an individual or an organisation
- Danger (probability) of arising of negative deviations from set levels of goals of an individual or of an organisation

Within the European Union, the Framework Directive 89/391 /EEC of 12 June 1989 on the introduction of measures to encourage improvements in the safety and health of workers at work presents an important basis for assessing the risks. On the EU level, there are no rigidly stipulated rules on how the risk assessment has to be performed and hence the Framework Directive has been transposed into the national legislation. Practical implementation varies from one worksite to another and is always adapted to their specific features [4, 5].

1.1 New Approach to the Occupational Safety Issue

A new approach to the occupational safety requires a methodical style targeted to prevent occurrence of negative aspects – prevention of consequences of accidents. Prioritised is participative management focusing on the human factor and on safety and health culture. Hence, within a company, the Occupational Safety and Health (OSH) is becoming an issue not only for the management but for the employees as well. The Slovak Republic intentions in the field are based on the OSH European strategy concept for years 2008–2012, which contains the following challenges [5]:

- *Holistic approach* with respect to individual classes of work and inclusive of psycho-social aspects and ease at work
- *Promoting the culture* of risk prevention and building partnerships in the OSH, improving the social dialogue, economic stimulation and social responsibilities
- *Advocating social policy* as an aspect of competitiveness

New approaches of the EU to protecting the employee have become the ground of the Slovak Republic state policy on occupational safety and health. Corporate occupational safety policy connects to the national policy, objectives of which are as follows:

- *Strengthening of the legal responsibility* relating to the OSH issues
- *Expanding positive attitudes* of the top-ranking political and state representatives, employers, employees and general public to the OSH
- *Advocating systematic solutions, planning, complexity* and introducing good practices in the OSH management
- *Ensuring consistent implementation of regulations* in everyday practice and compliance with the OSH principles
- *Supporting economic involvement of employers* so that safe work would become 'profitable'
- *Advocating holistic approach to* resolving the OSH issues, as well as securing satisfactory working conditions
- *Ensuring effective labour inspections* and continuous supervision over applying the OSH regulations [5]

1.2 Hazardous Work and Damaging Factors of the Working Environment

Hazardous work is the work with an increased risk of occurrence of occupational illness, professional poisoning and other health damage related to work. When defining the risk, it is important to identify risk factors first as these may adversely influence the health of the employee, which include the following [4]:

- Chemical factors (dust, chemical agents, allergens, toxic substances, carcinogens and mutagens)
- Biological factors (microorganisms and human endo-parasites)
- Physical factors
- Physical stress and psychical stress

The evaluation of the influence of a risk factor on health performed must be both qualitative and quantitative measurements, and results of these shall be compared against the permissible limit values. When the presence of a risk factor in the working environment is confirmed, the employer becomes obliged to elaborate and implement operating rules for manipulations with the risk factor, when integral parts of these shall be [3]:

- Risk assessment
- Evaluation of possibilities of eliminating the damaging factor effects
- Measures for diminishing the exposure
- The way of ensuring appropriate medical supervision for the concerned employees by way of occupational health service

The risk assessment results in inclusion of individual professions into risk factor categories. Developed for individual working environments, there were four categories of work activities. If the risk assessment result points to the 3rd or 4th category, the employer is obliged to introduce measures that are to cut down the hazardous factor effects to the lowest attainable level, as well as to submit a proposal to include the work task amongst risk works as classified by the pertinent regional public health authority [3].

The risk work tasks performing employee is, generally, affected by damage to his or her health by occupational disease. In case of employees performing work activities falling into categories 3 and 4, the employer is obliged to arrange for their regular medical preventive examinations (once in 2 years for category 3, and once a year for category 4) the objective of which is to early establish onset of the disease and to temporarily disclose the employee from the risk work up to his/her full recovery and reversal of his/her health condition to normal. The employer of those employees of who are performing risk work tasks is under the obligation to prepare and submit to the regional public health authority by 31 December of each year information on the results of evaluation of health risks and on the measures implemented for their reduction or elimination at workplaces where the employees are performing such risk tasks [4].

2 Damaging Factor 'Noise' and Risk Categories in a Mining Company

Noise is an undesired sound intensity (noisiness) which is quantified in decibels (dB). The decibel scale is a logarithmic one, which means that increase in the level of noise by three decibels presents in fact twofold increase in the noise intensity. Nonetheless, it is not only intensity that is a gauge of noise harmfulness; another and more important factor is the duration of exposure to noise. For that reason, used for measuring are time-weighted average levels of noise [6].

Noise at work presents a global issue identified in a host of branches of industry. Exposure to excessive noise may result in damages to hearing. Loss of hearing due to noise may result from either one-off exposure to an acoustic pulse in excess of 140 decibels (dB(C)) or from exposure to extreme intensity noise over 85 decibels (dB(A)) for several hours of each workday for a prolonged period of time. According to the estimates, exposed to noise in the EU countries are some 60 million workers – 30% of the labour force [7, 8].

According to the level and nature of the work and work environment factors that can potentially be of impact upon the health of the employees, and according to evaluation of health risks and of changes in the health status are work tasks included in four categories [3]:

1st category – includes work tasks not involving risk of damage to health of an employee due to work tasks and/or to the working environment

2nd category – includes work tasks at which, considering the risks, no damage to health can be presumed (un-exceeded limits governed by special regulations)

3rd category – includes work tasks in which:

(a) Exposure of the employee is not reduced by the work task factor and by the working environment to the level of limits stipulated by special regulations. To decrease the risk, organisational measures must be introduced inclusive of using personal protective work instruments (PPWI).
(b) Work tasks in which exposure of the employee is reduced by technical measures to the level of prescribed limits but in which mutual combination of the work factors can result in damages to health of the employee.
(c) Work tasks with no prescribed limits, though in which exposure to work factors can induce damage to health of the employee.
(d) Work tasks performed in a controlled zone.

4th category – included here are exceptional work tasks, and for a limited period of time not in excess of a period of 1 year:

(a) Works in which it is impossible to reduce, by introduced technical and/or organisational measure, exposure of the employee to the work factor and the working environment factor to the level of set limits; in which exposure to the work factor and the working environment exceeds limits; which may result in a change in the health of the worker and necessitate other specific protective measures, including the use of personal protective equipment.
(b) Works that by the rate of exposure to individual work factors and working environment should fall in the third category but where mutual combination of work factors and of working environment increases the health damage risk.

Deciding on framing a risk or hazard into a category is a Slovak Republic Regional public health authority based either on a proposal of the employer or at its own discretion.

3 The Risk Category Noise Evaluation Process for the Worker Profession in Processing Industrial Minerals

The Extent extent of work tasks of employees of the labourers in processing non-ferrous ores usually involves the below-outlined activities:

- Inspections of the worksite and machinery at the beginning and end of the shift
- Monitoring the operation of belt conveyors
- Operating the belts, feeding tables, belt conveyors, and pumping units
- Cleaning and removal of foreign objects off the belt conveyers
- Securing commencing, operating and shutting down of machinery and technologies, supervising the magnesite ore charging
- Monitoring proper run of devices when in operation
- Cleaning of floors and workplaces

- Evaluating appropriateness of the working environment conditions
- Managing the teamwork
- Recording technical data on progress and results of work tasks
- Securing dispensing of meals
- Keeping the working environment in proper order

Included amongst the risk factors having an impact upon the non-ferrous ores processing workmen can be *noise* and *the microclimate*. Further, difficult working conditions characterising factors are height, physical stress, night work, lifting and carrying loads [7].

3.1 Level, Type and Duration of Exposure to Noise

In question is processing of raw magnesite ore in heavy suspensions. Operations are controlled from the control room and are continuous (24/7).

Elaborated for the purpose of illustrating exposure of the workman at processing non-ferrous raw materials to noise in a selected company has been a table showing how was the work-time during a shift scheduled for the profession; the purpose was to determine how long is an employee performing his or her professional operations when working in noisy environment (Table 1).

Based on observing operations of workmen at processing industrial minerals in a selected processing plant, it has been determined that a workman is for some 6 h and 30 min a day performing 'excessively noisy' work tasks, as documented in the above table.

Table 1 Overview of exposure to noise in a plant for the profession of workman at treatment of industrial minerals

	Duration of individual activities		
	Time		Time
Description of performed activities	From	To	Total
Arrival to the workplace and scheduling of tasks	7:00	7:15	0:15
Inspection of the machinery	**7:15**	**7:45**	**0:30**
Work cycle	**7:45**	**11:00**	**3:15**
Lunch	11:00	11:30	0:30
Work cycle	**11:30**	**14:15**	**2:45**
Cleaning works; entering records in books	14:15	14:30	0:15
Preparations for leaving the workplace; leaving	14:30	15:00	0:30
Noisy operations – 6 h 30 min	8 h in aggregate		
Non-noisy operations – 1 h 30 min			

Source: original (personal) processing

3.2 Measuring Exposure to Noise

During measuring exposure to noise was the operator holding the position of workman at processing industrial minerals performing supervision of operations, pumps, washing units, vibration and resonance screens and shifting of materials and of separators. As shown in the table above, in case of the profession, exposure to noise continued for 6.5 h. Established during the measuring was equivalent noise level $L_{Aeq,T}$ of 100 dB and peak level $L_{CPk,T}$ C of acoustic pressure of 144.6 dB. Calculated from by measuring determined values was the standardised level of exposure to noise for an 8-h work-shift, $L_{AEX,8h,}$ of the workman processing industrial minerals of 99.1 dB, which was increased by measurement error of 3.6 dB(U). Simultaneously, assessed during measurements was efficiency of hearing protection safety products – so- called standard exposure to noise behind hearing protectors $L_{AEX,8h,R}$. Measured with actually used hearing protectors type Classic E.A.R. EN 352-2, SNR - 28 dB was, for the workman processing industrial minerals, equivalent level C ($L_{Ceq,T}$) of 99.1 dB.

Upon deducting the effect of hearing protectors: $99.1 + 3.6 - 28 = 74.7$ dB

Determined further, during measurements performed, was the peak level C behind protectors $L_{CPk,T,R}$ for: $144.6 + 3.6 - 28 = 120.2$ dB

3.3 Comparison of Measurement Results

Comparison of Correspondence and/or Non-correspondence of Results with the Standardised Level of Exposure to Noise Limit Level $L_{AEX,8h,L}$

- *Protectors on* → $102.7 - 28 = 74.7 < 87$ – limit value of exposure to noise *was not exceeded*
- *Without protectors* → $102.7 > 87$ – limit value of exposure to noise *was exceeded by* 15.7 dB

Determination of the hearing protection degree is shown in the below table (Table 2):

Table 2 Determination of the hearing protection degree

Standardised level of exposure to noise $L_{AEX,8h,R}$	Degree of protection
$L_{AEX,8h,R} > 87$ dB	Inadequate
(87–82 dB)	Acceptable
(83–77 dB)	Good
(78–72 dB)	Acceptable
$L_{AEX,8h,R} < 72$ dB	Too high (excessive)

Source: Guideline No. 30/2010 of the SR Ministry of Health of 07. 09. 2010 [9]

Whereas $L_{AEX,8h,R}$ 74.7 dB falls within the 72–78 dB interval, the degree of hearing protection proves to be *acceptable*.

Comparison of correspondence and/or non-correspondence of results with the exposure to noise operational values.

- Upper operational value: 85 dB
- Lower operational value: 80 dB

For the worker operating in processing industrial minerals, the standardised exposure to noise level increased by measurement error (U), reached the value of 102.7 dB > 85 dB, *and hence was exceeded by* 17.7 dB.

Comparison of Results with Peak Levels of C Acoustic Pressure:

- Comparison with limit value of peak level of C acoustic pressure – 140 dB:
  - For the workman workmen at processing industrial minerals, the $L_{CPk,T+U}$ amounts to

    148.2 > 140

    148.2 – 28 = 120.2 ($L_{CPk,T,R+U}$) < 140

- Comparison with upper performance value of the peak level of C acoustic pressure ($L_{CPk,h}$) – 137 dB
  - For the workman at processing industrial minerals, the $L_{CPk,T+U}$ 148.2 > 137 dB, which means that the level was *exceeded* by 11.2 dB

3.4 Inclusion of the Profession into the Risk Works Category

The governing *noise quantities* at workplaces are standardised level of exposure to noise and peak level of C acoustic pressure. For the purposes of the occupational health and safety, stipulated for the noise exposure risks are:

1. *Exposure to noise limit values*
2. *Exposure to noise operational values* $L_{AEX,\,8h,\,L} = 87$ dB a $L_{CPk} = 140$ dB
   (a) *Upper* operational values of exposure $L_{AEX,\,8h,\,a} = 85$ dB a $L_{CPk} = 137$ dB
   (b) *Lower* operational values of exposure $L_{AEX,\,8h,\,a} = 80$ dB a $L_{CPk} = 135$ dB

Whilst considered when exercising the limit value is also attenuation produced by hearing protectors the contrary, i.e. disregarding the attenuation, is the case when exercising operational values. The noise limit value may not be exceeded for the worker regardless if s/he uses hearing protectors or not.

Performed within the profession is being an activity (control room) necessitating mental concentration or hearing involving (verbal) communication – in question is an activity that may fall under individual categories of work.

Established for protection of health of employees against non-specific, especially disturbing or abusing effects of noise, are operational values of standardised

Table 3 Operational values of standardised exposure to noise $L_{AEX,8h}$ levels A for work categories

Work category	Activity	Noise at workplace $L_{AEX,8h,R}$ (dB)
I	Activity requiring continuous concentration or unhindered communication; creative activity	40
II	Activity, communication in which an important integral part of the work is being performed; activities characteristic with enormous demand on precision, speed and/or attention or alertness	50
III	Routinely performed activities communication in which a part of the work is being performed; activities performed based on partial auditory information	65
IV	Activities executed whilst using noisy machinery or instruments, performed in noisy environment that does not meet the requirement to be included into categories I, II or III	80

Source: SR Government regulation No. 115/2006 *Coll.* [10]

exposure to noise levels for individual groups of works, as categorised in the below table (Table 3).

Special groups of employees, per wording of the SR National Council Act No. 124/2006 *Coll.* on occupational safety and health, as amended, are:

- Pregnant women
- Mothers up to the end of the 9th month after childbirth
- Nursing mothers
- Juveniles
- Physically challenged employees

According to Decree no. 448/2007 *Coll.* regarding the details of factors of work and the work environment in relation to categorisation of works from the viewpoint of risks to health and regarding the requisites of the proposal for classification of works into categories, included into risk category 4 for the noise risk factor are:

(a) Works at which exceeded are upper operational values of exposure to noise and where the standardised level of exposure to noise $L_{AEX,8h}$ is exceeded by 10 dB or more, or where the peak level C of acoustic pressure L_{CPk} is exceeded by 3 dB or more.
(b) Works at which the standardised level of exposure to noise or peak level C of acoustic pressure corresponds with the criteria of category 3 but when at a time observed in the employee are changes in hearing capacity due to effect of noise.

However, in the profession of workman operating in processing industrial minerals, when the peak level C of acoustic pressure L_{CPk} was exceeded by 11.2 dB, the profession will be included into *category 4*. Exceeded by 17.7 dB was at a time the standardised level of exposure to noise $L_{AEX,8h}$.

In the selected plant, the employees performing the profession of workman operating in processing of industrial minerals will be pursuant to provisions of Section 30, Sub. 5, Lit. a) of Act No. 355/2007 *Coll.* for category 4 once a year participating on preventive medical examination in relation to their work tasks.

4 Measures for Elimination or Reduction of Exposure to Noise

Indirect influences upon health and safety of employees present influences on health and safety resulting from mutual effects of noise and warning acoustic signals, respectively, or other noises that must be monitored to reduce the risk of accidents.

4.1 Proposed Preventive and Safety Measures

Used at the control rooms of the evaluated workplaces are acoustic signals that are aired by the control unit at whatever deviation from the preset range of respective parameters, and also whenever an exceptional event occurs. Installed on the outside of the plant are acoustic alarms, whilst double-paned windows that are facing the plant make the control room soundproof. Appointed employees are supposed to regularly check general conditions at individual workplaces using a portable noise-level meter that is capable of recording also peak values (max measurable short-term or momentary values of noise at a workplace). If, based on the risk assessment, the employer detects a condition hazardous to the health of employees, he shall provide an appropriate medical supervision integral part of which are preventive medical check-ups.

Entitled to examination of hearing are employees exposed to noise where the upper operational values are exceeded, and also the employees exposed to noise where the lower operational values are exceeded, provided the assessment and measurement have proved the existence of possible hazard to their health.

Employer shall provide to the employee suitable and appropriate personal protective safety instruments for hearing protection, so-called hearing protectors in the case the risk of hearing damage of the employee cannot be eliminated by implementing other measures. Hearing protectors are selected so that the risk to hearing capacity would be eliminated or so that the possible risk would be reduced to the least possible extent (Tables 4 and 5).

Table 4 Measures for elimination or reduction of exposure to noise

Measures for elimination or reduction of exposure to noise according to Government Regulation 115/2006 *Coll* – so-called general principles of prevention	Performance of measures by the employer
The work methods designed to alleviate exposure to noise and selection of appropriate devices with minimum noise pollution	Replace resonance screen by less noisy ACH screens.
Structural and spatial arrangement of the workplace and workstations	Sources of noise to be situated in the external environment, and to be remote manipulated from the control room/station
Appropriate information for and practical training of employees focused on proper use of working devices	Regular maintenance – scheduled temporary stoppages of operations
Technical measures – e.g. noise curtains, covers, noise absorbing linings, dampers or insulation	Control room – double-paned windows – soundproofing
Proper maintenance of processing devices, of workstations and on-site systems	Importance of regular lubricating – to reduce noise coming from amongst friction surfaces
Work organisation focused on noise reduction: 1. Limited duration and level of noise the worker are exposed to 2. Appropriate schedule of work with adequate breaks for rest	Equipping the control room with PCs. Breaks – 1/2 h for lunch

Source: original (personal) processing

4.2 *Proposed Manner of Informing the Employees About Risks Associated with Performing Hazardous Work Tasks*

In accordance with the SR Government regulation No. 115/2006 *Coll.*, the employer shall make sure that employees exposed to noise that equals to or exceeds the lower operational values of exposure, as well as delegates of employees for the OSH were reasonably informed about and practically trained in possible hazards arising out of exposure to noise.

The practical training and information pertain especially to [11]:

- Natures of such risks
- Measures adopted in relation to elimination or mitigation of exposure to noise to the lowest possible level, inclusive of conditions under which such measures are being implemented
- Exposure to noise limit and operational values
- Results of assessing and measuring of noise along with substantiation of their importance and possible hazards
- Proper use of hearing protectors
- Reason for and manner of determining and publicising symptoms of damages hearing

Table 5 Measures resulting from exceeding the upper operational values

Measures necessitated by exceeding the upper operational values	*Measure to be adopted by the employer*
The employer shall forthwith adopt measures ensuring elimination or alleviation of exposure to noise	Through providing PPWI (personal protective work instruments)
Identified shall be workplaces or worksites at which employees are exposed to or at which it can be anticipated that they will be exposed to noise that exceeds the upper operational values of noise	Identify the workplaces that are hazardous as to the occurrence of excessive noise.
Such worksites or workplaces *shall be clearly delineated and marked by* safety health hazard signs and warning signs, and wherever technically practicable, they will be declared *restricted areas*	Caution and warning signs
If exposure to noise equals to or exceeds the upper operational values of exposure, the employees *must use hearing protectors.* The employer shall ensure and supervise usage of these protective instruments and will be held responsible for effectiveness of measures relating to providing PPWI (personal protective work instruments).	Plugs with mean attenuation SNR 28 dB Daily checked by the shift supervisor and the foreman Once or twice a week checked by qualified OSH technician

Source: original (personal) processing

- Circumstance under which the employees are entitled to appropriate preventive medical examinations
- Safe work procedures from the point of decreasing exposure to noise to the lowest level possible

5 Conclusions

We focused on noise, which is one of the underestimated risk factors in the workplace by the employees themselves. Employees frequently fail to acknowledge impacts of the factor upon their health, which is in contrast with factors such as dust or extreme physical load that are perceived as tangible or clearly visible. Hence, from the part of employers, it is necessary to pay sufficient attention to the risk factors discussed.

Based on the steps taken, we arrived at the results that within the profession of workman at the treatment of industrial minerals, the peak level C of acoustic pressure LCPk was exceeded by 11.2 dB, when at a time the standardised level of exposure to noise LAEX,8 h was exceeded by 17.7 dB, which means that the profession should be included in risk category 4. According to the provisions of

Act no. 355/2007 Coll., based on the proclaimed risk category 4, the employer must adopt measures for the elimination of these risks at the workplace without any delay and within a period of one year adopt any possible measures to degrade the category.

This contribution is a partial result of the project, solving VEGA MŠVVaŠ SR 1/0515/18 'The decision-making model of process of evaluating raw material policy of regions'.

References

1. Cehlár, M., Teplická, K., & Seňová, A. (2011). Risk management as instrument for financing projects in mining industry. In *SGEM 2011: 11th International multidisciplinary scientific geoconference* (Vol. 1, pp. 20–25, 913–920). STEF92 Technology Ltd., Bulgaria, Albena, Sofia.
2. Jurkasová, Z., Cehlár, M., & Khouri, S. (2016). Tools for organizational changes managing in companies with high qualified employees. In *Production management and engineering sciences* (pp. 409–412). Leiden: CRC Press/Balkema.
3. Act No. 355/2007 Coll. on protection, support and development of public health and on amendments and supplements to certain acts.
4. Seňová, A., & Čulková, K. (2012). Risk management in area of security and protection of health during the work 2012. In *New technologies – Trends, innovations and research* (pp. 347–376). Rijeka: InTech.
5. Šukalová, V. (2006). Riadenie OSH – moderný prístup. (OSH Management – a Modern Approach). In Práce a mzdy č. 1/2007. Žilina: Poradca s. r. o. http://www.poradca.sk/SubPages/OtvorDokument/Clanok.aspx?idclanok=59733&zor=1 2010/11/14.
6. OSHA European Agency, Historical background on occupational safety and health legislation. http://osha.europa.eu/sk/legislation/directives 2011/02/02.
7. Seňová, A., Teplická, K., & Pavolová, H. (2015). The process of risk assesment method RBMI in the mining enterprise. In *SGEM 2015* (pp. 705–712). Sofia: STEF92 Technology.
8. European risk observatory. Outlook 1 New and emerging risks in occupational safety and health, Luxemburg: The Publications Office of the European Union 2009. p. 24. http://osha.europa.eu/sk/publications/outlook/te8108475enc_osh_outlook 2013/28/01.
9. Guideline No. 30/2010 of the SR Ministry of Health of 07. 09. 2010 that adjusts the procedure used at assessing the exposure to noise level in employees using hearing protectors.
10. SR Government regulation No. 115/2006 Coll. on minimal health and safety requirements for protecting employees against the exposure to health related risks and hazards.
11. Hnilica, J., & Fotr, J. (2009). *Aplikovaná analýza rizika ve finančním managementu a investičním rozhodování. (Applied risk analysis in financial management and investment decision-making)* (p. 124). Praha: Grada Publishing.

Chemical Management in Automotive and Mass Industry

Jozef Zajac, Michal Hatala, Svetlana Radchenko, Darina Duplakova, and Frantisek Botko

1 Introduction

The main task of the processing medium is to allow the machining with the greatest efficiency. This is mainly to ensure tool life and machined surface quality at low energy consumption, the impact of cooling and lubricating and cleaning effect of the media used. Thus, these effects are the basic, most needed properties of the processing medium, and some of the functions may be more or less emphasized, depending on the conditions of the operation [1].

From the technological and operational point of view, these requirements can be compiled as follows [2]:

- Cooling effect
- Lubricant effect
- Cleaning effect
- Protective effect
- Cutting effect
- Operational stability
- Health safety
- Low cost
- Others

The necessary number of process media types is conditioned by the purpose of their use and the requirements put on them.

J. Zajac · M. Hatala · S. Radchenko · D. Duplakova · F. Botko (✉)
Technical University of Kosice, Faculty of Manufacturing Technologies with a Seat in Presov, Presov, Slovakia
e-mail: frantisek.botko@tuke.sk

L. Knapcikova et al. (eds.), *4th EAI International Conference on Management of Manufacturing Systems*, EAI/Springer Innovations in Communication and Computing, https://doi.org/10.1007/978-3-030-34272-2_15

Depending on the operating conditions, a greater or lesser emphasis is put on the individual basic effects of the media, giving rise to a basic classification of processing media to liquids with predominantly cooling effect (water-miscible) and liquids with predominantly lubricating effect (water-immiscible) [2].

The classification of cutting fluids according to STN 22 0131 is as follows [3]:

1. Aqueous solution
2. Oil products used for cutting fluids
   (a) Aqueous emulsions of emulsifying oils and lubricants
   (b) Oils without additives
   (c) Cutting oils with fatty substances (animal or vegetable origin)
   (d) Cutting oils with chemical additives or fatty substances
3. Concentrates of high-pressure additives
4. Fatty substances of animal or vegetable origin
5. Synthetic liquids

At present, the classification of processing media according to foreign sources leads to some simplification of classification. Manufacturers also adapt to this classification and offer their product range according to the following classification. For example, according to Leiseder, the classification of processing media is as follows:

Cutting fluids:

1. Water miscible: aqueous solutions, emulsion liquids
2. Water immiscible

Also, according to the same source of water-miscible processing media based on composition, the distribution is as follows:

1. Synthetic liquids
2. Semi-synthetic fluids
3. Emulsion fluids

During the application, ageing of processing media occurs due to their oxidation, thermal and mechanical stress, pollution and catalytic effect of metals.

Aqueous solutions and emulsions, i.e. water-based media, are subject to change most rapidly. The water has high evaporation and due to the temperatures resulting from the cutting process, there is increased evaporation which causes the electrolyte salts to fall out of the aqueous solutions and settle in the powder layer on the machine and its functional surfaces, making the machine run more difficult. Electrolytes disrupt machine coatings and can cause health problems by sticking to workers' skin. Loss of salts leads to a reduction of pH and a risk of corrosion. At the same time, impurities and/or lubricating oils penetrate into the solutions. In case of heavy contamination, the conditions for the reproduction of bacteria arise. Sudden changes also occur in classical emulsion fluids, although the consequences are not as severe in all respects as in aqueous solutions. Acidic and basic soaps are formed, the stability of the emulsion is disrupted by the aggregation of the dispersed phase particles and their sedimentation and attachment of the fatty components to the chips. This coarsens the emulsion (enlargement of the dispersion phase particles), depletes it (concentration reduction) and disrupts the stability of the emulsion. The consequences are in reducing pH and deteriorating against corrosive properties. At the same time, the microorganisms are multiplied in the emulsion if the emulsion is polluted and especially if the other conditions are favourable [4].

They are at least subject to changes in partially synthetic bioresistant type emulsions, the properties of which, provided that there is no excessive leakage of lubricating oil from the machine and impurities, are not significantly altered even with long operating times. Fatty oils and oily substances are considerably more stable in operation, and they are ageing, which is manifested in these types of oils mainly by the formation of sticky substances of a resinous nature which, due to the heavily aged oil and dirt, fall out of the oil and form sticky deposits on the machines. At the same time, their viscosity increases [5].

Petroleum oils and cutting oils with chemical additives are the most stable in operation. They are also changing, but they do not manifest as behaviour as fatty and oily oils. Their properties do not change within the prescribed period, provided that they are not heavily contaminated, in particular by moisture. The penetration of water greatly accelerates their ageing and creates the risk of corrosion. Fatty substances, petroleum and cutting oils do not carry bacteria unless they consistently contain at least 2% water by weight [6].

Various physical and chemical indicators are followed for assessing the status of process media:

- PH value
- Concentration
- The amount of bacteria
- Nitrite and nitrate content
- The content of foreign oils
- Viscosity
- Flash point
- Water content
- Contaminant content
- Ageing resistance
- Others

A wide range of processes for machining and forming metal materials are used in mechanical engineering. Greases that serve as lubricants and coolants are an indispensable aid. On the proper lubricant selection and properties, the quality of the surface of the workpiece, the life of machine tools and the like depend [7].

Material temperature reaches 200–500 °C according to its type, cutting speeds and chip cross-section, which is important from the point of view of material strength. Lubricating fluid effect is manifested in the formation of acceptable cut-off film in subtle erosions surface with plastic deformation of the material that facilitates deformation of the material delays in the creation of the various elements chip, thereby reducing deformation work [8].

To select the machining fluid, it is crucial to take into account the following:

- Machining operations at the beginning of the process requires fluid with the cooling and lubricating effect and the end of the process fluid with high lubricating effect. With increasing cutting speed and chip thickness is increasing pressures and temperatures in the cutting area and thus the demand for metal cutting fluids.

Nowadays, especially in companies producing components for automotive and industrial bearings, there is a trend towards services from non-domestic sources – outsourcing [9].

In the fluid management, it is possible to talk about several levels of outsourcing:

Level 1
Service company is servicing only one type of product (e.g. metalworking liquids) from a single source and purchase is a matter of logistics department of the final customer.

Level 2
Service company is servicing one type of products from multiple vendors and purchase is a matter of logistics department of the final customer.

Level 3
Service company is servicing one type of products from multiple vendors, including purchase

Level 4
Service company is servicing all types of products from multiple vendors, including purchase

Level 5
Service company is servicing one type of products from multiple vendors, including purchases and directly participates in optimizing the selection of individual products

Successful operation of the process media in the production process is the introduction of outsourcing to proceed in phases:

Phase 1
Creating a preparatory team for the implementation of the comprehensive care of process media. This team includes representatives of the service company and the employees of the final customer. The team includes staff from the departments of energy, technology, logistics, quality of production and technical offices, who are preparing to enter the project

Phase 2
The preparatory phase start-up fluid or chemical management in the production process in one module, where there is the quantification of individual partial tasks in a real production environment.

Phase 3
Implementation of fluid or chemical management in the production process in a single module. Duration of this phase depends on the nature of the project, subsequently the module takes 3 months and corrections to the initial entry requirements.
Phase 4
The preparatory phase start-up fluid or chemical management in the production process in other modules. On the basis of the facts of the deployment and operation of the first module, it can be done with prediction problems in other modules. This phase partly overlaps with phase 3.
Phase 5
Implementation and verification of fluid or chemical management in the production process in all modules with partial correction demands of both parties.

The results of the work of the team are [10]:

- Reduced costs per unit of production
- Optimization of process media in all modules
- Improving the quality indicators of individual media
- Increasing the life of the individual media and reducing downtime
- Raising awareness and increasing the qualifications of operating staff

2 Practical Approach to Chemical Management in Automotive and Mass Industry

Practical approach to chemical management in automotive and mass industry is presented by determination of the effect of the process media grease properties on the frictional wear. After the realization of the experiments, the worn areas were measured. There were determined the dependencies of the size of these areas on time and the dependence of the engine power on the size of worn area.

The experiments were performed at constant load force and constant speed. Only time intervals were changed. By recording the change in current, the dependence between the wear area and the engine power was determined. The experiment was performed at a constant load of 15 kg and a speed of 1450 rpm. The time interval of 2 min is followed, which gradually increases depending on the size of the worn area. The multimeter records the electrical current by which the motor power is calculated during each measurement. Finally, the dependence between this power and the size of the worn area is determined (Fig. 1).

The following requirements were taken into account during the experiments:

- Use of appropriate and available process media
- Workspace technical equipment
- Testing device design

Fig. 1 Clasp multimeter M97A

Table 1 Recommended cutting fluid concentrations

Material	Aluminium and aluminium/copper alloys (%)	Cast iron, carbon steel and low alloy steel (%)	Alloys, alloy steel and stainless steel (%)
Operation			
Grinding	3.0	3.0	3.5
Conventional machining	3.5	4.0	5.0
Other machining and thread cutting	5.0	5.0	6.0

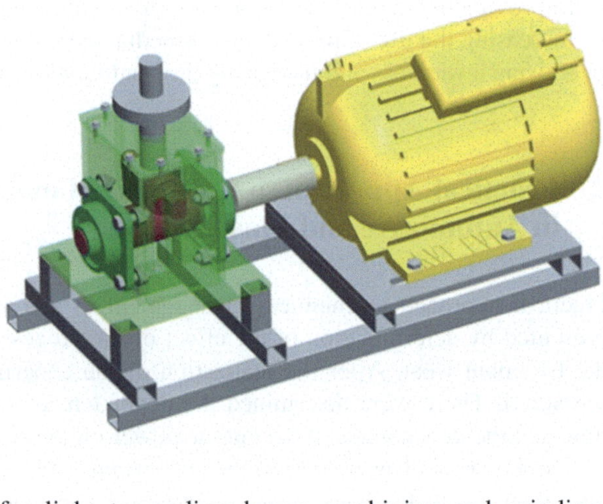

Fig. 2 Test apparatus

Emulsion cutting fluid for light to medium-heavy machining and grinding, Cimstar S 300 – a complex mixture of chemicals and mineral oil, which can be used in a wide range of machining operations – was chosen to carry out the experiments. It is very stable and resistant to bacteria and secondary contaminants, including oil, and maintains good working and cleaning conditions. A CIMCOOL T.A. kit was used to determine the cutting fluid concentration (Table 1).

The test apparatus (Fig. 2) was designed as an experimental device and consists of the following parts: electric motor, shafts, housing, bearing case, clutch, tank and pump.

Fig. 3 Mounting of the shafts

Transmission of the torque from the electric motor to the lower shaft is ensured by the coupling. The mutual mounting of the shafts is intermittent (Fig. 3). The loading force is exerted on the top shaft, which, unlike the lower shaft, does not rotate. When the shafts are in contact with each other, friction occurs at the point of contact, which, as a result of increasing wear, changes to surface contact. The greater the load force and the longer the measurement time, the greater the wear area.

The following table (Table 2) shows the values that were achieved in the implementation of the first experiment – measurement of wear during the dry friction.

There is a rapid increase in wear after the first measurement, and this wear increases by an average of only a few tenths of a millimetre. The friction surface is much larger after the first measurement than after the end measurements using the cutting fluid. Due to the high temperature, the worn surface is recrystallized and tempered (Figs. 4, 5 and 6).

Table 2 Achieved values – dry friction

	t (min)	a (mm)	b (mm)	S (mm^2)	I (A)	Pm (W)	m (kg)
1.	0	0	0	0	3,5	2889,6	0
2.	2	12,5	13	510,25	3,6	2972,16	15
3.	4	14,3	14	628,628	3,6	2972,16	15
4.	5	14,5	14	637,42	3,6	2972,16	15
5.	6	14,9	14,1	659,6826	3,6	2972,16	15
6.	7	15	14,2	668,82	3,6	2972,16	15
7.	8	15	14,5	682,95	3,5	2889,6	15
8.	10	15	14,6	687,66	3,5	2889,6	15
9.	12	15	14,8	697,08	3,6	2972,16	15
10.	15	15	14,9	701,79	3,6	2972,16	15
11.	18	15,1	15	711,21	3,4	2807,04	15
12.	21	15,1	15	711,21	3,5	2889,6	15
13.	26	15,2	15	715,92	3,5	2889,6	15
14.	31	15,3	15,2	730,2384	3,5	2889,6	15
15.	38	15,6	15,3	749,4552	3,5	2889,6	15

Fig. 4 Wear caused by dry friction

During the second experiment, a 100% semi synthetic cutting fluid concentrate was applied. The results are shown in the following table (Table 3).

The second of the three experiments carried out is purely experimental for the reason that 100% concentrate of this cutting fluid is not used in automotive industry. The measurement was found to be successful and at the end it was found that the resulting area was reduced to almost half and also that it did not recrystallize because temperature changes were minimized (Figs. 7, 8 and 9).

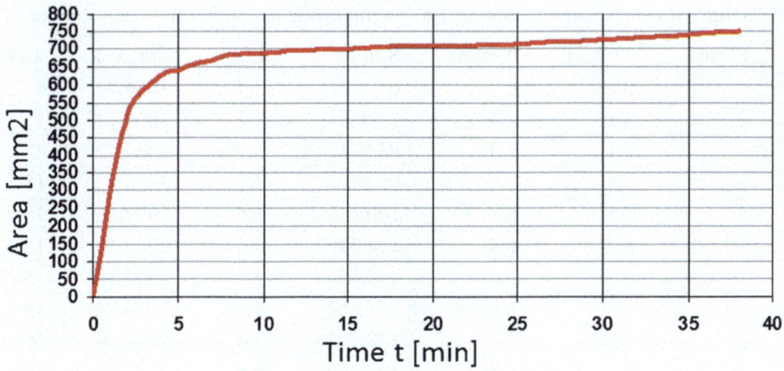

Fig. 5 The course of area and time dependence – dry friction

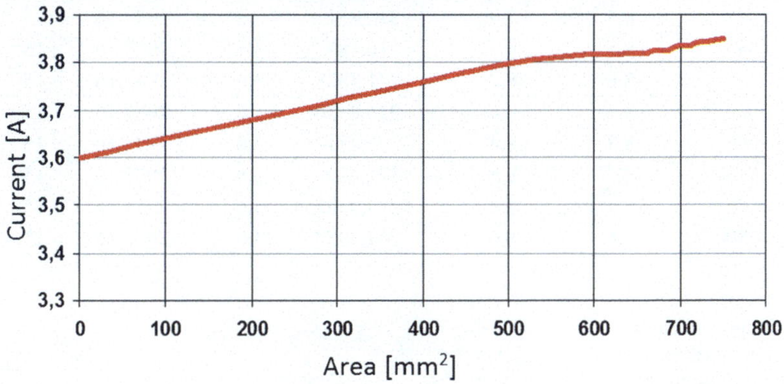

Fig. 6 The course of current and area dependence – dry friction

During the last experiment, concentrated cutting fluid (4% semi synthetic cutting fluid concentrate) was used. A refractometer was used to check the correct concentration, a table digital pH meter was used to determine the pH of the concentration obtained, and indicator strips were used to determine the hardness (water, mix) and the nitride and nitrate content of the sample. Cutting fluid samples were taken after each measurement followed by detection of impurities (Table 4).

4% concentration semi synthetic cutting fluid is used in engineering as a multi-purpose cutting fluid for machining and grinding of ferrous and most non-ferrous metals, including stainless steels and tool steels. From this reason, this experiment was the greatest importance in practical terms. The resulting worn surface is the smallest of all measurements, and temperature changes that could adversely affect the entire friction process have also not occurred (Figs. 10, 11 and 12).

Table 3 Achieved values – application of 100% semi synthetic cutting fluid concentrate

	t (min)	a (mm)	b (mm)	S (mm^2)	I (A)	Pm (W)	m (kg)
1.	0	0	0	0	3,7	3054,72	0
2.	2	4,1	4,5	57,933	3,7	3054,72	15
3.	4	5,1	5,3	84,8742	3,7	3054,72	15
4.	6	5,4	5,7	96,6492	3,7	3054,72	15
5.	8	5,5	5,8	100,166	3,6	2972,16	15
6.	10	6,2	6,4	124,5952	3,6	2972,16	15
7.	15	6,7	6,6	138,8508	3,6	2972,16	15
8.	20	7	6,9	151,662	3,7	3054,72	15
9.	25	7,8	7,3	178,7916	3,8	3137,28	15
10.	30	8	7,5	188,4	3,8	3137,28	15
11.	35	8,1	7,7	195,8418	3,7	3054,72	15
12.	40	8,2	7,8	200,8344	3,7	3054,72	15
13.	45	8,5	7,9	210,851	3,6	2972,16	15
14.	50	8,5	8,1	216,189	3,6	2972,16	15
15.	55	8,6	8,2	221,4328	3,6	2972,16	15
16.	60	8,6	8,4	226,8336	3,6	2972,16	15
17.	65	8,7	8,4	229,4712	3,6	2972,16	15
18.	70	8,7	8,5	232,203	3,6	2972,16	15
19.	75	8,7	8,6	234,9348	3,6	2972,16	15
20.	80	8,8	8,6	237,6352	3,6	2972,16	15

Fig. 7 Wear caused by application of 100% semi synthetic cutting fluid concentrate

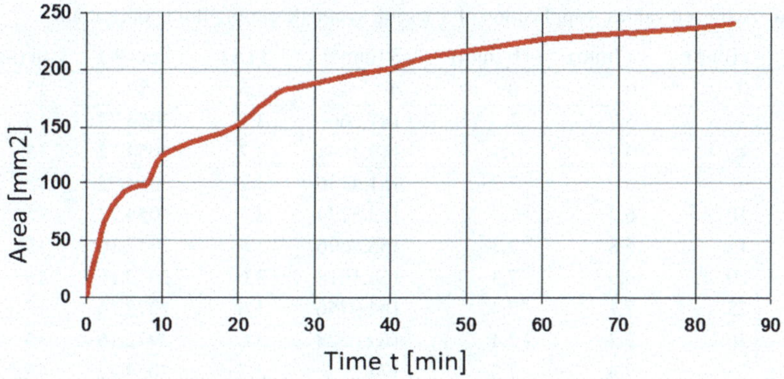

Fig. 8 The course of area and time dependence – application of 100% semi synthetic cutting fluid concentrate

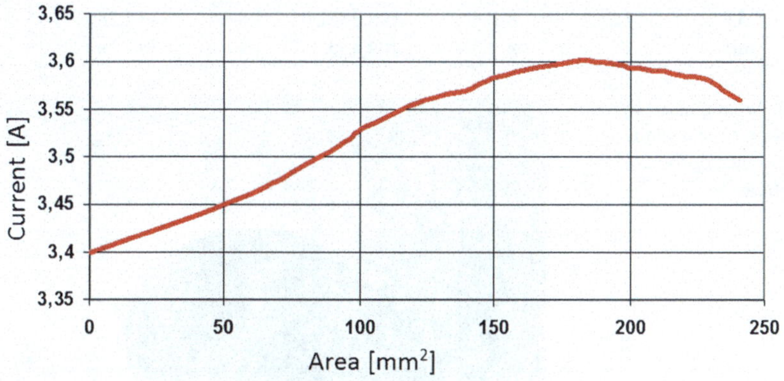

Fig. 9 The course of current and area dependence – application of 100% semi synthetic cutting fluid concentrate

3 Conclusion

Based on the experiments carried out, it was found that the method of lubrication and the type of process media significantly affect the entire course of friction surfaces. The biggest friction and therefore the biggest friction surfaces were created during dry friction. Significant temperature changes were observed in this friction method which caused recrystallization on the resulting friction surface. Fast wear growth after the first measurement was noted, with an average increase of only a few tenths of a millimetre. The friction surface is much larger after the first measurement than after the end measurements using the cutting fluid. The implementation of the second 100% semi synthetic cutting fluid concentrate experiment was purely

Table 4 Achieved values – application of 4% semi synthetic cutting fluid concentrate

	t (min)	a (mm)	b (mm)	S (mm^2)	I (A)	Pm (W)	m (kg)
1.	0	0	0	0	3,5	2889,6	15
2.	2	6,7	7	147,266	3,7	3054,72	15
3.	4	6,7	7,1	149,3698	3,7	3054,72	15
4.	6	6,7	7,2	151,4736	3,7	3054,72	15
5.	10	6,7	7,3	153,5774	3,7	3054,72	15
6.	15	6,8	7,3	155,8696	3,6	2972,16	15
7.	20	6,9	7,3	158,1618	3,6	2972,16	15
8.	25	7,1	7,4	164,9756	3,6	2972,16	15
9.	30	7,3	7,4	169,6228	3,6	2972,16	15
10.	35	7,4	7,5	174,27	3,7	3054,72	15
11.	40	7,5	7,6	178,98	3,7	3054,72	15
12.	45	7,6	7,8	186,1392	3,7	3054,72	15
13.	50	7,7	8	193,424	3,8	3137,28	15
14.	55	7,8	8,1	198,3852	3,8	3137,28	15
15.	60	8	8,1	203,472	3,8	3137,28	15

Fig. 10 Wear caused by application of 4% semi synthetic cutting fluid concentrate

experimental since 100% concentrate of this cutting fluid is not used in mechanical engineering. The measurement was judged to be successful with the conclusion that the resulting area was reduced to almost half and also not recrystallized because temperature changes were minimized. 4% concentration semi synthetic cutting fluid is used in engineering as a multi-purpose cutting fluid for machining and grinding of ferrous and most non-ferrous metals, including stainless steels and tool steels. Therefore, this measurement was of the utmost importance. The resulting worn surface is the smallest of all measurements, and temperature changes that could adversely affect the entire friction process have not been shown.

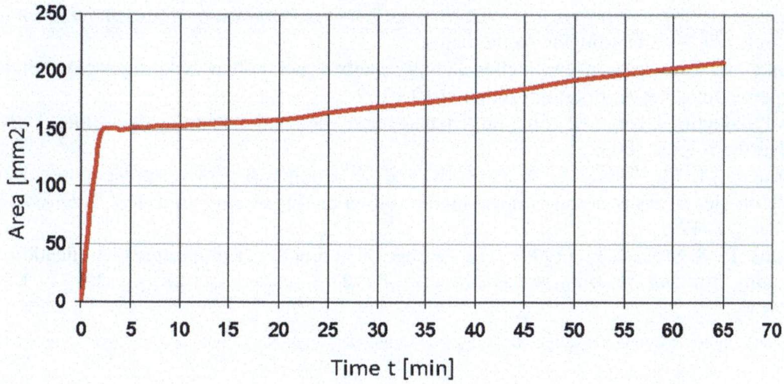

Fig. 11 The course of area and time dependence – application of 4% semi synthetic cutting fluid concentrate

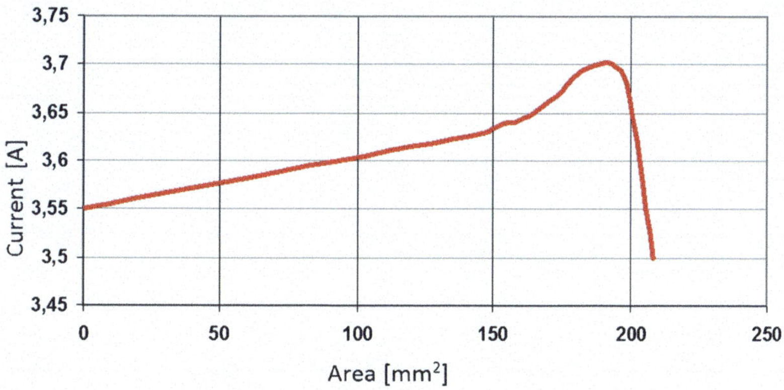

Fig. 12 The course of current and area dependence – application of 4% semi synthetic cutting fluid concentrate

Acknowledgement This work was supported by grants VEGA 1/0492/16 and KEGA 039TUKE-4/2017.

References

1. Brinksmeier, E., Heinzel, C., & Wittmann, M. (1999). Friction, cooling and lubrication in grinding. *CIRP Annals, 48*(2), 581–598.
2. Coughlin, R. W., Williams, D., Seveau, E., Veith, R., & Howes, T. (1992). Enumeration of microorganism in metalworking fluids using photometric methods. *CIRP Annals, 41*(1), 357–359.
3. STN 22 0131. Metal-working machines. Cutting compounds.
4. Zajac, J., & Paško, J. (2003). Verifikácia procesných kvapalín v ložiskovom priemysle. *Strojarstvo, 1/2003*, 23.
5. Pusavec, F., Stoic, A., & Kopac, J. (2009). The role of cryogenics in machining processes. *Tehnički vjesnik, 16*(4), 3–10.

6. Zajac, J. (1998). *Perspektywiczne trendy w szlifowaniu kształtowym z uwzglednieniem efektywnosci. TPI 98*. Vlaardingen, Netherlands.
7. Minke, E. Einsatz konventioneller Schleifscheiben bei hohen Schnittge-schwindigkeiten, Schleiftechnisches Kolloquium Achen 9/10.10.97.
8. Iowa Reduction Center – Cutting fluid management for small machining operations, University of Northern Iowa 2003.
9. Harnicarova, M., Zajac, J., & Stoic, A. (2010). Comparison of different material cutting technologies in terms of their impact on the cutting quality of structural steel. *Tehnicki vjesnik, 17*(3), 371–376.
10. Bojtoš, P., & Štekláč, D. (1999). The cutting fluids functional properties identification, when grinding. In: *2nd International conference of PhD Students*, 8.–14.8 (pp. 29–34). Miskolc, Hungary.

Impact of BIM Technology on Development of Digital and Managerial Competencies of Project Managers in Construction Industry

Peter Mesaros, Tomas Mandicak, Annamaria Behunova, Jana Smetankova, and Katarina Krajnikova

1 Introduction

The construction project presents a series of relative processes. The success of the project is primarily influenced by the order of these processes. The primary aim is to execute individual processes in the right order and in the right way. The project has temporary character and results of these processes are a specific product. The construction project is a one-time organized process of construction, renovation, etc. An important factor in the project implementation process is the project team from which the success of the project depends [1]. Construction projects are strictly defined by result requirements, the cost and time constraints and are bounded by the environment in which they are implemented [2].

Project management is a basic element of any construction project. The construction project manager must acquire several skills and competences through which he manages the whole team. Construction projects are characterized by many unpredictable changes that need to be addressed promptly, which is the key to the stability of process [1]. Construction project management is the direction, regulation and supervision of the project from development to its completion. The aim of

P. Mesaros · T. Mandicak (✉) · J. Smetankova · K. Krajnikova
Technical University of Košice, Faculty of Civil Engineering, Košice, Slovakia
e-mail: peter.mesaros@tuke.sk; tomas.mandicak@tuke.sk; jana.smetankova@tuke.sk; katarina.krajnikova@tuke.sk

A. Behunova
Technical University of Košice, Department of Industrial Engineering and Informatics, Prešov, Slovakia
e-mail: annmaria.behunova@tuke.sk

L. Knapcikova et al. (eds.), *4th EAI International Conference on Management of Manufacturing Systems*, EAI/Springer Innovations in Communication and Computing, https://doi.org/10.1007/978-3-030-34272-2_16

the project is to fully satisfy the client's requirements for functionality and budget [1]. Decision-making in construction project is influenced by many factors and its difficult process [3, 4].

A construction project manager is a responsible person for planning, coordinating, budgeting and overseeing the project. A construction project manager is responsible for the following activities [1]:

- Putting together the budget and negotiate cost estimates
- Choosing the most efficient construction method and strategies
- Staying in touch with the clients for works- or budget-related issues
- Arranging the work timetables
- Discussing about technical and contract details with workers and other professional parties
- Cooperating with building and construction specialist
- Keeping an eye on the personnel in the construction site

Construction project management refers to project management tasks and managers based on their past experience. It is more useful than following a theoretical approach in new case [5]. The primary objective of construction project management is to manage construction through the specification of project objectives and plans, including developing the scope, schedule, budget, decision-making and selection of project participants, increasing resource efficiency (worker, equipment), supporting coordination and contracting, planning, estimating, design and construction and effective communication development [6].

Learning process is a first step and presumption for better results of every construction manager. The achievement of higher level of competencies increases the precondition for its successful results in the management of construction projects [7]. Increasing of information and communication technology impacts productivity levels and developing of managerial competencies of managers [8].

El-Baz and El-Sayegh described four groups of managerial competencies for construction industry: concretely technical competencies, management competencies, leadership management and financial competencies. Technical competencies include statistical analysis, decision analysis, resource optimization, information technology control, SCM and so on. Management competencies include typical planning and managerial competencies as strategic and operation planning, human resource planning, change management and so on. Leadership competencies are ability of managers to have effective communication, vision, responsibility and motivation. The last one represents financial competencies. It includes cost management ability, financial analysis, accounting and so on [9].

Kerzner in his study mentioned leadership abilities, creativity, ability to make decisions, ability to identify problems, ability to organize work to subordinates, effectiveness, ability to persuade and having ambition, vision and leadership abilities [10]. Project Manager Competency Development (PMCD) by Project Management Institute (PMI) standard defines three project managers' competence areas. They are knowledge competencies, performance competencies and personal competencies. International Project Management Association defined 20 technical competency elements and 15 behavioural competency elements. It includes 11 contextual competency elements too [11].

Building information modelling (BIM) is a digital representation of the physical and functional characteristics of a device that connects project information databases [12]. BIM is defined as the use of ICT technology to make building more efficient across a project life cycle to provide safer and more productive environment, eliminate environmental impact and promote efficient operational management and use of the building across its life cycle [13].

BIM is an intelligent 3D model-based process that gives architecture, engineering and construction professionals the insight and tools to more efficiently plan, design, construct and manage buildings and infrastructure. Application of BIM is steadily gaining popularity in construction industry, and as such, BIM knowledge represents an important learning outcome in higher students' education. The main problem in this area, which is related to BIM training in building management, is to define the learning outcomes, course curricula and specific knowledge needed for future building and project managers [14].

In the context of BIM and learning competencies of projects was set the most used and mentioned project competencies and their main groups. These were set on literature review and mentioned sources (Table 1).

Based on these facts, BIM is a tool that helps eliminate costs and time load across the building's life cycle, and probably, it has a positive effect on development of digital and managerial competencies. Of course, on productivity and economic parameters impact, there are other factors like materials and other aspects with them (e.g. waste management and so on.) [15, 16]. In spite of this fact, BIM technology includes many points of view. Currently, the industry is undergoing rapid changes that will also affect the population and one of the reasons why BIM technology is here [17].

Table 1 Groups of competencies and competencies of project managers

Technical competencies	Behavioural and personal competencies	Financial and economic competencies
Ability to organize work to subordinates Ability to make decisions Ability to formulate goals Ability to use project management software Ability to design project	Ability to communicate Ability to motivate team members Help in solving problems Ability to resolve conflicts Ability to humans resource planning Ability to motivate Ability to take a responsibility	Ability to manage the scope, time and cost of the project Ability to cost management Budgeting and accounting

Source: own processing based on literature review

2 Methodology

The data collection was done by a questionnaire survey form. The questionnaire is one of the most common methods of research use. It is used for mass and faster detection of facts, attitudes, values, opinions, etc. The questionnaire contained simple and comprehensible questions about the learning competencies of project manager's development by use of BIM technology. The structure of the questionnaire was basic information about the project managers and construction projects. Another part of the questionnaire had direct questions about the research problem. Project managers used the Likert scale for the quantification of impact in the main research questions (1, low impact on development of digital and managerial competencies; 5, high impact on development of digital and managerial competencies).

The research sample consisted of project managers who participated in construction projects in Slovakia and Czech Republic. Project managers and construction projects were selected randomly. The return rate was 4.89%, which means 147 project managers. The research sample therefore incudes 2% of project managers of large companies, 19% of project managers who work in medium-sized companies and 79% of project managers who work in small and microenterprises.

A similar research sample was in Czech Republic. Totally, participants (project managers) in Czech Republic were 142.

For determining the weights and the resulting model, the Analytic Hierarchy Process (AHP) method was used. The AHP method to support multi-criteria decision-making was originally developed by Prof. Thomas L. Saaty. The basis for decision-making is empirical decision-making criteria and using AHP is to outline the whole decision problem as a hierarchical structure [18]. AHP is based on the value of the information obtained, derives ratio scales from paired comparisons of criteria to discover and correct logical contradictions. Questionnaire surveys in general contain subjective opinions of respondents. These opinions were taken as inputs and divided according to alternatives $A_1,...,A_m$ (competencies) and specific criteria $K_1,...,K_n$ (abilities of project managers). This process allows to translate these opinions into measurable numeric relations.

Internationally, AHP is used in a wide range of applications, for example, for the evaluation of suppliers, in project management, or for our selection of the best alternative in the decision tree. As a result, the main model represents the intensity of each competency acquired through the use of BIM technology, where priorities (weightings) and a consistency ratio could be calculated. AHP helped us to make decisions in a more rational way and to make them more transparent and better understandable [19]. Using AHP as a supporting tool for decision-making helps to gain a better insight in complex decision problems. It means the selection of the best alternative in the decision tree. As you need to structure the problem as a hierarchy, it forces you to think through the problem, consider possible alternatives (decision criteria) and select the most significant criteria with respect to the decision objective.

The criterial matrix for the evaluation of alternatives according to criteria is represented by $Y = (y_{ij})$, where rows are alternatives and columns are criteria. For every criterion, it was necessary to calculate its weight (number from 0 to 1). The more important the criterion, the higher the number of its weight (denoted v_j for criterion $K_j, j = 1,...,n$). Some of the methods of criteria weights:

Method of entropy: no preferences, weights of criteria are equal ($w_j = 1/n$)

Method of order and Fuller's method: ordinal information about criteria are known

The Scoring Method and the AHP Method (Saaty's method): ordinal information and distances between criteria are known. Scale of points for criteria are 1, 3, 5, 7, 9, where 1 = criteria are equal, 3 = the first criterion is more significant than the second one, 9 = the first criterion is absolutely more significant than the second one [20].

Mathematically, the AHP method is based on the solution of an Eigenvalue problem. The results of the pair-wise comparisons are arranged in a matrix. Saaty's matrix is squared, consists of estimated elements $s_{ij} \approx \frac{w_i}{w_j}$ which are ratios of weights i-th and j-th criterion. Elements on the main diagonal are equal to 1. In ideal case, for every $i, j, k = 1,\ldots,n$ should apply

$$s_{ij} = \frac{w_i}{w_j} = \frac{w_i}{w_k} . \frac{w_k}{w_j} = s_{ik} . s_{kj} \tag{1}$$

and then, this matrix is called consistent and reciprocal. So, it could be multiplied from right side by normalized right Eigenvector $(w_1, w_2, \ldots, w_n)^T$ (n is Eigenvalue) gives the ratio scale (weighting), the Eigenvalue determines the consistency ratio:

$$\begin{pmatrix} 1 & \frac{w_1}{w_2} & \cdots & \frac{w_1}{w_n} \\ \frac{w_2}{w_1} & 1 & \cdots & \frac{w_2}{w_n} \\ \cdots & \cdots & \cdots & \cdots \\ \frac{w_n}{w_1} & \frac{w_n}{w_2} & \cdots & 1 \end{pmatrix} \cdot \begin{pmatrix} w_1 \\ w_2 \\ \cdots \\ w_n \end{pmatrix} = \begin{pmatrix} w_1 \\ w_2 \\ \cdots \\ w_n \end{pmatrix} \cdot n$$

In practice, it is very rare to obtain fully consistent matrix. Then, the maximal Eigenvalue of matrix is taken and the index of consistency $CI = \frac{\lambda_{max} - n}{n-1}$. As an appropriate value for approximation of weights, Saaty suggested a normalized geometrical mean of row's numbers in Saaty's matrix:

$$w_i = \frac{\sqrt[n]{\prod_{j=1}^{n} s_{ij}}}{\sum_{k=1}^{n} \sqrt[n]{\prod_{j=1}^{n} s_{kj}}}. \tag{2}$$

From these normalized weights for alternatives, the Saaty's normalized matrix was generated where in i-th row and j-th was normalized weight for i-th alternative was derived in Saaty's analysis of alternatives for j-th criterion. For this matrix, the weighted row's sums were determined (according to normalized weights of each criterion). The best alternative was then the one with the largest weighted row's sum. The results were arranged to tree.

3 Results and Discussion

Results of research are divided to the tree hierarchy and show impact rate or level on development of digital and managerial competencies. The main element has a weight value of one. This is then divided into elements at level 2. The formulation of the main objective emerged from the research questions determined on the basis of the theoretical analysis of the issue and the following basic research questions for the area.

First level: Main objective of the AHP model – Analyse the impact of development of digital and managerial competencies by BIM technology.

Second level: Partial objectives of the AHP model – Quantify the impact of the use of BIM technology on selected groups of project manager's competencies.

Third level: Quantify the impact of the use of BIM technology on digital and managerial competencies of project managers.

For more details, BIM technology impacts especially technical competencies. This group of manager's competencies achieved value 0.54, what represents the biggest value on the learning competencies of project managers. The next group was set as financial and economic competencies. The last one was the so-called group of behavioural and personal competencies with a value of 0.17. Separately, BIM is the most positive in the development of ability to design project. It achieved a value 0.19 and it is the biggest value. Interestingly, more than 50% of respondents find this answer more important. The ability to manage the scope and time of the project is the second in this case (Fig. 1).

Project managers achieved this competency when using BIM technology. Ability to use project management software was set as the third important and more positive learning competency by BIM technology. Similar situation is in Czech Republic (Fig. 2).

BIM has the highest impact on development of ability to organize work to subordinates, use project manager software and design project. The most important group to development of competencies were technical competencies. They achieved 0.48 impact rate according to the AHP method. Generally, there are small differences, but globally the results are very similar. It can be from more reasons. First of all, BIM technology is used in similar rate and level as in Slovak construction industry. Next, it has very nearest culture and market conditions of these countries. And the last one, a lot of projects and construction companies are in the same markets in

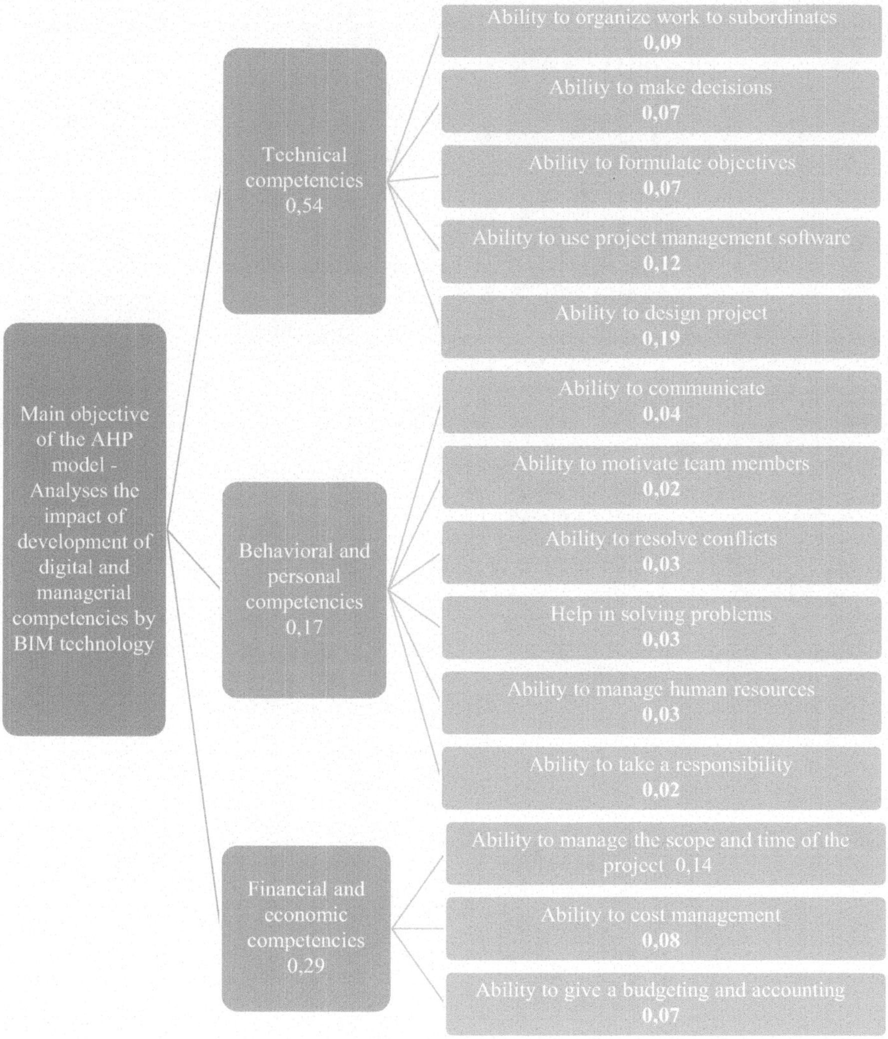

Fig. 1 Development of digital and managerial competencies by BIM technology in Slovak construction industry

Slovakia and Czech Republic. Actually, this model was applied and tested in a large construction company in Slovakia. It is currently exploring whether the theoretical model discovered in this research is applicable in practice and whether it brings real benefits. This is another phase of research that cannot be accelerated.

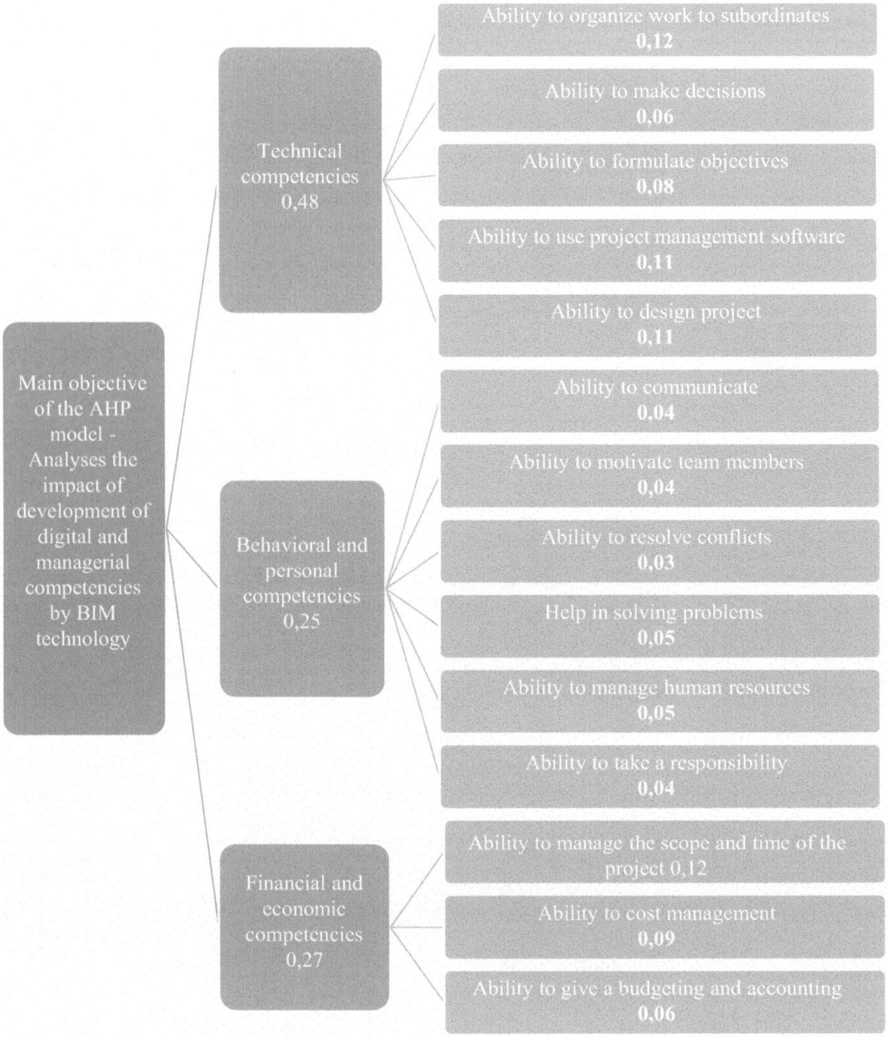

Fig. 2 Development of digital and managerial competencies by BIM technology in construction industry in Czech Republic

4 Conclusion

Management of construction projects is difficult and involves competencies-requirement processes. Project managers present an important issue of construction project management. Currently, digital age is a thankful space for the implementation of technology that helps in the management processes. BIM technology presents one of them. BIM technology can be considered as one of the most effective and powerful tools in construction project management. Research problem and objectives were identified based on theoretical background and market demand. The impact of BIM technology on the development of digital and managerial competencies was analysed and based on AHP method quantified. There were described main impact groups and digital and managerial competencies. Results compare two very similar construction markets in two countries. Results in both countries are very similar. It takes into account a little difference between culture and industry conditions, and it is not a surprise. But some differences were recorded, e.g., in the perception of soft managerial or behavioural competencies. BIM impacts more in Czech Republic than Slovakia on this issue. The survey sample and the similarity of countries showed a strong link between the construction markets and market constraints. Next research questions are oriented to extend research countries. It is worth considering moving research in this direction as well. That means comparison of different countries, e.g., country from West Europe or different continent. This reasoning is supported by the claim that these countries are generally pioneers in the use and implementation of BIM technology. It is important for extending it to the whole world context.

Another area for research is one that is already described in the results. This means that research has progressed to a stage where the model is tested under real conditions. One of the construction companies in Slovakia is in the stage of testing the research conclusions, and therefore in the near future, it will be possible to verify the theoretical model in real conditions. Generalized research results in individual points are as follows main points of results.

Technical competencies are the most important work of the development of competencies of project managers. Ability to design is other individual with the highest impact on the development of manager's competencies.

Acknowledgements This work was supported by the Slovak Research and Development Agency under contract no. APVV-17-0549.

References

1. Designing Building Wiki. (2019). *Construction project.* [Online] Available at: https://www.designingbuildings.co.uk/wiki/Construction_project. Accessed 12 May 2019.
2. Mandičák, T. & Mesároš, P. (2017). Impact of ERP systems on cost reducing in construction project management in Slovakia. *OTMC 2017.* – Zagreb: Croatian Association for Construction Management. (pp. 345–355). – ISBN 978-953-8168-21-5.
3. Radziszewska-Zielina, E., & Szewczyk, B. (2016). Supporting partnering relations management in construction projects' implementation using AHP and fuzzy AHP, WMCAUS 2016. *Procedia Engineering – Journal – Elsevier, 161*, 1096–1100.
4. Radziszewska-Zielina, E., Śladowski, G., & Sibielak, M. (2017). Planning the reconstruction of a historic building by using a fuzzy stochastic network. *Automation in Construction, 84*, 242–257.
5. Maqsood, A. (2006). Applying project histories and project learning through knowledge management in an Australian construction company. *The Learning Organization, 13*(1), 80–95.

6. Geniebelt, P. *Construction project management processes: Everything you need to know*. [Online] Available at: https://geniebelt.com/blog/construction-project-management-processes. Accessed 12 May 2019.
7. Mesároš, P., et al. (2017). Impact of education and experiences of managers on performance of Slovak construction enterprises. *ICETA 2017*. – Danvers: IEEE, (201, pp. 263–268). – ISBN 978-1-5386-3294-9.
8. Bolek, V., et al. (2016). Factors affecting information security focused on SME and agricultural enterprises. *Agris On-line Papers in Economics and Informatics, 8*(4), 37–50.
9. El-Baz, H. & El-Sayegh, S.M. (2007). *Developing engineering management core competencies*. In Fifth LACCEI International Latin American and Caribbean Conference for Engineering and Technology "Developing entrepreneurial engineers for the sustainable growth of Latin America and the Caribbean: Education, innovation, technology and practice.
10. Kerzner, H. (2005). *Project management* (9th ed.). Hoboken: John Wiley & Sons.
11. Caupin, G., et al. (2009). *IPMA, ICB IPMA competence baseline version 3.0*. Nijkerk: International Project Management Association, June 2006
12. Ahmad Latiffi, A., Brahim, J., Mohd, S., & Fathi, M.S. (2014). The Malaysian government's initiatives in using Building Information Modeling (BIM) in construction projects. *Sustain Solut Struct Eng Constr*. no. Cream, 767–772.
13. Ariyaci, Y., Egbu, C., & Coates, P. (2012). Building information modelling (BIM) implementation and remote construction projects: Issues, challenges, and critiques. *Journal of Information Technology in Construction, 17*(May), 75–92.
14. Kolarić, S., et al. (2018). *BIM training in construction management educational practices in Croatia and Slovakia*. In Proceedings of the Creative Construction Conference. https://doi.org/10.3311/CCC2018-130. Creative Construction Conference 2018, CCC 2018, 30 June – 3 July 2018, Ljubljana, Slovenia Sonja.
15. Knapčíková, L., et al. (2018). Advanced materials based on the recycled Polyvinyl Butyral (PVB). *MMS Conference 2017*. – Ghent: EAI (pp. 1–9).
16. Knapčíková, L., et al. (2016). Material recycling of some automobile plastics waste. *Przemysl Chemiczny, 95*(9), 1716–1720.
17. Behún, M., Gavurova, B., Tkacova, A., & Kotaskova, A. (2018). The impact of the manufacturing industry on the economic cycle of European union countries. *Journal of Competitiveness, 10*(1), 23–39. – ISSN 1804-171X.
18. Saaty, R. (2009). *Participant names and papers*. In Proceedings of the 10th International Symposium on the Analytic Hierarchy/Network Process. Pittsburgh, Pennsylvania: ISAHP.
19. Goepel, K. D. (2018). Implementation of an online software tool for the analytic hierarchy process (AHP-OS). *International Journal of the Analytic Hierarchy Process, 10*(3), 469–487.
20. Saaty, H., et al. (2008). *Group decision making: Drawing out and reconciling differences*. Pittsburgh: RWS Publications. ISBN 978-1-888603-08-8.

Impact of BIM Technology on Time-Reducing in Conditions of Slovak Construction Industry

Peter Mesaros, Tomas Mandicak, Annamaria Behunova, Katarina Krajnikova, Jana Smetankova, and Marcel Behun

1 Introduction

Information and communication technology are important in every field. On-time knowledge and information in construction field can involve project managers and engineers moderate problems on a construction worksite and reduce the time and cost of construction project [1]. Requirements on data and information are globally more and more every day. Currently, construction projects are becoming more complex and difficult to manage [2]. The construction sector is still more making attempts to increase productivity through the use of information and communication technologies (ICT), which certainly include building information modelling [3]. BIM is an intelligent 3D model-based process that provides architect construction experts insight and tools to more effectively plan, design, picture, engineering build and manage buildings and infrastructure [4]. In the minding of construction industry, it using well-structured digital information [5]. Every investment in technology should bring some benefits [6]. Currently, the industry is undergoing rapid changes that will also affect the population [7]. The problem

P. Mesaros (✉) · T. Mandicak · K. Krajnikova · J. Smetankova
Technical University of Košice, Faculty of Civil Engineering, Košice, Slovakia
e-mail: peter.mesaros@tuke.sk; tomas.mandicak@tuke.sk; katarina.krajnikova@tuke.sk; jana.smetankova@tuke.sk

A. Behunova
Technical University of Košice, Department of Industrial Engineering and Informatics, Prešov, Slovakia
e-mail: annmaria.behunova@tuke.sk

M. Behun
Technical University of Košice, Institute of Earth Resources, Košice, Slovakia
e-mail: marcel.behun@tuke.sk

L. Knapcikova et al. (eds.), *4th EAI International Conference on Management of Manufacturing Systems*, EAI/Springer Innovations in Communication and Computing, https://doi.org/10.1007/978-3-030-34272-2_17

statement is focused on examining the impact of BIM technology on the benefit that is expected after its successful implementation. That is especially time-reducing in construction process. This can be quantified and measured by several methods [8]. These can also be measured and explored when reconstructing old and historic buildings [9]. It is general assumption that BIM implementation of BIM technology brings time-reducing in construction process. This issue is key for understanding a next statement of trends and implementation of BIM technology in construction industry – in other words, the analysis and quantification of the impact of BIM technologies on time-reducing.

2 Building Information Modelling (BIM)

The construction sector is still making more attempts to increase productivity through the use of information and communication technologies (ICT), which certainly include building information modelling [10]. The topic about building information modelling is not new in construction industry. Many participants of construction industry understand of the term building information modelling than software application but construction design process. For some participants, this technology is a completely new method that requires the implementation of new practices, methods, contracts, relationships, etc.

There are many definitions of the term building information modelling (BIM). BIM presents a digital, three-dimensional virtual project presentation that helps in designing, planning, building and cost planning and building project maintenance [11]. The National Building Information Model Standard Project Committee defined the BIM as follows: BIM presents a digital representation of the physical and functional characteristics of building. It is used as general source of knowledge, information on equipment, construction. BIM is a helpful tool to make decisions throughout the building's life cycle [12].

BIM is 'a repository' that offers to user an object in n-dimensions: 3D model, 4D model with information of time, 5D model with cost information, 6D model with information of sustainability and 7D model with information of facility management.

The pre-investment phase is an important part of the building's life cycle. The main aim at the pre-investment phase is to ensure the development of the documentation which is needed to provide for the functionality, sustainability, operation, safety and security of the building as well as the operating phase itself. One of the benefits of implementing BIM at this stage is to improve coordination and cooperation. BIM is a virtual knowledge database that facilitates easy information exchange. Other benefits include system conflict detection, constructivist assessment and risk elimination, a high level of customization of building systems and building activities, and optimization of time schedule and cost estimation. Cost estimating and setting a time schedule is an integral part of the traditional building process. Poor collaboration and coordination, late detection of building system collisions, inefficient assessment of construction due to 2D information, lack of information to effectively adapt building systems and activities all affect the optimization of cost and time schedule estimates. BIM provides tools that enable early identification and reduction, or complete elimination of collisions in building implementation [13].

Productivity in the investment and operating phase is very important [14]. In the construction industry, we sense productivity as the amount of construction output produced from a set of sources (inputs), such as materials, workers, equipment, etc. [15]. These resources affect costs. A construction is considered productive only if it is effective with respect to the generated inputs. Early identification of needs leads to improve time scheduling. Improving the time schedule brings early delivery of materials, its storage and material handling. These measures will increase construction efficiency. BIM provides a lot of benefits of management process, namely data collection and production management, work efficiency, security and motivation, as well as building maintenance and operation benefits. A lack of security could lead to a delay in the construction schedule and to a reduction in efficiency. In the building maintenance and operation phase, BIM delivers all available information that has been collected from design to commissioning. This information is a valuable data that will help to effectively manage the operation and maintenance of buildings. Maintenance and operation efficiency is demonstrated primarily by reducing the time and cost of repair and maintenance [12, 14].

There are a lot of researches showing the benefits of using building information modelling tools across the building's life cycle. Furneaux and Kivvits recognized the impact of using BIM at cost. Their research has shown significant cost savings through increased interoperability. The use of software applications reduces the user's total cost of up to two thirds, which is $ 15.8 billion a year in the survey (Australia). The research also showed considerable time savings that had to be spent on project preparation itself [15, 16]. Based on these facts, BIM is a tool that helps eliminate costs and time load across the building's life cycle. Of course, on productivity and economic parameters impact others factors like materials and other aspects with them (e.g. waste management and so on.) [17, 18]. In spite of this fact, BIM technology includes many points of view.

3 Methodology

3.1 Research Aim and Hypotheses

The research is focused on the issue of time-reducing and technology in life cycle building and construction process. Based on the theoretical background, the main research aim was set to analyse and quantify the impact of BIM implementation on time-reducing in construction projects.

Based on research questions, research hypotheses were set as follows:

Hypotheses for Research Area: Impact of BIM Implementation on Time-Reducing in Construction Project

H_0: The mean ranks of two groups are equal.

H_1: The mean ranks of two groups are not equal (assume that mean of 'time-reducing before implementation BIM technology' < mean of 'time-reducing after building information modelling implementing').

3.2 *Data Collection and Research Sample*

All data in research were conducted by online form questionnaire. The research sample was approached by e-mail with the request to participate in the research that included Slovak construction enterprises which participated in projects whose structure had several phases.

A total of 111 respondents were interviewed (construction enterprises/developers in Slovakia), which implemented their projects divided into at least two phases. Their constructions also consisted of realization of consistent objects (building objects whose individual types of decisive structures in the production area of the object are evenly distributed and also their laboriousness is evenly distributed, e.g., residential with construction and technologically comparable floors). The companies had information about the time-consuming planning using traditional methods of organizing, managing and construction planning and also information using innovative ways of building, specifically building information modelling. Based on these facts, 24 respondents participated in the questionnaire survey. It represents a good return of 21,62%. Some of respondents did not have information about time-consuming after implementation of BIM technology, or they were only in the first phase of implementing BIM technology and they did not realize any project with this technology. These respondents had to be excluded from our sample.

To determine the answer, we used Likert scale ranging from 1 to 5 on the basis of fixed values that had been done based on the arithmetic average of the values for the selected area under consideration: 1, long-time duration of project (respondent's consideration that project(s) had worse time duration), and 5, short-time duration of project (respondent's consideration that project(s) had better time duration). The extent and interval values for each stage of the scale were thoroughly explained to the respondents.

3.3 *Research Steps and Methodology*

First of all, it was set problem statement based on theoretical background. It was set research questions and hypotheses in this issue. The next step was data collection. All information about the research sample and the data collection process was collected and recorded.

The next step was testing normality. The normality assumption needs to be considered for validation of data presented in the literature as it shows whether correct statistical tests have been used [19]. For this, a very common test called Shapiro-Wilk test was used. It is a test for normal distribution exhibiting high power, leading to good results even with a small number of observations (respondents). The basic idea is to estimate the variance of the sample in two ways: (a) the regression line in the QQ-Plot allows to estimate the variance and (b) the variance of the sample can also be regarded as an estimator of the population variance. This test provides better power than the K-S test even after the Lilliefors correction [19]. The desired significance level alpha in all used tests was 5%. The desired hypotheses: the research sample does have normal distribution (with unspecified mean and variance) or the research sample does not have normal distribution. Computation of this test was done in MATLAB. The research sample did not have normal distribution. Based on this result, Mann–Whitney U test was used. All results of this test are shown in the results section.

To confirm our hypotheses about mean ranks, the Mann–Whitney U test was used. It is nonparametric alternative to t-test and it does not assume any assumptions related to the distribution of scores [20]. This test is used to compare two sample means that come from the same population, and used to test whether two sample means are equal or not. It gives the most accurate estimates of significance, especially when sample sizes are small.

Equation (1) for calculation of the Mann–Whitney U:

$$U = n_1 n_2 + \frac{n_2 (n_2 + 1)}{2} - \sum_{i=n_1+1}^{n_2} R_i \tag{1}$$

where

U = Mann–Whitney U test
n_1 = sample size one
n_2 = sample size two
R_i = rank of the sample size

In the last step, the index of point's change was calculated (Eq. 2):

$$\text{PCH} = \frac{\sum_{i=1}^{n} O_A - O_B}{20n}, \tag{2}$$

where

O_B = respondent's opinion before implementation BIM
O_A = respondent's opinion after implementation BIM
n = sample size

It represents the change of respondent's opinion in researched sample, either increase (positive number) or decrease (negative number), or no change (number close to zero).

4 Results

In the first step, the time-reducing was monitored before and after the implementation of the building information modelling in construction projects. Shapiro–Wilk test was used to determine if the null hypothesis of normality is a reasonable assumption regarding the population distribution of a random sample. The desired hypotheses:

(H_0) Research sample does have normal distribution (with unspecified mean and variance) or (H_1) research sample does not have normal distribution.

Computation of this test was done in MATLAB. We have tested samples individually and also differences between each measurement (duration before implementing – duration after implementing) were tested. These results were obtained (Table 1)

It could be said that research data were not normally distributed. Probability that respondent's perception of improvement time-reducing (duration) of construction projects is in our sample 0.8 (pink coloured area) which means that there could be 80% of respondents who considered BIM implementation in their project(s) as useful – less or more (see Fig. 1). But it should be noted that our sample size is too small for proving this high percentage.

Table 1 Results of research sample distribution

	Duration before implementing	Duration after implementing	Duration differences
Mean	2.125	3.208	1.083
Standard deviation	1.116	1.382	1.283
Number of respondents	24	24	24
Skewness	0.964	−0.084	0.911
Kurtosis	0.509	−1.187	0.549
p-value			0.916

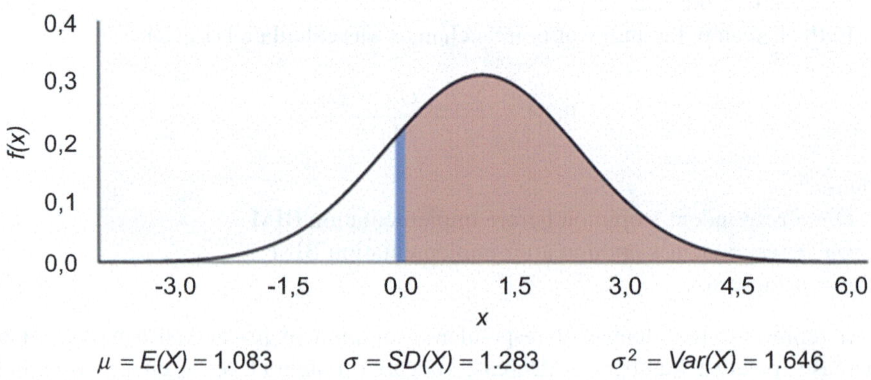

Fig. 1 The graph of distribution of time duration differences

Table 2 Results of Mann–Whitney U test for time duration

	Duration before implementing	Duration after implementing
Sum of ranks	458.5	717.5
Mean of ranks	19.1	29.9
Expected mean of ranks	24.5	24.5
U-value	417.5	158.5
Z-score	−2.65994	
p-value	0.00782	
Expected U-value	288	
Expected sum of ranks	588	

Subsequently, the nonparametric Mann–Whitney test was used. As we mentioned previously, the null and two-sided research hypotheses were stated as follows:

H_0: The mean ranks of two groups are equal.
H_1: The mean ranks of two groups are not equal (assume that mean of 'time-reducing before implementation BIM technology' < mean of 'time-reducing after implementation BIM technology') (Table 2).

H_0 was rejected, so means of samples were not equal. In addition, mean of ranks were numerically compared so we could assume that really 'Duration before implementation of BIM technology' < 'Duration after implementation of BIM technology.' We were also interested in linear correlation between our groups (samples) and it was computed in Excel additionally. The correlation coefficient $r = 0.489$ so it could be talked about soft positive correlation.

The main aim of the research was to show at least a few respondents that implementation of BIM brought a reduction in the time-consuming preparation and in the realization of buildings. The companies which were included in the results had their own experiences of planning the construction process in their project without using BIM and with using BIM.

The index of point's change PCH was equal to 1042 which means that implementation of BIM technology had a positive effect on time-reducing in construction projects. The duration of construction projects is shortened when BIM technology is used. Generally, it means recommendation for implementation of BIM technology as a progressive ICT in construction projects. According to PCH, the duration of construction is shortened. It is an important information for construction sector.

5 Conclusion

ICTs bring many advantages in construction sector. Implementation of ICT brings assumption of positive effects on economic parameters. Project construction management is a difficult process with many stakeholders. BIM technology represents a progressive tool in construction industry. It is beneficial in every stage of construction project. There exist a lot of points of view on the advantage of implementation and use of this technology. Productivity is an important indicator in analyses. Productivity can be perceived through more parameters. Construction industry and construction project bring many specifics. Time or duration is one of the most important parameters in construction project. The time-reducing parameter was investigated in this research in correlation with use and implementation of BIM technology. Research points out the positive impact of BIM technology on time-reducing in construction projects. Generally, time-reducing achieved was about 20% in surveyed respondents and projects. Of course, taking into account the size of the research sample, it has to be stated that this is a trend, but it cannot make clear conclusions for the whole industry. In other site, the use of BIM technology in the Slovak construction industry is on the rise, but it still represents a relatively low utilization rate. From this point of view, the sample includes the real state of construction in Slovakia and thus the findings of this research represent important information and indications for the industry. There are still a few questions from a scientific perspective. What impact would this technology achieve if the BIM technology were used to a greater extent and thus the research sample was larger? Next, it is an international question and view on this. How is it abroad? There are many researches on the use of BIM technology abroad. Our research also addressed the issue of comparing the use of BIM technology from a global perspective. There are also several studies to reduce costs through this technology. The time of construction and the overall reduction in time is a question that is not much solved on a global scale. There is also scope for expanding our research in this area. Therefore, the aim of further research in this area will be a pan-European or worldwide view of the issue.

Acknowledgements The paper presents a partial research results of project VEGA 1/0828/17 'Research and application of knowledge-based systems for modeling cost and economic parameters in Building Information Modeling'.

This work was supported by the Slovak Research and Development Agency under contract no. APVV-17-0549.

References

1. Shu-Hui, J., Shih-Ping, H., & Hui-Ping, T. (2013). Applications of building information modelling (BIM) in knowledge sharing and management in construction. *International Journal of Civil Engineering and Environmental Engineering, 7*(11), 828–832.
2. Travaglini, A., Radujković, M., & Mancini, M. (2014). Building information modelling (BIM) and project management: A stakeholder's perspective. *Organization, Technology and Management in Construction Journal, 6*(2), 1058–1065.
3. Azhar, S. (2011). Building information modeling (BIM): Trends, benefits, risks, and challenges for the AEC industry. *Leadership and Management in Engineering, 11*(3), 241–252.
4. Mesároš, P., et al. (2018). *Applications of knowledge technology in construction industry*. In ICETA 2018: Proceedings – 16th IEEE International Conference on Emerging eLearning Technologies and Applications, pp. 367–372. New Jersey (USA): Institute of Electrical and Electronics Engineers. ISBN 978–1–5386-7912-8.

5. Maqsood, A. (2006). Applying project histories and project learning through knowledge management in an Australian construction company. *The Learning Organization, 13*(1), 80–95.
6. Radziszewska-Zielina, E., & Rumin, R. (2016). *Analysis of the profitability of investment in renewable energy sources on the example of a semi-detached house*. In International Conference on the Sustainable Energy and Environment Development, SEED 2016, E3S Web Of Conferences, 10, 00079.
7. Behún, M., Gavurova, B., Tkacova, A., & Kotaskova, A. (2018). The impact of the manufacturing industry on the economic cycle of European union countries. *Journal of Competitiveness, 10*(1), 23–39. ISSN 1804-171X.
8. Radziszewska-Zielina, E., & Szewczyk, B. (2016). Supporting partnering relations management in construction projects' implementation using AHP and Fuzzy AHP, WMCAUS 2016. *Procedia Engineering – Journal – Elsevier, 161*, 1096–1100.
9. Radziszewska-Zielina, E., Śladowski, G., & Sibielak, M. (2017). Planning the reconstruction of a historic building by using a fuzzy stochastic network. *Automation in Construction, 84*, 242–257.
10. Words & Images. (2009). *Building information modeling: Understanding and operating in a new paradigm*. Falls Church: The Foundation of Wall and Ceiling Industry.
11. National BIM Standard-United States. National Building Information Model Standard Project Committee. Retrieved on 8/6/13. http://www.nationalbimstandard.org/about.php. Last accessed 2019/05/13.
12. Fadeyi, M.O. The role of building information modeling (BIM) in delivering the sustainable building value. *International Journal of Sustainable Built Environment, 6*(2), 711–722. https://www.sciencedirect.com/science/article/pii/S2212609016302047. Last accessed 2019/05/12.
13. Richardson, D. (2014). *Productivity in the construction industry*. Canberra: Australia Institute.
14. Ying, Z. (2005). *Significant factors affecting construction productivity* (Doctoral dissertation), National University of Singapore.
15. Furneaux, C., & Kivvits, R. *BIM — Implications For Government, CRC for Construction Innovation, Brisbane*. http://eprints.qut.edu.au/26997/. Last accessed 2019/05/12.
16. Yan, H., & Damian, P. *Benefits and barriers of Building Information Modelling*. http://homepages.lboro.ac.uk/cvpd2/PDFs/294_Benefits%20and%20Barriers%20of%20Building%20Information%20Modelling.pdf. Last accessed 2019/05/12.
17. Knapčíková, L., et al. (2018). *Advanced materials based on the recycled Polyvinyl Butyral (PVB)*. In MMS Conference 2017, Ghent: EAI (pp. 1–9).
18. Knapčíková, L., et al. (2016). Material recycling of some automobile plastics waste. *Przemysl Chemiczny, 95*(9), 1716–1720.
19. Ghasemi, A., & Zahedias, S. (2012). Normality tests for statistical analysis: A guide for non-statisticians. *Journal Endocrinology Metabolism, 10*, 486–489.
20. MacFarland, T. W., & Yates, J. M. (2016). Mann–Whitney U test. In *Introduction to nonparametric statistics for the biological sciences using R*. Cham: Springer.

Management of Time Structure with Respect to a Production Process Efficiency

Michaela Ocenasova, Jana Ocilkova, and Martin Ocilka

1 Management of Time

Production consists of the performance of each manufacturing plant; therefore, it is a key area in order to succeed in the ever-growing competitive environment. Given this, it is in the interest of manufacturing plants to monitor the organization and management of production processes in order to identify the negative aspects that affect their efficiency, and ensure their subsequent elimination. One of the essential aspects that directly affect the economic efficiency of manufacturing plants is also the time structure of their production process, which basically reflects the flow of products through different workplaces, depending on the time factor, and addresses compliance of deadlines in line with the plans, the capacity utilization, and determines the time necessary for performance, control, and handling [1]. The time structure of production is given, irrespective of the type of production, by aggregating the time distribution of the production of products within a particular manufacturing plant. In general, it can be stated that the time structure of production is determined by a particular structure of the continuous production time of a particular product, given by a summary of the times needed for the work, downtimes, and the transformation process [2, 3].

M. Ocenasova (✉) · J. Ocilkova · M. Ocilka
Technical University of Košice, Faculty of Mining, Ecology, Process Control and Geotechnologies, Institute of Earth Resources, Kosice, Slovakia
e-mail: michaela.ocenasova@tuke.sk; jana.ocilkova@tuke.sk; martin.ocilka@tuke.sk

L. Knapcikova et al. (eds.), *4th EAI International Conference on Management of Manufacturing Systems*, EAI/Springer Innovations in Communication and Computing, https://doi.org/10.1007/978-3-030-34272-2_18

1.1 General Characteristics of the Time Structure of Production Process

The material flow through the workplace shall be such as to ensure smooth production of the final product, which consists of individual components. Therefore, this step is regulated by the time ties. A particular, pre-planned flow of unfinished parts through the plant creates a temporal structure of the production process [4]. The time structure of the production process takes into account, in addition to the length of time of the product flow through the manufacturing process, also the time of human work and the time used by machinery and equipment. The time structure of the production process is expressed by continuous time of production. Continuous time of production is a time indicator reflecting duration of the production process [5], which is formed by cycles indicated below [6]:

- Pre-production cycle, i.e., the time of preparatory work to the start-up of the product manufacture (the structural preparation and development work on a new product, technological preparation of production, including the preparation of tools, material, the economic and planning preparation of new manufacturing and so on)
- The production cycle, i.e., the time of basic manufacturing operations, time of transport, control and other operations in the manufacturing process, and durations of breaks (times of interruptions and delays) occurring during the product manufacturing.

A particular problem in the context of production timing is in idle and waiting times, i.e., times of necessary breaks in the production process, which are given by the mode of work and a complicated combination of partial production operations [7].

2 Analysis of the Temporal Structure of the Production Process

When analyzing the time structure of the production process, we consider the time factor as a most significant factor of the work. This parameter directly affects commercial and economic results of each manufacturing plant; it is based on direct measurement of partial working times, which are commonly used for the determination of wage costs or development of work schedules [8]. One of the analytical methods of the time structure of production processes is the method of continuous study of the time consumption, which we used in tracking the time structure in the production of aluminum castings. This particular method can detect, record, and evaluate the real time consumption of a particular employee or manager, as well as production facility. This method is based on measurements of time consumptions forming the image of [8, 9]:

- Operation, which is used to assess the workplace or production cycle organization
- Working day, which is used to assess the organization of the workplace, working hours used losses, and the like.

3 Methodology and Methods of Assessment of the Continuous Study of Production Time Consumption

To obtain results of the real-time consumption in the production of a production batch, we used the following indicators [5, 10]:

Occupancy rate (U_1)

$$U1 = \frac{T1 + T2}{T} * 100 \tag{1}$$

A share of breaks conditionally necessary (U_2)

$$U2 = \frac{T3}{T} * 100 \tag{2}$$

A share of needless time consumption caused by the worker (U_3)

$$U3 = \frac{T'2 - T2 + TD}{T} * 100 \tag{3}$$

A share of needless time consumption due to technical and organizational losses (U_4)

$$U4 = \frac{TE}{T} * 100 \tag{4}$$

A percentage of the possible labor efficiency increase by eliminating unnecessary time consumption caused by workers:

$$U5 = \frac{T'2 - T2 + TD}{T - (T'2 - T2 + TD + TE)} * 100 \tag{5}$$

A percentage of the possible labor efficiency increase by eliminating unnecessary time consumption caused by technical and organizational losses:

$$U6 = \frac{TE}{T - (T'2 - T2 + TD + TE)} * 100 \tag{6}$$

Percentage of the total possible increase in labor efficiency:

$$U7 = U5 + U6 \tag{7}$$

where

T1 – time of work

T2 – the time of generally required breaks

T3 – the time of conditionally required breaks

TD – personal loss of time

TE – technical and organizational loss of time

T – time of the shift

In view of the foregoing, the results of the above observations and quantifiers can be used to [5]:

- Quantify various activities expressed by the time consumption.
- Analyze the structure of the work time consumption.
- Analyze the lost times by causes.
- Develop the performance curves throughout the shift, especially if also monitoring the amount of the production output (performance).

4 Analysis of the Temporal Structure of Aluminum Casting Production

For the purpose of construction of the production cycle, we compiled partial time consumptions of partial production operations through the personal observation as given in Table 1.

In view of the above detailed analysis, we concluded that the most critical point of the production process of aluminum castings was in the operations of mechanical machining of finished products that gave rise to the most losses of the batch-production consumption time. Therefore, we have decided to dedicate more to the image analysis of the staff performance from the beginning to the end of the production shift and keep time records of all mechanical operations with the time data rounded to whole minutes (Table 2).

The final stage of observation was to evaluate the image of the work shift. At this stage, we compute the unit time from the consecutive time. Each unit time is evaluated according to the activity carried out, or inactivity. The same activities are summarized into an actual balance of time consumption. The actual balance reflects the amount of time in minutes and percentage per each category of the time of the work shift observed. A developed balance of the real time consumption is given in Table 3.

Development of the actual balance of the time resulted in the values needed to calculate the utilization of an employee, time of breaks, losses of time and increase in efficiency. Partial calculations leading to the increased efficiency of production,

Table 1 Time consumptions in partial operations in production of a production batch of aluminum castings

Workplace	Working operation	Consumption of time (min./production cycle)	Time of downtime (min/production cycle)	Percentage of downtime out of the production cycle time
Melting plant	Transportation (as part of melting)	1:37	0	
	Raw material charged into the furnace	0:54	0	
	Melting	9:00	0	
	Transport to further processing	1:59	0	
For the workplace		**11:53**	**0**	**0%**
Casthouse	Cleaning, lubrication of cavities	0:19	0	
	Casting	1:05	0	
	Transport to the warehouse (melter as part of melting)	1:10	0	
For the workplace		**1:24**	**0**	**0%**
Hydraulic press	Manipulation with parts	105	0	
	Transport	2	2	1,87%
For the workplace		**107**	**2**	**1,87%**
Grinding plant	Delivery of parts	1:29	1:29	3,67%
	Grinding	37:30	0	
	Transport	1:28	1:28	3,63%
For the workplace		**40:27**	**2:57**	**7,30%**
Mechanical machining	Delivery of parts	10	10	2,23%
	Takeover of grinding rolls	10	10	2,23%
	Machining of castings	379	0	
	Furnishing of the area to store the production batch to	10	10	2,23%
	Placement of castings	20	20	4,45%
	Removal	20	20	4,45%

(continued)

Table 1 (continued)

Workplace	Working operation	Consumption of time (min./production cycle)	Time of downtime (min/production cycle)	Percentage of downtime out of the production cycle time
For the workplace		**449**	**70**	**15,59%**
Tumbling	Tumbling and drying	25:00	0	
	Transfer to storage	1:28	0	
For the workplace		**26:28**	**0**	**0%**

Source: Processed by Author

which were quantified upon formulas (1)–(7), are indicated in Table 4. Results of the analysis made us to conclude that the efficiency could be increased by 25.33% in the mechanical machining of aluminum castings while, due to the elimination of unnecessary time consumption due to technical and organizational reasons, it would be possible to increase efficiency of product process by 22.72% and the needless consumption of time caused by employee by 2.61%. The most significant factor determining the actual efficiency was the level of utilization of a particular employee that was 79.79% and the unnecessary consumption of time, which was negligible, accounting only for 2.08% of the total consumption of time (Fig. 1).

The analyses carried so far resulted in measures increasing the efficiency of the production process, designed to eliminate the consumption of time in the mechanical stage of actual production of aluminum castings in the context of addressing any shortcomings of technical and organizational losses, e.g., arrival and/or removal of castings as well as arrival and/or removal of the transport box, including the transfer of finished aluminum castings to the warehouse or takeover of necessary tools and the like. In this way, consumption of time would be reduced at the workplace by 70 minutes per employee; if having four employees a shift, it would be 280 minutes per shift. Given the standard production output of 24 pieces per hour, then, if eliminating the consumption of that time, it could be possible to increase production by 112 pieces per hour, i.e., by about 2240 pieces per month.

By summarizing the results, we concluded that the production of aluminum castings should be constantly monitored and regularly assessed through the indicators of consumption of time indicated above, as presented in the model proposed for managing the partial production operations forming the time images of partial time consumption in the production of aluminum castings (Fig. 2).

Table 2 Observation sheet of the work shift image

Number	Consecutive time	Unit time	Time symbol	Title of time consumption
1	5:45			Start of the shift
2	6:00	0:15	TE	Recognizing difficulties the night shift
3	6:10	0:10	TE	Bringing-in a pallet with castings
4	6:20	0:10	TE	Takeover of grinding rolls
5	8:20	2:00	TB1	Continuous work – product machining
6	8:30	0:10	TD	Products fell down due to improper storage – stacking
7	8:40	0:10	TE	Furnishing an empty transport box for the products machined
8	8:50	0:10	TB1	Placing products into the transport box
9	10:03	1:13	TB1	Continuous work – product machining
10	10:06	0:03	TE	Defects found on products
11	10:20	0:14	TE	Consultation with the foreman
12	10:25	0:05	TE	Defective product taken away
13	11:35	1:10	TB1	Work assumed
14	11.40	0:05	TB1	Placing products into the transport box
15	11:45	0:05	T2	Leaving the workplace – toilet
16	11:50	0:05	TB1	Work assumed
17	12:20	0:30	T2	Lunch break
18	13:20	1:00	TB1	Completion of the production batch
19	13:25	0:05	TB1	Placing products into the transport box
20	13:45	0:20	TE	Removal of the transport box with products machined
21	13:45			End of the shift – the end of the observation

Source: Processed by Author

5 Conclusion

In the article, we focused our attention on the issue of time consumption management in the production process of aluminum castings and its impact on efficiency of production process. Using the method of the continuous time consumption, which consisted in the construction of a time image of partial operations in the comprehensive process of that production, we identified the most critical location of production, which was mechanical machining of aluminum castings. Upon the balance of the actual time consumption at work in this workshop, we quantified

Table 3 Balance of the actual time consumption

Type of the time	Time symbol	Minutes	Percentage of the shift time
Time of unit labor	TA1	0	0
Time of batch labor	TB1	348	72.5
Time of shift labor	TC1	0	0
Working time	**T1**	**348**	**72.5**
Time of generally required breaks	T2	35	7.29
Time of conditionally required breaks	T3	0	0
Personal losses of time	TD	10	2.08
Technical and organizational losses of time	TE	87	18.13
Time of the shift	**T**	**480**	**100**

Source: Processed by Author

Table 4 Quantification of the real time consumption indicators

1. Time consumption indicator	**Symbol**	[%]
Level of occupancy	U_1	79.79
Share of conditionally required breaks	U_2	0
Share of needless consumption of time caused by employee	U_3	2.08
Share of needless consumption of time due to technical and organizational losses	U_4	18.13
Percentage of possible increase in labor efficiency by eliminating the unnecessary time consumption caused by employee	U_5	2.61
Percentage of possible increase in labor efficiency by eliminating the unnecessary time consumption due to technical and organizational losses	U_6	22.72
2. The overall percentage of possible increase in labor efficiency	U_7	**25.33**

Source: Processed by Author

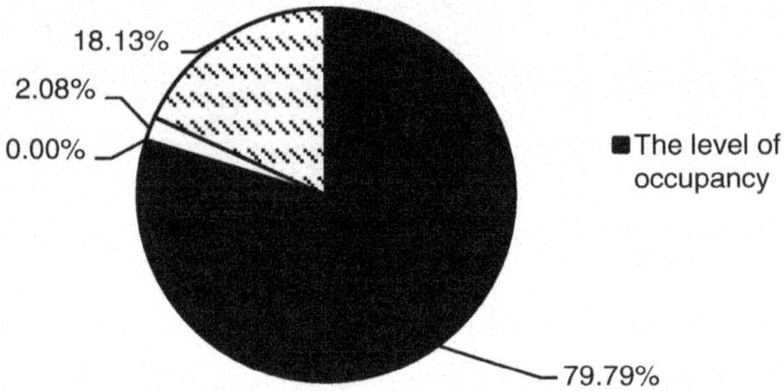

Fig. 1 Share of the actual use of the shift time in the production process of aluminum castings. (Source: Processed by Author)

individual indicators to point to a working occupancy of an employee, the time of breaks and time losses, which directly determine the possible increase in efficiency, which could be increased in that workshop by 25.33%; this would mean ultimately an increase in production by about 2240 pieces per month. This fact would have a positive impact on the development of the overall economy, i.e., the economic efficiency, and thus the competitiveness of the enterprise.

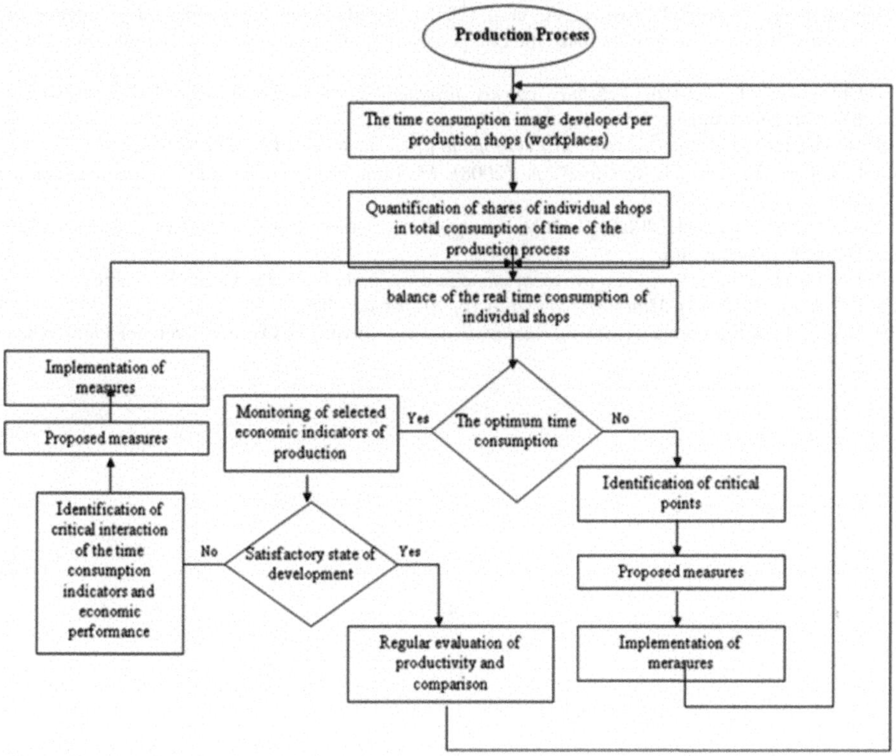

Fig. 2 Model proposed for the management of the time consumption in the production process. (Source: Processed by Author)

In view of the foregoing, we are of the opinion that the balance of the real-time consumption in any production plant is in close interaction with the possibilities of optimizing all areas of production and thus also gaining some competitive advantage and defining a further future development strategy in order to remain competitive.

This contribution is a partial result of the project VEGA MŠVVaŠ SR 1/0515/18 "The decision-making model of process of evaluating raw material policy of regions."

References

1. Pribulová, A., Futas, P., Rosová, A., et al. (2013). Influence of foundry dust on moulding mixtures quality. *Metalurgija, 52*, 51–54.
2. Pavolová, H., Cehlár, M., Seňová, A., & Dugas, J. (2015). Evaluation of the time structure of the manufacturing process using the method of continuous study of the time consumption in the context of productivity increase. In *Engineering sciences and production management 2015* (p. 68). Košice: Petit.

3. Straka, M., Trebuna, P., Rosová, A., et al. (2016). Simulation of the process for production of plastics films as a way to increase the competitiveness of the company. *Przemysl Chemiczny, 95*, 37–41.
4. Dubovická, L. (2010). *Vybrané aspekty produkčného manažmentu* (p. 102). Bratislava: Ekonóm (Economist).
5. Ďudák, J. (2012). *Časová štruktúra výrobného procesu: teoretické základy*. Nitra: SPU.
6. Leščišin, M., Stern, J., & Dupaľ, A. (2008). *Manažment výroby* (p. 328). Bratislava: Sprint víra.
7. Kukpkovič, M., et al. (2003). *Podnikové hospodárstvo: Komplexný pohľad na podnik* (p. 452). Bratislava: Sprint víra.
8. Kavan, M. (2002). *Výrobní a provozní management* (p. 424). Praha: Grada Publishing.
9. Pavolová, H. (2014). *Manažérstvo výroby* (p. 90). Košice: TU.
10. Novák, J., & Šlampová, P. (2007). *Racionalizace výroby* (p. 75). Ostrava: Technická Univerzita Ostrava.

On Statistical Method-Based Information System for Decision Support of Municipalities

B. Hamernikova, V. Fabera, V. Benes, M. Jetmar, M. Jerabek, J. Kubat, J. Pancik, and M. Placek

1 Introduction

Analysis of public administration and public sector efficiency is an important area of public economics and public finance theory. The issue of the effectiveness of local authorities is also a question of the theory of regional development and management. The analysis of the effectiveness of municipalities and their comparison is an important tool of public policies and activities of public administration bodies (ministries, regions, municipalities). The statistical methods such as DEA (Data Envelopment Analysis) and FDH (Free Disposable Hull) are used to analyze the effectiveness of municipalities and their comparison. These methods are used to evaluate the effectiveness of units based on the size of inputs and outputs in the business and public sectors. Units are compared with each other.

We designed and obtained a grant for a project that uses the statistical methods mentioned above [5, 7]. The project will thus improve the quality of decision-making and management processes in public administration and increase the transparency of its activities at the level of municipalities, regions, and ministries. In next chapters, we briefly describe the basis of our project, its objectives, and first results. The main topics described in this contribution are the web application and associated database description.

B. Hamernikova · V. Benes · M. Jetmar · J. Kubat · J. Pancik (✉) · M. Placek
AMBIS a.s., Prague, Czech Republic

V. Fabera · M. Jerabek
Czech Technical University in Prague, Faculty of Transportation Sciences, Prague, Czech Republic

L. Knapcikova et al. (eds.), *4th EAI International Conference on Management of Manufacturing Systems*, EAI/Springer Innovations in Communication and Computing, https://doi.org/10.1007/978-3-030-34272-2_19

2 Project of Information System for Decision Support of Municipalities

2.1 Performance Management in Public Sector

According to the last bibliographical analyses, e.g., [2, 10], performance management is still one of the most popular topics of research focusing on public management and public administration. Performance management is action, based on performance measures and reporting, which results in improved behavior, motivation, and processes, and which promotes innovation. The key aspects of performance management are deciding what to measure and how to measure it, interpreting the data, and communicating the results [5]. In the last decade, the potential of big data for performance management has also been discussed as well as the related problems of data misinterpretation and attribution problems [8]. Following the performance management literature, a general performance management model for local authorities has been developed. The performance management tools play an important role in this model (top of Fig. 1). Developing such "user-friendly" software tool dedicated for common users from municipalities was the goal of our project.

2.2 Performance Management in Czech Municipalities

Local government revenues in the Czech Republic amounted to 11.2% of gross domestic product (GDP). The EU average is 10.9% of GDP and 9.9% of GDP in the Eurozone [4]. Local government revenues in the Czech Republic consist mainly of shared taxes. The highest share is the value-added tax. According to the Eurostat data of 2016, local governments spend 10.2% of GDP on public spending. The EU-wide average was 10.8% of GDP in 2016; the share of the Eurozone countries was 9.7% [4]. These expenditures are realized in order to carry out the mission of the local government, which is defined in the Municipalities Act No. 128/2000 Coll. as follows: "The municipality takes care of the comprehensive development of its territory and the needs of its citizens; in the performance of its tasks it also protects the public interest."

2.3 Performance Analysis with DEA

For our performance analysis, the Data Envelopment Analysis was chosen. We are currently experiencing a rapid increase in the use of this performance analysis tool in municipalities [3, 11]. Data Envelopment Analysis (DEA) determines the most efficient municipalities in the sample of our research. These form the "best-practice

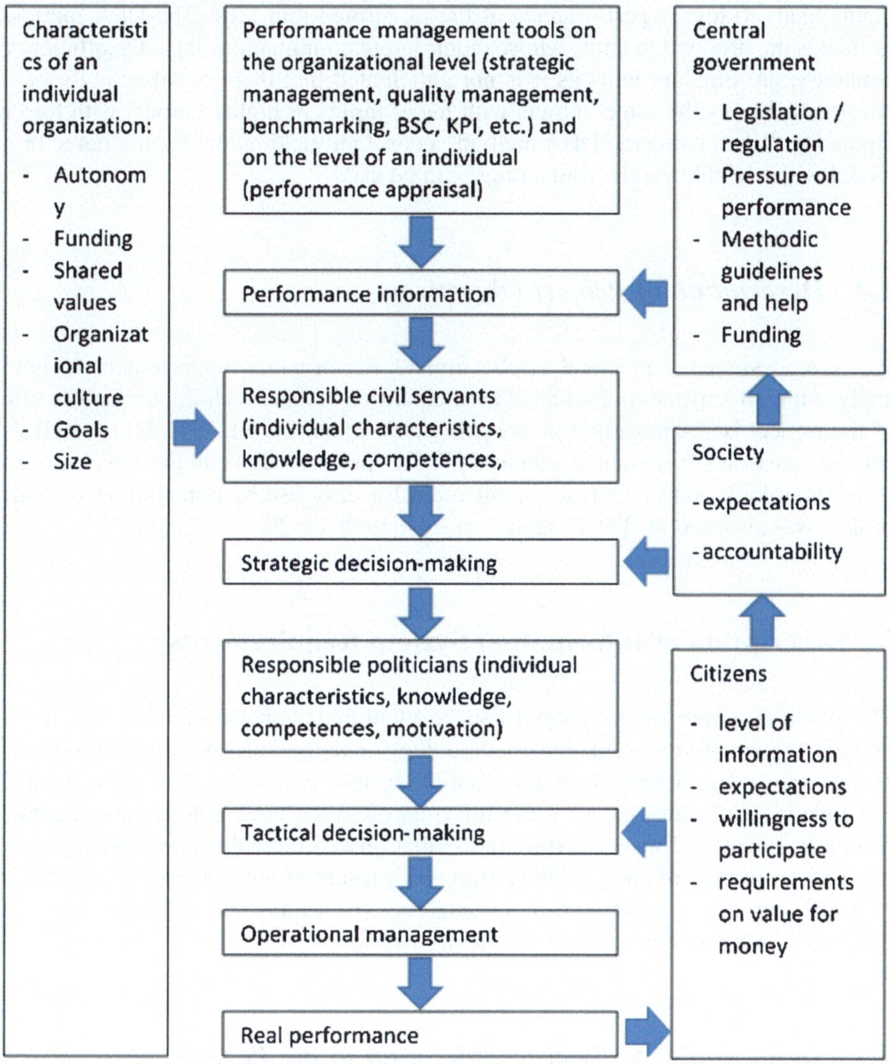

Fig. 1 General performance management model for local authorities

frontier" in a multidimensional space defined by inputs and outputs. The relative efficiency of municipalities lying under this best-practice frontier is computed by their deviations from the frontier [12]. This method is suitable for evaluating the efficiency, performance, or productivity of homogeneous production units – i.e., units that produce identical or equivalent effects, which we will refer to as outputs of this unit. Outputs are, by their nature, maximizing; their higher value results in higher performance of the tracking unit. To produce effects, the production unit utilizes inputs which are minimized by their nature; the lower value of these inputs leads to higher performance of the monitored unit [10]. The DEA method estimates the production units, whose input/output combinations lie at the efficiency boundary, are efficient units as it is not anticipated that there could actually be a unit that achieves the same outputs with lower inputs or higher outputs with lower inputs [1]. When using the DEA method, a constant yield model from a range or a model with variable yields from a range can be used.

2.4 Description of Project Objectives

We have designed a practical application of performance management to help analyze the effectiveness of selected municipalities in the Czech Republic. The title of the project is: "Application of nonparametric methods (DEA, FDH) to analysis and comparison of municipal efficiency." The project output is the development of methodology and IT software application for easy use by potential users. This project was awarded by TACR agency to solve in 2018–2021 years [7].

3 Description of Information System Requirements

The planned output of the project can be identified with the creation of a fully fledged information system and methodology and the software application will be part of it. The definition of information system is broader than many people can imagine and states that "... the information system is a collection of people, technical means and methods (programs) that ensure the collection, transmission, processing, storage of data, for the purpose of presenting information for the needs of users operating in management systems ... " [6]. Figure 2 defines the nature of the project outputs as parts of an information system (IS).

3.1 Definition of Software Application in the Project

Software application for information system in general is created by [6]:

- Users with different roles
- Presentation layer for users
- Data/information processing layer
- Database Management System (Database)
- Data
- Documentation
- User access to data

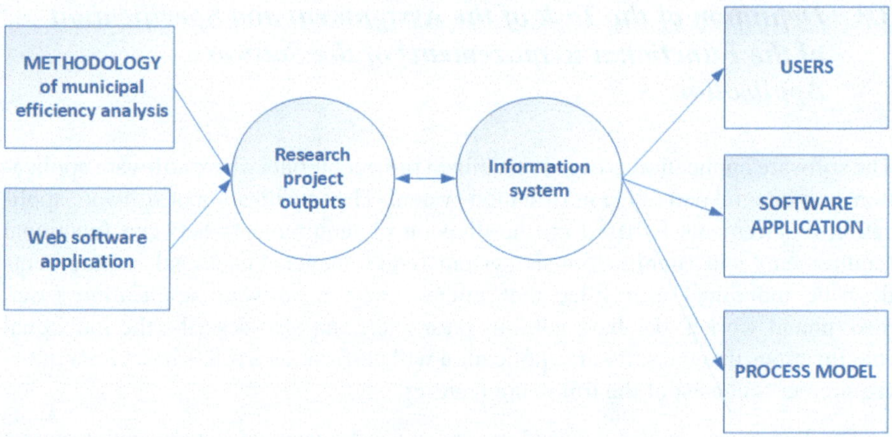

Fig. 2 Nature of the project outputs as parts of an information system

The planned software application will serve for needs of public administration. Within the meaning of Act No. 365/2000 Collection of Laws [9, 13] which defines the concept of public administration information system (IS), the developed software application will form the basis for the public administration agenda information system. The requirements for the public administration agenda information system are not so strictly viewed as requirements for public information systems from the state regulator side.

3.2 *Defining the Procedures for Creating a Software Application*

It has been determined that the software application is part of the IS (which it creates together with the methodology). Standard IS development methods will be used in the development of the software application. These methods define the individual stages of software application development. The following stages are known: specification of requirements, analysis, design, implementation, testing, deployment, using, changing, and their implementation. The development of the software application by the research team, which would be solved without the implementation of the whole IS as a project output, represents a risk of alignment of project objectives. All mentioned stages of software application development can be repeated and versions of the information system can be created in iteration from the simpler variant to the final version of the IS, namely, functional IS model, IS prototype, and final IS version.

3.3 Definition of the Task of the Assignment and Specification of the Functional Requirements of the Software Application

The software application is based on defined project outputs where software application is an integral part of the information system. The specification of software application requirements is based on the division of requirements between functional requirements and (nonfunctional) system requirements. Functional requirements describe individual user roles that interact with a software application (user, government worker, database administrator, etc.) and also describe the individual role functions that the software application will perform on a role's base. Functional requirements consist of the following features:

- The IS process model is based on the methodology that is part of the project output. The process model includes a description of individual processes, process owners and process inputs and outputs.
- A list of roles and a list of their assigned use cases (and usage scenario for each use case) can be created based on a process model.
- Individual functions of the software application can be assigned to individual use cases.

Software application system (nonfunctional) requirements describe the system from the perspective of software architecture, applied technologies, security, standards used, and so on. Individual functional and system (nonfunctional) requirements will be gradually filled in the system and software versions of the functional model, e.g., in functional prototype and final version. All the above-mentioned requirement documents will be part of the IS documentation and software application.

Figure 3 shows the hierarchy of requirements as was described in the text above.

3.4 WebDEAr Software Application Architecture

In the case of a software application for a project, one will be kind of a so-called web application. The user of the output of the project (the output is an information system that consists of methodology and software application) will be communicated with the software application via a web browser. The working name of the web application is WebDEAr. The WebDEAr web app is like any web app and consists of (Fig. 4):

1. Presentation layer, a part of software application called the front-end WebDEAr, which is the application's user interface and allows user interaction in one of the defined roles.

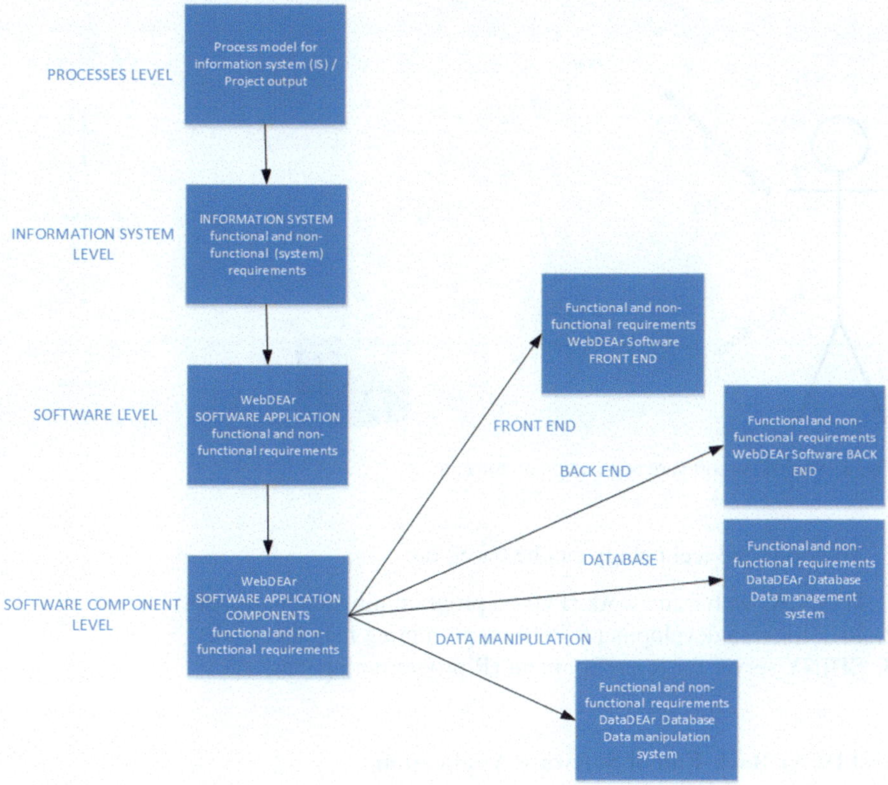

Fig. 3 Hierarchy of requirements

2. Information layer of software application, the so-called back-end WebDEAr, which represents the program implementation within the project developed algorithms DEA and FDH and represents also other activities necessary for running the software application.
3. Database (DataDEAr) that represents a database management system (DBMS).
4. The user data itself.

WebDEAr-Front-End of Software Application

The WebDEAr web application runs on a dedicated server that provides all the necessary services. The front-end user interface has user inputs (such as buttons) and graphical visualization outputs. The user interface will be specified in the request document in the Front-end section.

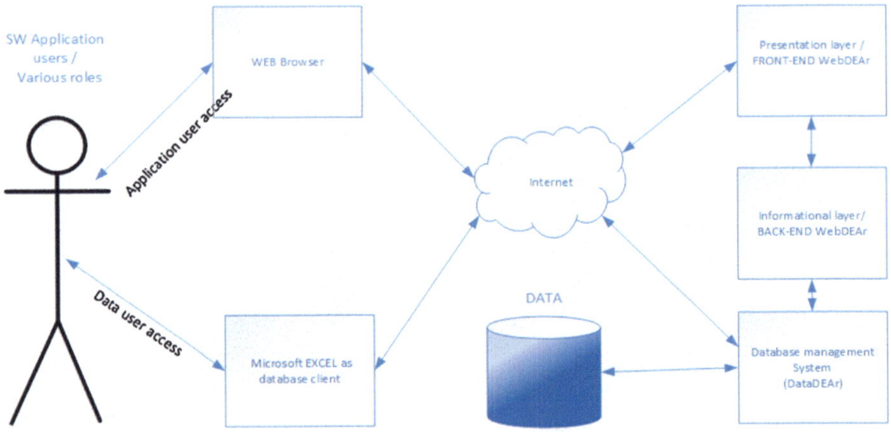

Fig. 4 WebDEAr software application architecture

The front-end technology can be based on:

1. DJANGO web framework (Python programming language path)
2. PHP for web development (PHP programming language path)
3. SHINY server for R environment (R programming language)

WebDEAr Back-End of Software Application

The WebDEAr back-end is in detail specified in the requirement document. Back-end technology used can be based on:

1. Programming language Python with libraries for DEA and FDH implementation, libraries for visualization (SciPHY), numerical calculations (NumPHY), and libraries for communication with database (DataDEAr) (path of programming language Python)
2. Based on PHP programming language (PHP programming language path)
3. Based on the R programming language (R programming language)

The DEA and FDH algorithms are the outputs of the project economist team and are ported in to the back-end of WebDEAr application by being delivered to the IT team in the form of R-written functions along with a validation set of data. The functions in R representing DEA and FDH algorithms are validated on a set of data that is provided along with the functions written in the R language to the IT team, which also verifies the correctness of porting into by team used language (Python or PHP). The DEA and FDH algorithms used in the WebDEAr back-end are specified in detail in the back-end requirement document, including their verification method.

DataDEAr Database Solution Architecture: Relational Data Management System

Database management system (DBMS) is based on relational database SQL server (Microsoft MS SQL Server Express, Oracle SQL Server Express or MySQL). Express version DBMS is a free version and is used for commercial and noncommercial purposes – it is limited to a few GB of data needed for the project. WebDEAr Back-End communicates with DBMS. The DataDEAr is specified in detail in the requirement document in the Database section.

4 Description of Developed Functional Prototype of Information System

4.1 Description of Implementation of Functional Prototype Architecture

The requirements for a functional prototype of a software application are:

- Access to the application without authentication
- Displaying information about the municipality or type of municipality
- Displaying the results of the municipality, resp. categories of municipalities including inputs and outputs
- Calculation according to the chosen method with entered data (inputs/outputs) for a particular municipality
- Logging user requests, collecting information about their work with the application

It is a classic three-tier architecture. The data layer is represented by the Oracle relational database management system. The application layer that performs data verification and processing is made up of several modules. The lowest level of PL/SQL performs procedures and functions that are part of the database system. The calculation using DEA and FDH methods is solved by "plug-in software modules," which are available as extensions directly to the Oracle database system. Because it is necessary to perform other statistical calculations, the research team decided to use the R language, which is directly designed for statistical analysis of data and their graphic representation; the module is also available for Oracle. The presentation layer is provided by the Apache web server with dynamically generated pages in PHP; PHP's scripts also include partial data processing that interferes with the application layer.

A virtual server with the Oracle Linux 7.5 operating system was put into operation at the Faculty of Transportation Sciences Czech Technical University in Prague. The Oracle Database Version 18c and Apache HTTP Server were installed.

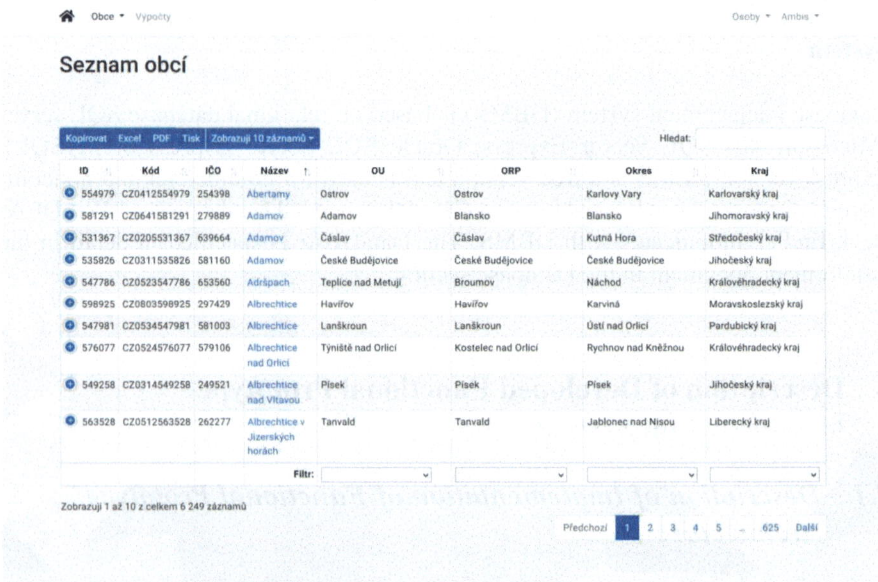

ID	Kód	IČO	Název	OU	ORP	Okres	Kraj
554979	CZ0412554979	254398	Abertamy	Ostrov	Ostrov	Karlovy Vary	Karlovarský kraj
581291	CZ0641581291	279889	Adamov	Adamov	Blansko	Blansko	Jihomoravský kraj
531367	CZ0205531367	639664	Adamov	Čáslav	Čáslav	Kutná Hora	Středočeský kraj
535826	CZ0311535826	581160	Adamov	České Budějovice	České Budějovice	České Budějovice	Jihočeský kraj
547786	CZ0523547786	653560	Adršpach	Teplice nad Metují	Broumov	Náchod	Královéhradecký kraj
598925	CZ0803598925	297429	Albrechtice	Havířov	Havířov	Karviná	Moravskoslezský kraj
547981	CZ0534547981	581003	Albrechtice	Lanškroun	Lanškroun	Ústí nad Orlicí	Pardubický kraj
576077	CZ0524576077	579106	Albrechtice nad Orlicí	Týniště nad Orlicí	Kostelec nad Orlicí	Rychnov nad Kněžnou	Královéhradecký kraj
549258	CZ0314549258	249521	Albrechtice nad Vltavou	Písek	Písek	Písek	Jihočeský kraj
563528	CZ0512563528	262277	Albrechtice v Jizerských horách	Tanvald	Tanvald	Jablonec nad Nisou	Liberecký kraj

Fig. 5 List of municipalities

The WebDEAr.fd.cvut.cz domain has been set up to access the website and is publicly available. Basic data has already been imported into the database system.

Screenshots (Figs. 5, 6, 7 and 8) represent various outputs of front-end of web application including the DEA algorithm implementation in the back-end.

4.2 *Description of Relational Database Model*

Based on the analysis of the requirements for a software application for the analysis of municipal efficiency (the output of the project TL01000463-V002), a relational scheme of the database was created, which forms the core of data structures for storing the necessary data. The scheme currently contains 23 tables (Fig. 9).

The central table is the MUNICIPALITY table, containing basic information about municipalities, such as the code according to the CZSO, code NUTS5, IČO, name, GPS position (also in WGS84 format), altitude, catastral area, type of municipality (TYPE_MUNICIPALITY), reference to district (district municipality), municipality with extended competence, and municipality with authorized municipal office. There are currently 6249 entries in the MUNICIPALITY table.

The REGION table is a codebook that contains all regions (their code and name) of the Czech Republic. The REGION_DISTRICT table links the region codebook to district municipalities.

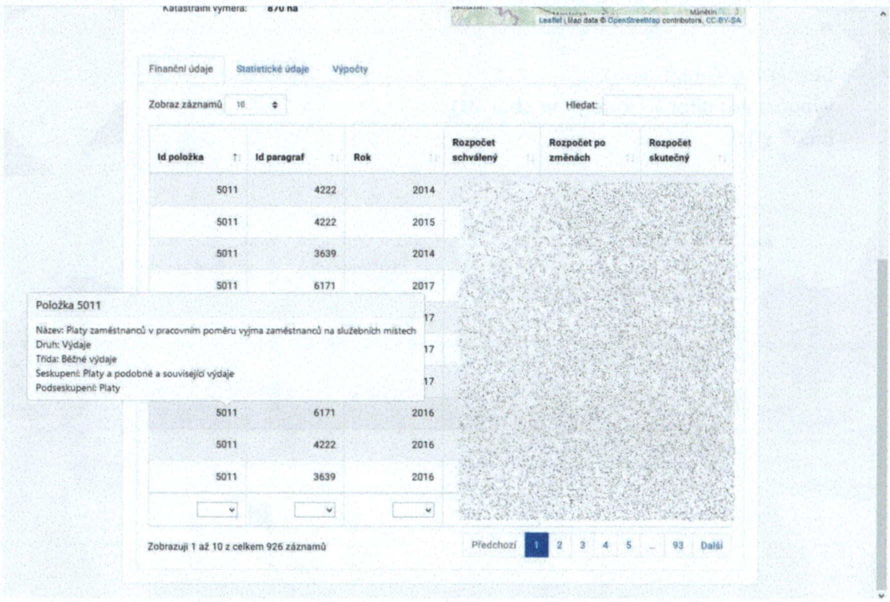

Fig. 6 Municipality financial details (anonymized data)

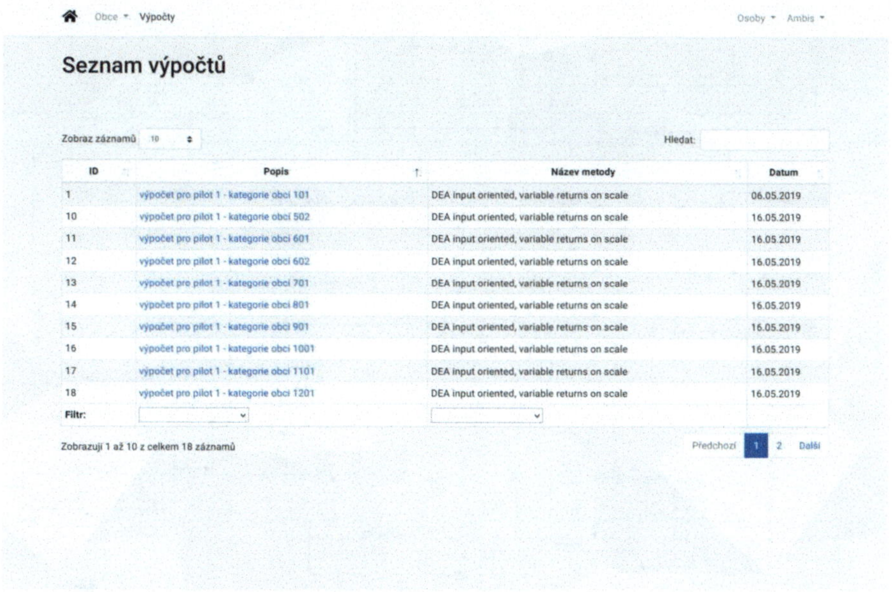

Fig. 7 List of statistical calculation with DEA algorithm

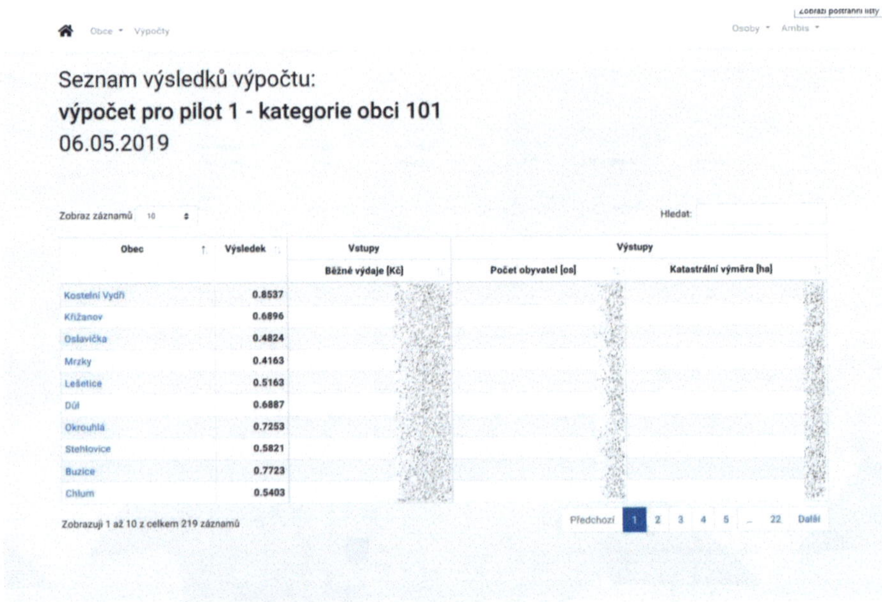

Fig. 8 Statistical calculation with DEA algorithm – detail (anonymized data)

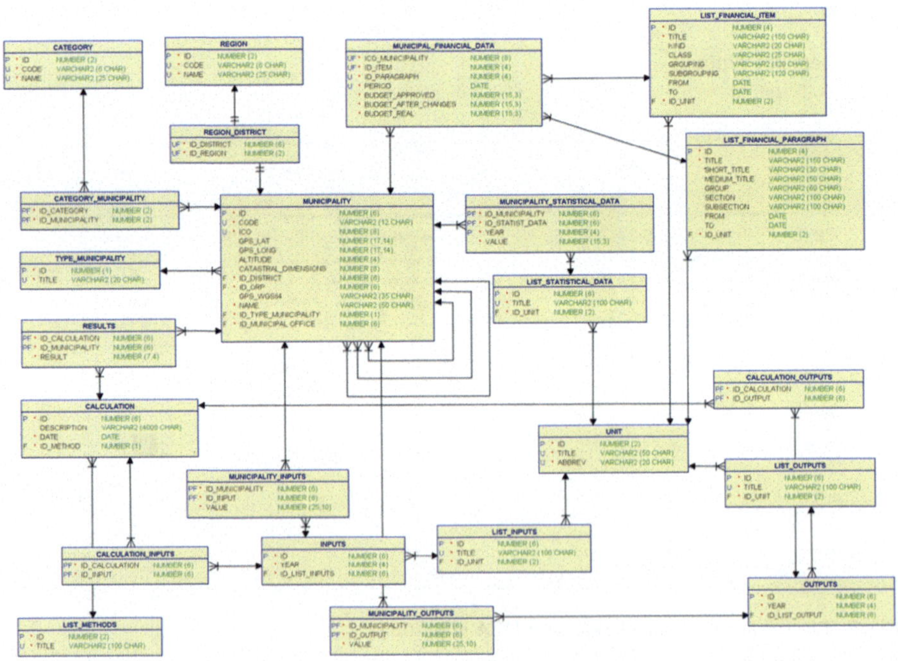

Fig. 9 DataDEAr database model

For the purposes of comparing the relevant municipalities, the enumeration table (List of Value table) CATEGORY is created for the purpose of categorizing municipalities that is linked to the MUNICIPALITY table using the reference table CATEGORY_MUNICIPALITY.

Important data for analyzing the effectiveness of municipalities are statistical and financial data. The statistical data list (code list) is in the LIST_STATISTICAL_DATA table, which also contains a reference to the UNIT code list. Individual statistical data on submunicipalities are linked in the MUNICIPALITY_STATISTICAL_DATA table.

For financial data, we distinguish paragraphs and items that have a code list in the LIST_FINANCIAL_PARAGRAPH table, respectively. LIST_FINANCIAL_ITEM. Both dials are linked to the UNIT table. The financial data of individual municipalities are linked to the period and the municipality in the MUNICIPAL_FINANCIAL_DATA table, where three data are available – budget approved, after changes, and actual. Currently, there are a total of 5,263,845 entries in the MUNICIPAL_FINANCIAL_DATA table for 524 entries in the LIST_FINANCIAL_PARAGRAPH table and 527 entries in the LIST_FINANCIAL_ITEM table. Financial data is from 2012 to 2017.

Various inputs and outputs of statistical and financial data are used to analyze the effectiveness of municipalities using the DEA and FDH methods. The INPUTS and OUTPUTS tables referring to LIST_INPUTS and LIST_OUTPUTS are used to record which data was used as inputs and which as outputs. Specific data are captured in the MUNICIPALITY_INPUTS and MUNICIPALITY_OUTPUTS tables, with a view to a possible later change of data.

Database structure design allows for the retention of all results in recalculations. The basic data in the CALCULATION table, including reference to the method used (table LIST_METHODS), are stored for each calculation. And the results from each calculation for each selected village are stored in the OUTPUTS table.

When processing data from various sources, we encountered not entirely 100% functional integrity constraints on financial data on the code list of municipalities, sections, and items.

5 Conclusion

In the frame of the project dedicated to develop a performance management tool for municipalities, a functional prototype of information system was designed. The requirement document for describing functional and nonfunctional requirements was prepared. Database model and database itself were developed and were filled with data from the public sector. Functional prototype of web software application which consists of front-end and back-end layers was developed. Back-end of developed application is filled with database connectivity functions and with DEA and FDH statistical functions. Porting of developed DEA and FDH algorithms from statistical level to back-end web application level was tested. Next time we will devote our concern to the development of process model of comparison of effectiveness of local authorities and the following comparison methodology. We have to optimize data structure of database model and data cleaning next time. Importing data mechanism and data manipulate based on Microsoft EXCEL as client for database management system should be prepared.

References

1. Boruvkova, J., & Kuncova, M. (2012). Porovnání očních oddělení kraje Vysočina pomocí DEA modelů. *Acta Oeconomica Pragensia, 5*, 75–84.
2. Corrado, C., Massiomo, A., & Sarto, F. (2016). A twenty-five years bibliometric analysis in business and public administration domains. 201. Foundations and trends in performance management. *Scientometrics, 108*(2), 595–611.
3. Emrouznejad, E., & Yang, G. (2018). A survey and analysis of the first 40 years of scholarly literature in DEA: 1978–2016. *Socio-Economic Planning Sciences, 61*, 4–8.
4. Eurostat. (2016). [Online]. [Cited: 07 15, 2017]. http://ec.europa.eu/eurostat/web/government-finance-statistics/overview.
5. Fryer, K., Jiju, A., & Ogden, S. (2009). Performance management in the public sector. *International Journal of Public Sector Management, 22*(6), 478–498.
6. Gála, L., Šedivá, Z., & Pour, J. (2016). *Počítačové aplikace v podnikové a mezipodnikové praxi – 3., aktualizované vydání* (p. 2016). Praha: Grada.
7. Hamerníkova, B. (2018). Application of nonparametric methods (DEA, FDH) to analyse and to compare the efficiency of municipalities. Project proposal TL01000463, 02/2018 – 01/2021. s.l.: TAČR, Technological Agentury of Czech Republic, program ETA, 2018.
8. Lavertu, S. (2016). We all need help: "Big Data" and the mismeasure of public administration. *Public Administration Review, 76*(6), 864–872.
9. Lunter, J. (2018). *Co je a co není informační systém verejné správy*. [Online]. [Cited: 5 19, 2019]. http://www.mvcr.cz/soubor/co-je-a-co-neni-isvs.aspx.
10. Osborne, S. P. (2017). Public management research over the decades: What are we writing about? *Public Management Review, 19*(2), 109–113.
11. Perpiňa-Narbon, I., & De Witte, K. (2017). Local governments' efficiency: A systematic literature review – Part 1, Part 2. *International Transaction in Operational Research, 00, 01*, 1–38, 1–30.
12. Šťastná, L., & Gregor, M. (2011). *Local government efficiency: Evidence from the Czech*. IES Working Paper. Vol. 14.
13. Zákon 365/2000, Zb. (2000). *Zákon o informačních systémech verejnej správy*. [Online]. [Cited: 5 19, 2019]. http://www.mvcr.cz/clanek/legislativa-zakon-c-365-2000-sb-o-informacnich-systemech-verejne-spravy.aspx.

Possibilities of Data Analysis Using Data Model

Stella Hrehova

1 Introduction

Data is the foundation of the decisions, knowledge and ultimately the wisdom for correct decisions and actions [1]. If this data is relevant, complete, accurate, meaningful and actionable, it will help. If not, it can prove to be useless and even harmful [2]. The following figure shows the division for well-managed data (Fig. 1).

Data analytics allows users to convert raw data into a comprehensive format. The converted details are used to cleanse, convert or model the data, so that it supports the decision-making process, derives conclusions and sets up predictive analytics [4].

A significant part of the process information is transmitted mainly by SCADA (Supervisory Control and Data Acquisition) systems, which enable to monitor the process of technology control on the screen (visualization), display and archive the state of observed variables and errors in the system or change some parameters of the regulatory process at the user level. These systems can also collect information on consumed and produced quantities of energy. Data obtained from the procedural level of management can be statistically processed and used to evaluate the efficiency of the production and supply of heat from both supplier and consumer perspectives [5].

The aim is to provide information about the average power consumption from multiple sources in graphical form and so to select data dynamically on requirements.

S. Hrehova (✉)
Faculty of Manufacturing Technologies with seat in Presov, Technical University in Kosice, Presov, Slovakia
e-mail: stella.hrehova@tuke.sk

L. Knapcikova et al. (eds.), *4th EAI International Conference on Management of Manufacturing Systems*, EAI/Springer Innovations in Communication and Computing, https://doi.org/10.1007/978-3-030-34272-2_20

Fig. 1 Well-managed data [3]

2 Creation of Data Model in MS Excel

Using the SCADA web user interface, it is possible to import data in an appropriate format and then analyse it (see Fig. 2).

Before the data model is created, it is necessary to assess the source data. Some of the potential problems are [6]:

- Inadequate and/or out of date
- Faulty data
- Improper analysis
- Inappropriate presentation
- Time lag

From the point of view of creating a data model, it is possible to work with either one large table where all data will be included or by creating a data model with linked tables. The presented paper shows the process of creating a data model by creating relations among separate tables.

2.1 Data Model Preparation

After data collection, it is important to prepare the data to be analysed. Organizing the data correctly can save a lot of time and prevent mistakes [7]. One of the most common source data errors are duplicate or missing items. This is especially in the case when the data is recorded at very short time intervals. To find out if there are duplicate items in the data file, we simply use the basic MS Excel tool – conditional formatting (see Fig. 3).

To remove duplicate items, the query is used, where the columns (parameters) are selected in a simple way through the wizard table that will be used in the analysis (see Fig. 4).

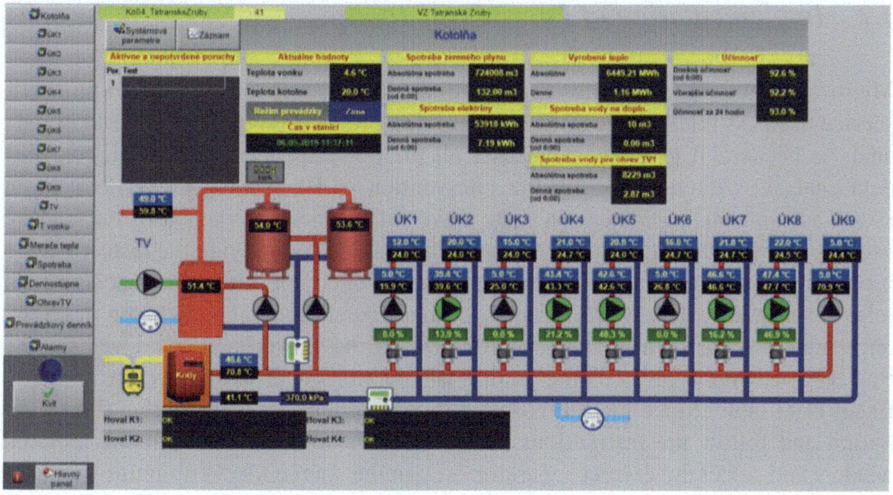

Fig. 2 The user interface of SCADA system

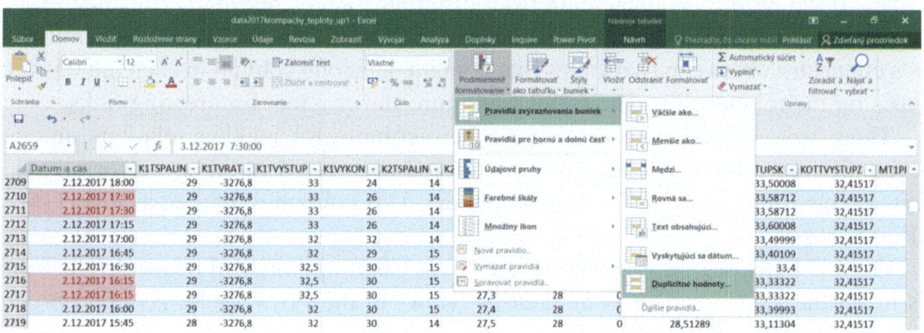

Fig. 3 Marked duplicate items

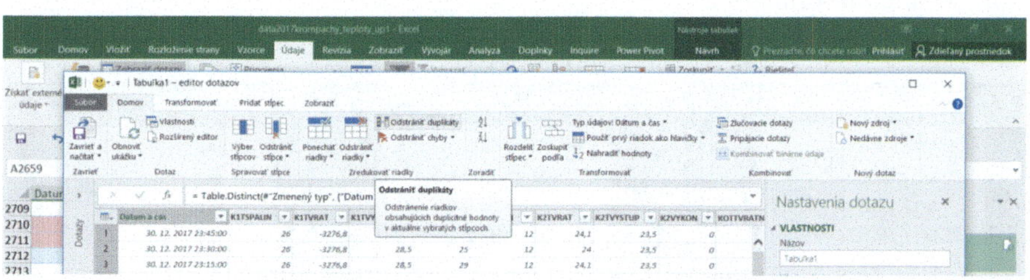

Fig. 4 Query to remove the duplicate items

2.2 Implementation of PowerPivot Tools

Microsoft PowerPivot for Microsoft Excel is a technology aimed at providing self-service business intelligence (BI), which is a real revolution inside the world of data analysis because it gives the final user all the power needed to perform complex data analysis without requiring the intervention of BI experts. PowerPivot is an Excel add-in that implements a fast, powerful, in-memory database that can be used to organize data, detect interesting relationships and provide the fastest way to browse information [8]. Data model becomes a collection of tables connected with relationships. Any data which is imported into Excel or are already given in Excel, once added to the data model, are available in the PowerPivot window [9]. The PowerPivot Ribbon gives us the additional functions over and above the standard Excel Data table. Since the file is saved in csv format, we will use the 'Text File' option and follow the instructions. It is important to set the first line as column headers and select the correct separator used in the source file. Within this view, we can simply click the checkboxes next to each column to determine which column will be included in the data model and which not.

The paper will analyse the consumption of energy from three different source tables. If we want to evaluate data from different sources relevantly, it is necessary to create a relationship among them to ensure associated related data. A relationship is a connection between two tables that contain data: one column in each table is the basis for the relationship. The following figure shows a tool for creating links.

There was no necessity to create a new column for making relations among three different tables. All data was monitored and stored in the same format and at the same time (every day at the same hour). This column is marked as PM_TIME (Fig. 5).

For this purpose, the Pivot Table and Pivot Chart options will be used. The base is to set the x-axis, which is marked as 'Row' in the Pivot Table. As appropriate column, the date column PM_TIME was selected.

We can see in Fig. 6 the tools of Pivot Table divide single date column into two parts. It is very useful for the user. It gets the possibilities using filters and so gains more detailed information.

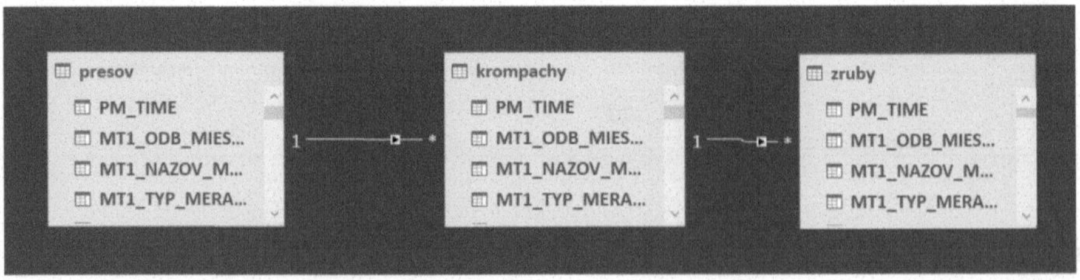

Fig. 5 Create relationship among three tables

Fig. 6 Part of Pivot Table

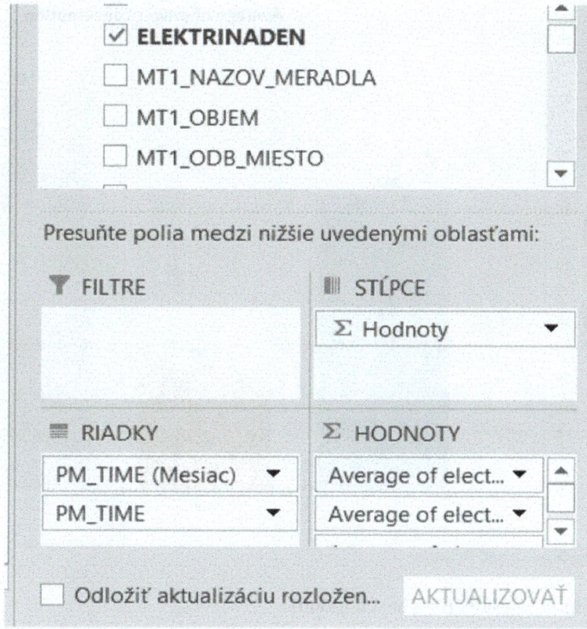

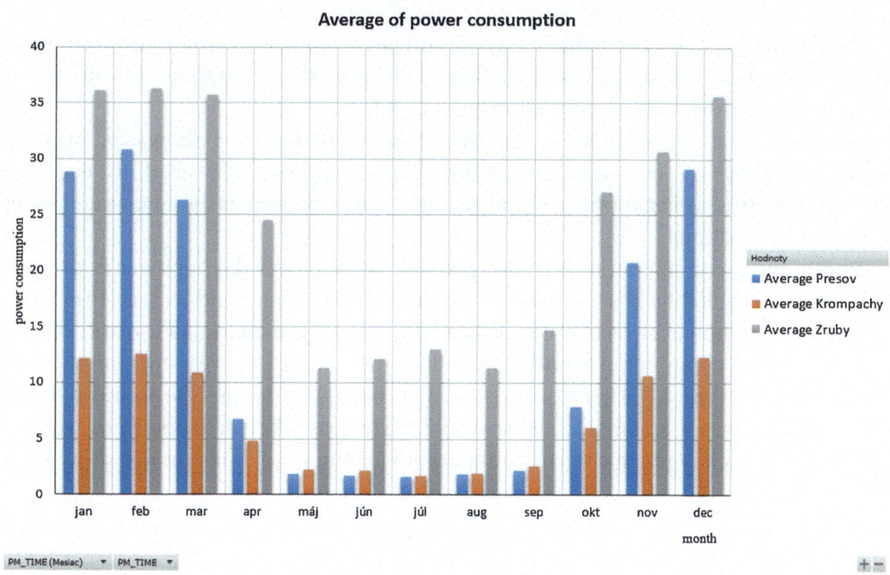

Fig. 7 Graphical results

Because we want to display the average of values, we have to change the target function and also the function description. Figure 7 shows the graphical results according to the months.

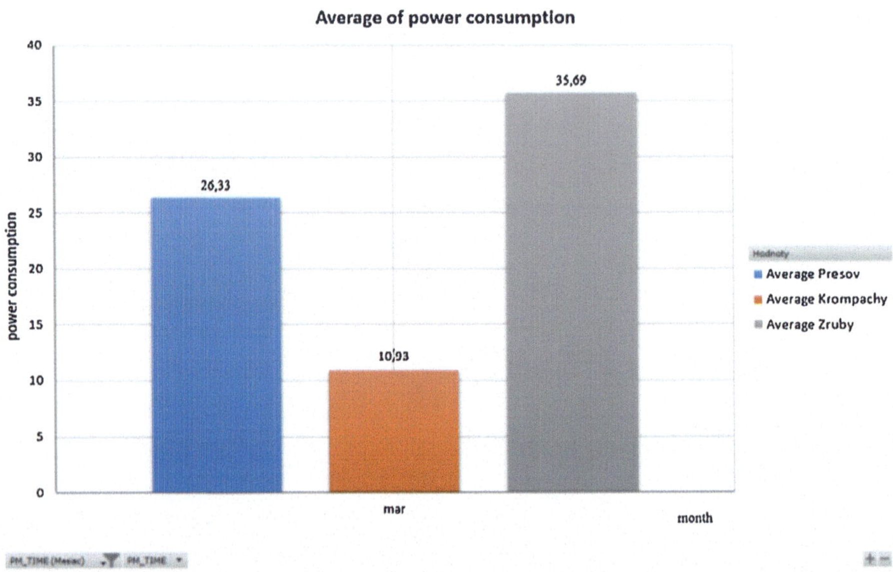

Fig. 8 Results per chosen month

Based on Fig. 7, it is possible to see significantly higher electricity consumption for the workplace 'Zruby'. This difference is given by the geographical location – it is the northernmost workplace with respect to others.

It is very easy to use filters using PivotChart tools. Figure 8 shows the average power consumption per selected month. The user does not need to use any other procedures to get more detailed results, which is especially advantageous in reporting or meetings.

In this narrowed selection, it is also possible to display 1 day's power consumption. This is enabled by the '+' tool in the PivotChart graph window (see Fig. 9).

In this case, however, it is preferable to use a line type chart. As a result of this visualization, we can see a significant fall in the average energy consumption in the locality of 'Presov'. Therefore it would be necessary to analyse in more detail whether there was a technical failure of the system or whether the system was set too sensitive to changes in external temperatures. Other locations do not show such behaviour.

3 Conclusion and Discussion

Data is only valuable when it is actually used or may be useful in the future [10]. This requirement is very important in terms of data efficiency and therefore data needs to be analysed as soon as possible. The paper describes the design of a data

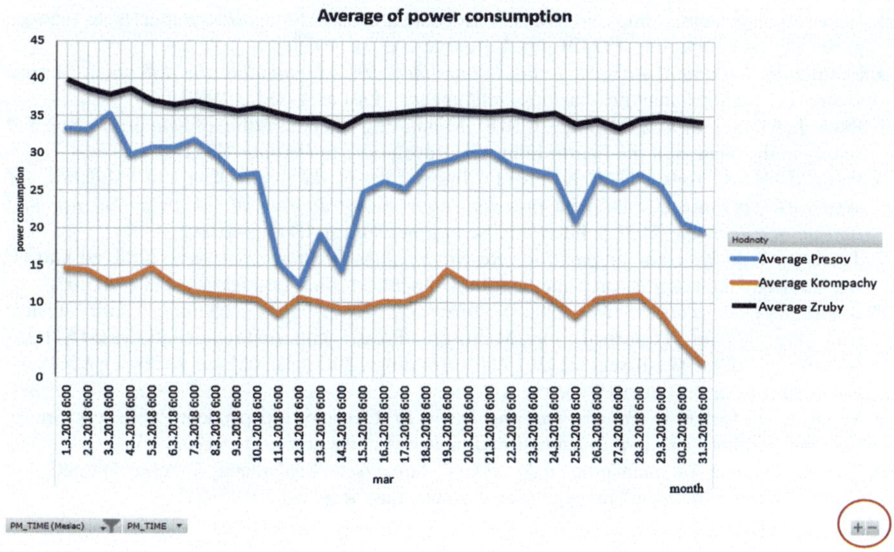

Fig. 9 Power consumption per day

model for data analysis obtained by monitoring the heating process and using MS Excel tools. The choice of the appropriate computer application plays a key role in each data processing. The advantages of the presented environment are its simplicity and sufficient content of the necessary tools for analysis. Another advantage is that the common users are able to analyse data, so data processing becomes really effective. On the other hand, it should be pointed out that the use of this application is based on the definition of analysis aims. As described in the paper, although we have a large amount of data, analysis and visualization have not been presented with the real values but cumulated (average values). These results have not to be correct so they could lead to incorrect conclusions. Therefore, it is necessary to assess whether the conclusions are sufficient and relevant. The next step could be to use the trendline selection menu to find out the most appropriate mathematical equation for predicting electricity consumption in the future.

Acknowledgement This work was supported by the Slovak Research and Development Agency under Contract No. APVV-15-0602.

References

1. Galeto, M. *What is data management*. https://www.ngdata.com/what-is-data-management/. Last accessed 2019/04/27.
2. The importance of data management in companies. https://www.ringlead.com/blog/the-importance-of-data-management-in-companies/. Last accessed 2019/04/27.

3. Data management. https://www.sas.com/content/dam/SAS/en_us/doc/other1/data-management-why-is-it-important-107421.pdf. Last accessed 2019/05/05.
4. Sharma, R. *Data analytics and business intelligence: Understanding the differences*. http://techgenix.com/data-analytics-business-intelligence/. Last accessed 2019/05/05.
5. Piteľ, J., Mižáková, J., & Mižák, J. (2018). Monitoring and controlling of heat production and supply in the context of the Fourth Industrial Revolution. In *Heating*. SSTP, Slovakia.
6. Patil, M. V., & Yogi, A. M. N. (2011). Importance of data collection and validation for systematic software development process. *International Journal of Computer Science and Information Technology, 3*(2). http://airccse.org/journal/jcsit/0411csit20.pdf.
7. Trochim, W.M.K. Data preparation. *Research methods knowledge base* (2nd ed.).http://www.uniteforsight.org/research-methodology/module5#_ftn1. Last accessed 2019/05/05.
8. Russo, A., & Ferrari, M. (2013). *Microsoft Excel 2013: Building data models with PowerPivot*, Published by Microsoft Press. http://index-of.co.uk/OFIMATICA/Microsoft%20Press%20Excel%202013,%20Building%20Data%20Models%20with%20PowerPivot.pdf. Last accessed 2019/05/05.
9. Singh, A., & Collie, R. *Power Pivot and Power BI*. Chicago: Independent Publishers Group. http://www.allitebooks.in/power-pivot-and-power-bi/. Last accessed 2019/05/05.
10. The importance of managing data assets. https://searchdatamanagement.techtarget.com/feature/The-importance-of-managing-data-assets. Last accessed 2019/04/27.

Regression Model by the Management for Preparation of the Permalloy Sample

Dusan Knezo and Denisa Oleksakova

1 Introduction

Nickel and iron alloys are called permalloy. These alloys are produced, for example in the form of wound rings, laminated sheets or other thin sheets [1]. Nickel and iron alloys have excellent properties of soft magnetic material and are applied in the area of engineering, material research, and business [2].

The goal of the researcher is the management and discovery of methods for the preparation of the material in the requested and suitable form for the application. One of these methods is the compaction of the powder (obtained by mechanical milling or mechanical alloying) in the requested shape. Methods of mechanical milling and mechanical alloying have recently been extensively developed and used to produce alloys containing various equilibrium and non-equilibrium phases. These methods of preparing alloys also have other possibilities for research work and permalloy application. Their advantage is the preparation of powder in large quantities and also the simple regulation of material processing parameters [3].

This article is aimed at the resulting magnetic properties of the compacted permalloy (permalloy – iron and nickel alloy with 81 percentage weight of nickel and 19 percentage weight of iron) using statistical and experimental methods.

Supported by the Faculty of Manufacturing Technologies with the seat in Prešov, Technical University of Košice.

D. Knezo (✉) · D. Oleksakova
Department of Natural Sciences and Humanities, Faculty of Manufacturing Technologies with the seat in Prešov, Technical University of Košice, Prešov, Slovakia
e-mail: dusan.knezo@tuke.sk

L. Knapcikova et al. (eds.), *4th EAI International Conference on Management of Manufacturing Systems*, EAI/Springer Innovations in Communication and Computing, https://doi.org/10.1007/978-3-030-34272-2_21

2 Methods of the Experiment

2.1 Experimental Samples

Compacted permalloy samples in the form of small roll (diameter about 10 mm, height approximately 3 mm and weight around 2 g) were prepared by pressing (at 800 MPa and a temperature of 600 °C) of milled powder. The fine powder was prepared by the mechanical milling of the iron and nickel mixture in the planetary ball mill for different lengths of time (10, 15, 20, 25, 30 h). Mechanical milling is a useful powder-processing technique that can produce a variety of equilibrium and non-equilibrium alloy phases [3]. We have prepared samples that were annealed at different temperatures (500, 1000 °C). After the annealing we have the resulting coercivity, which depends on milling time and annealing temperature [4].

2.2 Coercivity

The coercivity (the coercive force) was measured by the werzimat. Coercivity is the properties of the magnetic field to resist demagnetization of the permanent magnet in the external magnetic field [4]. This magnetic field property can also withstand the demagnetizing field [5]. We know two types of coercivity: real and internal. Real coercivity is the value at which the total induction in the magnet is equal to zero. Internal coercivity is the value at which the polarizing vectors of individual magnetic domains are abolished [2].

2.3 Statistical Methods

The comparison of the influence of milling time and annealing temperature of compacted permalloy to the value of the coercivity by experimental and statistical methods was detected by the multivariate regression model. The general purpose of multivariate regression is to learn more about the relationship between several independent (predictor) variables and a dependent (criterion) variable. The simplest form has one dependent and two independent variables. The aim of the multivariate linear regression model analysis is the task of fitting a single line through a scatter plot [6].

Multivariate linear regression is used in various analyses. It can be used to identify the force that expresses the effect of independent variables on dependent variables. It is also used to predict the effects or impacts on changes. Multivariate linear regression analysis helps us to understand how much a dependent variable changes when we change independent variables. Multiple linear regression analysis also predicts trends and values for the future [7].

3 Discussion

In Table 1, values of coercivity for annealed samples at different temperatures (independent variable) and different milling times (independent variable) are displayed. The coercivity (dependent variable) depends on the annealing temperature and milling time. It examines that changes (caused by the change milling time) modify the coercivity caused only by the temperature annealing and change of coercivity caused by two factors, i.e., the duration of the milling and the temperature of the annealing [4, 8].

Figure 1 shows the values of the coercivity for five samples (compacted from the powder milled another milling time) at a different annealing temperature. We can see that milled time does not modify the coercivity. The coercivity (its values) of permalloy decreases for the sample annealed at a higher temperature (1000 °C).

In Fig. 2 we can see the value of the coercivity for three samples (annealed at the different temperature) prepared from the powder obtained by mechanical milling for 10 to 30 h.

Table 1 The coercivity of permalloy

Milling time (h)	$Hc(A \cdot m^{-1})$ (0 °C)	$Hc(A \cdot m^{-1})$ (500 °C)	$Hc(A \cdot m^{-1})$ (1000 °C)
10	452	329	88
15	578	564	141
20	453	456	124
25	912	905	232
30	563	568	165

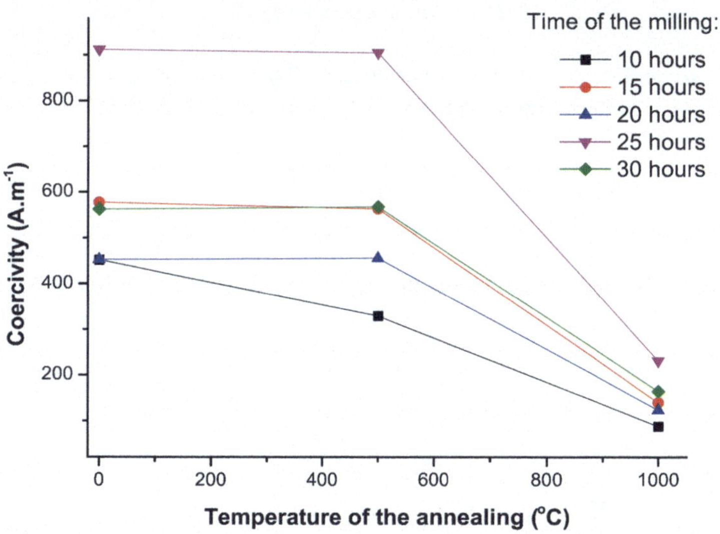

Fig. 1 The values of coercivity for samples compacted from the powder milled for 10, 15, 20, 25 and 30 h and their dependence on the annealing temperature

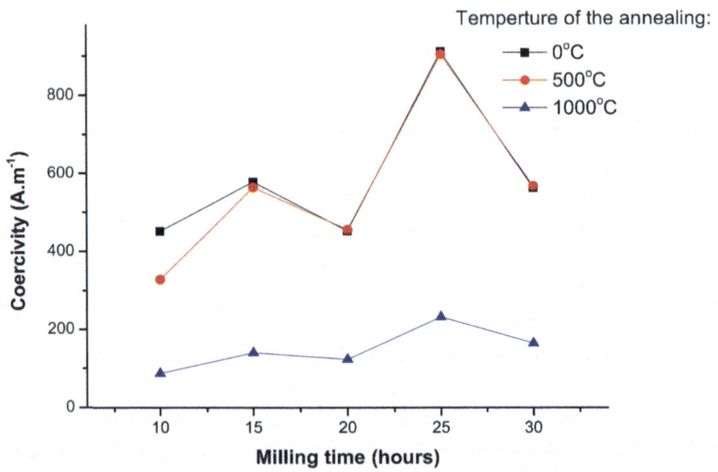

Fig. 2 The comparison of the value of the coercivity for compacted samples non-annealed and annealed at 500 and 1000 °C dependence on the time of the mechanical milling

The displayed values of the coercivity of permalloy in the dependence of annealed temperature results that the modification of the temperature from 0 to 500 °C does not vary the coercivity, but the modification of the temperature from 500 to 1000 °C leads to a decrease in the values of the coercivity.

The coercivity (Hc) is dependent on milling time (the variable x) and on annealing temperature (variable y):

$$Hc = a_0 + a_1 \cdot x + a_2 \cdot y. \tag{1}$$

where a_0, a_1 and a_2 are unknown parameters. These parameters can be estimated by the valid formula (by the using the least squares method) [9]:

$$S(a_0, a_1, a_2) = \sum_{i=1}^{n} (Hc_i - a_0 - a_1 x_i - a_2 y_i)^2. \tag{2}$$

The unknown parameters a_0, a_1 and a_2 can be the resulted of the solution of the system [9]:

$$a_0 n + a_1 \sum_{i=1}^{n} x_i + a_2 \sum_{i=1}^{n} y_i = \sum_{i=1}^{n} Hc_i, \tag{3}$$

$$a_0 \sum_{i=1}^{n} x_i + a_1 \sum_{i=1}^{n} (x_i)^2 + a_2 \sum_{i=1}^{n} x_i y_i = \sum_{i=1}^{n} x_i Hc_i, \tag{4}$$

$$a_0 \sum_{i=1}^{n} y_i + a_1 \sum_{i=1}^{n} x_i y_i + a_2 \sum_{i=1}^{n} (y_i)^2 = \sum_{i=1}^{n} y_i Hc_i. \tag{5}$$

We calculate the coefficients and the dependence of the coercivity on the milling time (hours) and on the temperature of the annealing ($°C$) is:

$$Hc = 440,133 + 10,800 \cdot x - 0,442 \cdot y. \tag{6}$$

The positive parameter a_1 is in the accordance with the fact that the coercivity increases with increasing milling time. Parameter a_2 is negative. It leads to decreasing of the coercivity with increasing annealing temperature. The results show that the ideal combination for the preparation of permalloy is a shorter duration of mechanical milling and a higher temperature of the annealing samples.

4 Conclusions

The aim of the investigation was to find out the correlation of the annealing temperature of compacted permalloy and the milling time of powdered permalloy with the magnetic properties of this material (in our case it was coercivity).

We used the method of multivariate regression, which showed that the coercivity of the permalloy is influenced by the annealing temperature and the milling time too. This also is in accordance with experiments. It may be caused because, owing to the mechanical milling, smaller and smaller particles of powder were created (with increased milling time). Rotation of the vector of magnetization is the more dominant process of the magnetization of smaller particles. After the pressure of the milled powder, the more dominant process of magnetization is the displacement of domain walls, and the coercivity of the sample decreases.

The multivariate regression analysis has exhibited that the increasing annealing temperature leads to decreasing values of coercivity. The same results were reported by the experiments. We suppose that the annealing leads to the creation of better magnetic contact between powders and the domain wall can move better. We assume that the pores between the particles of powder create pinning centers for domain wall movement and determines the value of the coercivity, so that coercivity decreases with increasing annealed temperature.

Acknowledgements This work was supported by the Scientific Grant Agency of the Ministry of Education of the Slovak Republic and the Slovak Academy of Sciences KEGA 002TUKE-4/2019.

References

1. Arnold, H. D., & Elmer, G. W. (1923). Permalloy, an alloy of remarkable magnetic properties. *Journal of the Franklin Institute, 195*, 621–632.
2. Coey, J. M. D. (2009). *Magnetism and magnetic materials*. Cambridge: Cambridge University Press.
3. Suryanarayana, C. (2001). Mechanical alloying and milling. *Progress in Materials Science, 46*, 1–184.
4. Ižaríková, G., & Olekšáková, D. (2012). Influence of the time of the milling and the temperature of the annealing to the coercivity of compacted samples investigated by the statistical methods. *Transfér inovácií, 23*, 52–54.
5. Luo, Ch., Fu, Y., et al. (2015). Temperature dependent coercivity and magnetization of light rare-earth Nd doped permalloy thin films. *Journal of Magnetism and Magnetic Materials, 374*, 711–715.
6. Budíková, M., Králová, M., & Maroš, B. (2010). *Pruvodce zakladnimi statistickymi metodami*. Praha: Grada Publishing.
7. Morrison, S. J. (2009). *Statistics for engineers: An introduction*. Chichester: Wiley.
8. Zhang, B., & Coddet, Ch. (2012). Studies of magnetic properties of permalloy prepared by SLM technology. *Journal of Magnetism and Magnetic Materials, 324*, 495–500
9. Hines, W. W., & Montgomery, D. C. (1990) *Probability and statistics in engineering and management science*. New York: John Wiley and Sons.

Statistical Evaluation of Market Potential Location of a New Product that Uses Renewable Energy Sources

Lucia Domaracka, Barbora Bencoova, Marcela Tausova, Daniel Slosar, and Marian Albert

1 Introduction

Today, the word "startup" is used very often, but many people do not exactly know what it means. The word "startup" is often connected with the new common company. But actually the word "startup" can refer to the company which has brought something new.

Mobile phones, tablets and notebooks, which are part of our everyday life and many people cannot imagine life without them, lead to addiction for people.

But what can we do, if mobile phones or camera will discharge through long-lasting hiking or summer festival. In this case, we have a few solutions, for example, to take a pack-set with us or devices which are able to generate, for example: generators, portable photovoltaic chargers or furnace which make energy from wood. According to the device which we use or conditions where we are, each one of these possibilities has advantages and disadvantages, too.

Requirements of users have increased and the offer of products as well. The offer also extends the invention called Chargebrella. It brings solutions to people, who have to be sure that they will be able to recharge batteries of their mobile devices if they will be somewhere without electrical power.

The principle of Chargebrella is to recharge devices with the assistance of solar energy or thermal sources, for example, fire. Because of it, Chargebrella is the unique device and there is no competition in Slovak or world market.

L. Domaracka · B. Bencoova · M. Tausova · D. Slosar · M. Albert (✉)
Institute of Earth Resources, Faculty of Mining, Ecology, Process Control and Geotechnologies
Technical University of Košice, Košice, Slovakia
e-mail: lucia.domaracka@tuke.sk; barbora.bencoova@tuke.sk; marcela.tausova@tuke.sk; daniel.slosar@tuke.sk

L. Knapcikova et al. (eds.), *4th EAI International Conference on Management of Manufacturing Systems*, EAI/Springer Innovations in Communication and Computing, https://doi.org/10.1007/978-3-030-34272-2_22

The usage of these devices can be more effective for poorer countries because the usage of them is more economical. On the other hand, the recharging of these devices at the places where a recharging of them is impossible can be interesting for people who are actively interested in outdoor sports.

2 Literature Review and Marketing Review

John mentioned "technological innovation since the time of the first Industrial Revolution is the proximate cause of global warming. Further innovation in technical and social systems is the necessary route to mitigation. Always and everywhere, innovation is messy, complicated, and contingent" [1]. And while prospective technologies have been reasonably well mapped, policy choices – dependent on political coalitions that solidify and dissolve unpredictably – cannot be similarly mapped. For example, governments must find ways to reduce emissions of greenhouse gases (GHGs) while at the same time providing ample supplies of low-cost energy for those who cannot afford high-cost energy, a difficult task in poor parts of the world and impoverished enclaves, even in the wealthiest countries.

Based on the Thomson effect in conjunction with Peltier, Joule and Fourier heat conductors, the energy and exergy analysis of the thermoelectric solar cell generator (SATEG) was presented. The cross-sectional area of the ring thermoelectric generator increases along the radial direction of origin of the flat thermoelectric generator. Compared to a flat thermoelectric generator, the circular thermoelectric generator has a higher overall heat transfer area.

The technology of direct energy conversion in the solid state for the conversion of heat to electricity is the production of thermoelectric energy [2–4]. The production of thermoelectric energy works on the Seebeck effect. The Seebeck effect is a phenomenon in which a temperature difference between two dissimilar electric conductors or semiconductors produces a voltage difference between the two substances. When the heat is applied to one of the two conductors or semiconductors, the heated electrons flow to the cooler one. The voltages produced by the Seebeck effect are small, usually only a few microvolts (millionths of a volt) per kelvin of temperature difference at the junction. If the temperature difference is large enough, some Seebeck-effect devices can produce a few millivolts (thousandths of a volt).

According to Vining, thermoelectric devices have better efficiency at lower performance compared to conventional thermodynamic power generation equipment and space conditioning. The most suitable for low-power applications are thermodynamic devices [5]. Electricity availability is a problem for building sustainable communications networks in evolving areas. We can say that low-cost energy supply encourages people to use communication services [6].

There were two ways to capture solar energy: photovoltaics that convert solar radiation to electricity or solar thermal systems that concentrate solar heat and use it to heat water or domestic heating. There is also another way of acquiring solar energy whose potential was recorded decades ago. However, this method was discontinued, for the reason that it could not be used in a practical and especially advantageous economic way. It is a thermochemical method where the solar energy is captured in the configuration of certain molecules, which can then release energy to produce usable heat [7].

A key role for successful service management is innovation, marketing and purchasing. Much remains to be done to understand how to cooperate with these three functions [8]. Marketing and innovation cannot be considered separately because today they are very closely related. A strong innovative marketing interface facilitates collaboration between development processes, so that marketing can focus on customer service innovation [9].

Managers, who are trying to get into a challenging area of internationalization, should create a competitive edge that will help them showcase their company's superior abilities [10, 11]. Managers discourage the spread of the unknown from entering the foreign market and from the use of international activities, as the growing foreign markets tend to be very difficult [12].

The decisive driving force of a company's business strategy is marketing resources that help companies gain competitive advantage over competitors (direct or latent) and lead to better brand performance and innovation and long-term marketing effectiveness [13, 14].

A key supporting role for the success of a new product is the business capacity of companies, as promotion and advertising influence new product awareness, which increases interest in buying a new product [15]. The leading role in developing a new product also displays a strong marketing ability for the company [16, 17].

3 The Used Research Methodology

We used the method of questionnaire in this thesis. According to Švec [18], this method is often used to get a lot of information very quickly through written questions about knowledge and opinions of people to the actual theme.

The methodology of questionnaire's processing includes:

(a) The creation of electronic questionnaire

The questionnaire was made through Google Docs form and was made up of 12 questions. These questions were made of open and closed questions.

(b) The filtration of target group

According to what we want to know, the questionnaire can be interpreted by more ways. The most important is to get back most of the sent questionnaires. In this case, it is 70%.

The goal of the questionnaire was to find out a target group of Chargebrella. Then, it was to find out the usage of electronic devices at the places where there is no electric power for the progress of Chargebrella. This questionnaire is the key point for Chargebrella's progress.

(c) The overall and graphic evaluation of the questionnaire

A graphic scheme was made to each question. Statistical data were composed in statistical program JMP, also called 'jump'. Each question also had a short objective description. Data were statistically evaluated and processed through histogram and contingency tables. The verification of statistical hypotheses was implemented then by using:

- Tests of correspondence between the share of the population with a fixed value – Pearson's test
- Confidence interval for testing Gaussian, standard normal distribution

$$p_v - u_{1-\alpha/2}\sqrt{\frac{p_v\,(1 - p_v)}{n}} < p < p_v + u_{1-\alpha/2}\sqrt{\frac{p_v\,(1 - p_v)}{n}} \tag{1}$$

- Analysis of the dependence through Fisher–Yates's exact test

(d) The interpretation of questionnaire's results

The results of questionnaire should not only be in the service of internal cases which would help Chargebrella's progress but it should also help investors. In this case, the investors can be understood like a target group of Chargebrella.

(e) Data output of questionnaire

The evaluation and interpretation's result of questionnaires should be a document whose content and extent correspond with time and interpretation's difficulty of this activity. It should indicate future solutions of this project.

4 Chargebrella

Briefly, Chargebrella is a portable generator, which is able to make more electric power than commonly using chargers. In technical respects, it is a thermo-generator which can make electric power from alternative and renewable sources of energy – concretely, from different sources of heat. During a sunny day, it uses solar radiation to generate electric power. In case there is not enough solar radiation or it is a night, it uses other sources of heat above 250 °C – common fire, a portable stove or a small oven.

The main advantage of this device is its ability to generate electric power not only during the day but also at night. It means that it is not dependent on one source of energy like the already mentioned devices. According to that, Chargebrella is portable and space-saving, and it is obvious it will help us very much. What is more, it uses renewable sources of power. It means, we can save quantity of electric power supplied to networks mostly from non-renewable sources (Fig. 1).

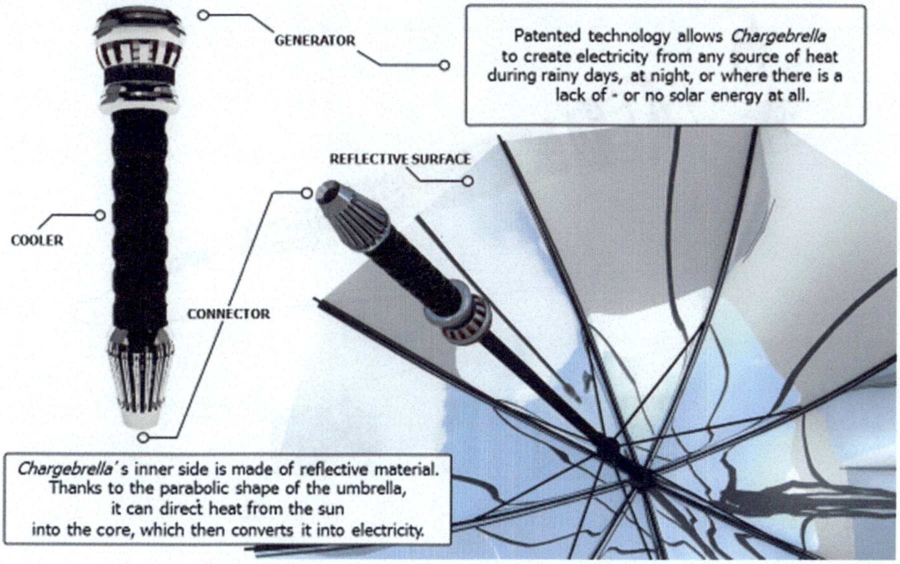

Fig. 1 Schematic of Chargebrella

For visualizing better, you can visualize a common umbrella or a little sunshade. It is enough to set the inner side of umbrella to Sun and a reflective part of the inner side will concentrate the required amount of solar radiation to handle. The handle will transform it into electric power. At night or during a rainy day, you can detach a parabolic part and use only the handle – to put it next to fire and electricity will be made directly from heat.

On the ground of that, this device is like an umbrella or a sunshade by shape, the title of the device was made by connecting the English words "charge" and "umbrella," which created the title Chargebrella.

Chargebrella has solution for people who are at places or in situations where there is no access to source of electric power to charge mobile devices. In regard to a small offer of finance available and portable devices for needs of individual users, the product fills an existing gap in the market (Fig. 2).

Devices, nowadays at the market, using renewable sources of power, are dependent on concrete renewable source, what limits their adaptation to different conditions. After 6 years of development, the device was made and it does not have any restrictions because it is able to use different sources of heat during any part of the day.

Chargebrella is a patent technology. Thanks to used materials, the lifetime of a thermo-generator is at least 30 years (or up to 300 years according to the quality of materials). The device is light, technically simple and has minimal faultiness.

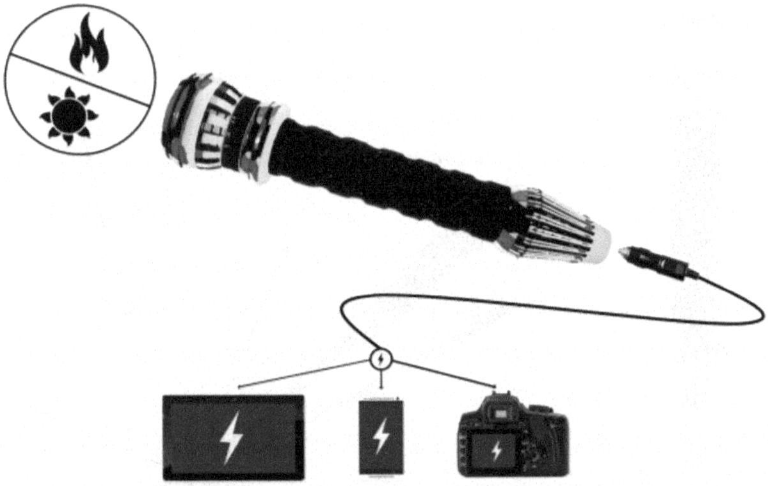

Fig. 2 Chargebrella

5 Results and Discussion

In this section, the evaluation of questionnaire is described. These results were processed by the statistical program JMP. The goal of this questionnaire was to find out information about this project, about its usage by usual customers and also to find out the prize, which they would be willing to pay for.

Discussing Problems

1. Age, gender and knowledge about project (H1, H2, H3)

Hypothesis 1: "More than 30% people have heard about project called 'Chargebrella'".

Hypothesis 2: "There are existing statistically relevant differences in the area of knowledge about project in regard to gender".

Hypothesis 3: "There are existing statistically relevant differences in the area of knowledge about project in regard to age" (Fig. 3).

It would be good to test the hypothesis:

Hypothesis 1 "we suggest that more than 30% people have already heard about project called 'Chargebrella'".

By probability testing by using unilateral correspondence test of percentage, it was found out that null hypothesis H0 cannot be declined and because of that, it was necessary to count the interval of reliability, which shows the extent of positive answers in values from 25.8% to 35.3%.

Next, it was good to test the answers, according to the gender and age of respondents (Tables 1 and 2).

Have you ever heard about project Chargebrella?

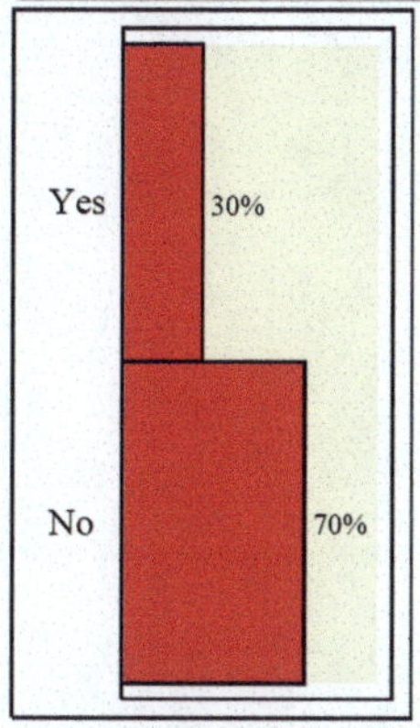

Frequencies

Level	Count	Prob
No	248	0,69663
Yes	108	0,30337
Total	356	1,00000

N Missing 5

2 Levels

Test Probabilities

Level	Estim Prob	Hypoth Prob
No	0,69663	0,70000
Yes	0,30337	0,30000

Test	ChiSquare	DF	Prob>Chisq
Likelihood Ratio	0,0192	1	0,8897
Pearson	0,0193	1	0,8896

Method: Fix hypothesized values, rescale omitted

Confidence Intervals

Level	Count	Prob	Lower CI	Upper CI	1-Alpha
No	248	0,69663	0,646985	0,742075	0,950
Yes	108	0,30337	0,257925	0,353015	0,950
Total	356				

Note: Computed using score confidence intervals.

Fig. 3 Graphic evaluation of the first question from the questionnaire "Have you ever heard about project Chargebrella?"

Table 1 Statistical test hypotheses 1 through compliance test proportion of the population with a specified value

k	N	p_0	TK	Kh	Conclusion
108	356	0,3	0,14	1,96	H0 cannot be declined

Table 2 Estimate of the confidence interval for the proportion of occurrence

pd	p	ph
0,252	0,300	0,348

Confidence intervals with alfa = 0,05

Fig. 4 Graphic evaluation of the second question from the questionnaire "What is the gender?"

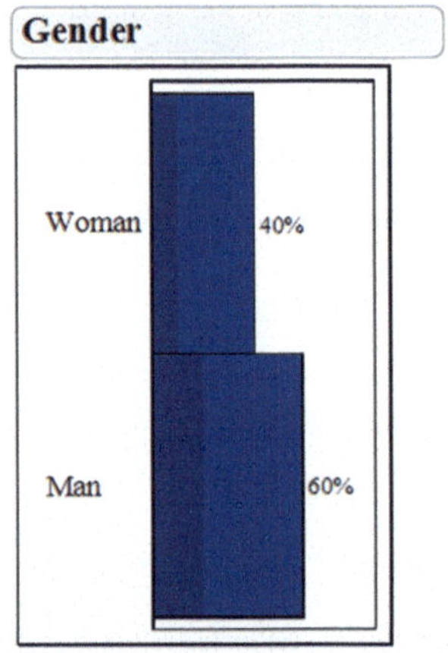

Frequencies

Level	Count	Prob
Man	212	0,59887
Woman	142	0,40113
Total	354	1,00000
N Missing	7	

2 Levels

$$p_v - u_{1-\alpha/2}\sqrt{\frac{p_v(1-p_v)}{n}} < p < p_v + u_{1-\alpha/2}\sqrt{\frac{p_v(1-p_v)}{n}}$$

Conclusion 1: On grounds of statistical testing, we can say that nowadays knowledge about project is around 25–35%.

Here, we see where we can find respondents answering positively to question "Have you ever heard about project Chargebrella?" – according to age (see Fig. 5) and gender (see Fig. 4).

From first sight:

A) There is a little difference in opinions according to gender – The project is more known for men than women (Fig. 4).
B) There is a bigger difference in opinions according to age up to 30 years and more than 30 years, people who are more than 30 years are more awake about project (Fig. 5).

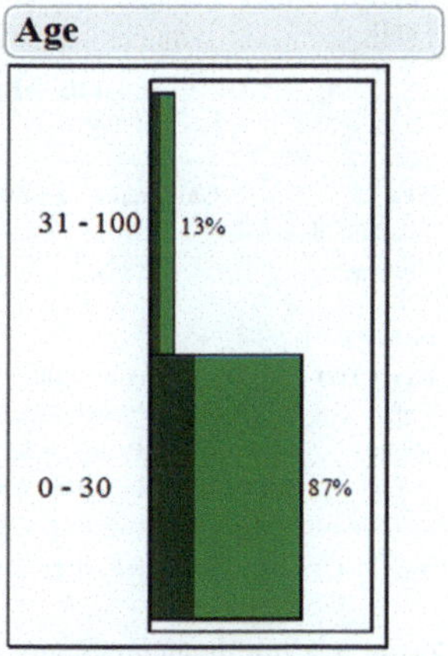

Frequencies

Level	Count	Prob
0 - 30	311	0,87360
31 - 100	45	0,12640
Total	356	1,00000
N Missing	5	
2 Levels		

Fig. 5 Graphic evaluation of the third question from the questionnaire "How old are you?"

Fisher's exact test

```
-----------------------------------------
TABLE = [ 71 , 36 , 139 , 106 ]
Left   : p-value = 0.9655659179758306
Right  : p-value = 0.05707753497735833
2-Tail : p-value = 0.0990537705011956
-----------------------------------------
```

Fig. 6 Fisher's exact test

From first sight:

A) There is a little difference in opinions according to gender – The project is more known for men than women (Fig. 4).

Hypothesis 2: "There are existing statistical significant differences in the area of knowledge about project according to age."

Verifying was done by using Pearson's chi-squared test (goodness-of-fit test) and Fisher's exact test.

Interesting: Fisher's exact test has confirmed the statistically significant difference in the opinions according to gender (Fig. 6), but Pearson's chi-squared test has not confirmed the statistical significance in the level of significance alfa = 0,05. (3,84 – critical value of chi square).

In the level of significance alfa = 0,1, differences were confirmed in both – Pearson's chi-squared test and Fisher's test. (2,71 – critical value of the chi square) (Fig. 7).

Tests

N	DF	-LogLike	RSquare (U)
352	1	1,4505770	0,0061

Test	ChiSquare	Prob>ChiSq
Likelihood Ratio	2,901	0,0885
Pearson	2,864	0,0906

Fisher's Exact Test	Prob	Alternative Hypothesis
Left	0,9656	Prob(Gender=Man) is greater for Have you ever heard about project Chargebrella?=No than Yes
Right	0,0571	Prob(Gender=Man) is greater for Have you ever heard about project Chargebrella?=Yes than No
2-Tail	0,0991	Prob(Gender=Man) is different across Have you ever heard about project Chargebrella?

Fig. 7 Comparison of Fisher's exact test and Pearson's chi-squared test

Table 3 Contingency table by gender

Have you ever heard about project?	Count Total % Col % Row %	Woman	Man	
	No	10630,1174,6543,27	13939,4966,1956,73	24569,60
	Yes	3610,2325,3533,64	7120,1733,8166,36	10730,40
		14240,34	21059,65	352

Conclusion 2: On grounds of statistical testing, we can say that Hypothesis 2 was confirmed and it means there are statistically significant differences in the area of knowledge about project according to gender. The project is more known for men than women.

During analysis, we used:

- the contingency table (Table 3),
- an analysis of the average value – examination of significance value's diversion from average (Figs. 8 and 9).

Here, we see where we can find respondents answering positively to the question "Have you ever heard about project Chargebrella?" – according to age and gender.

From first sight:

B) There is a bigger difference in opinions according to age up to 30 years and more than 30 years, people who are more than 30 years are more awake about project (see Fig. 5).

Hypothesis 3 "There are existing statistical significant differences in the area of knowledge about project according to age."

Data processing age was required to be reformed because the group of respondents up to 18 years was too small and we were not able to evaluate it. Because of that, we have made two categories of age (up to 30 years, over 30 years).

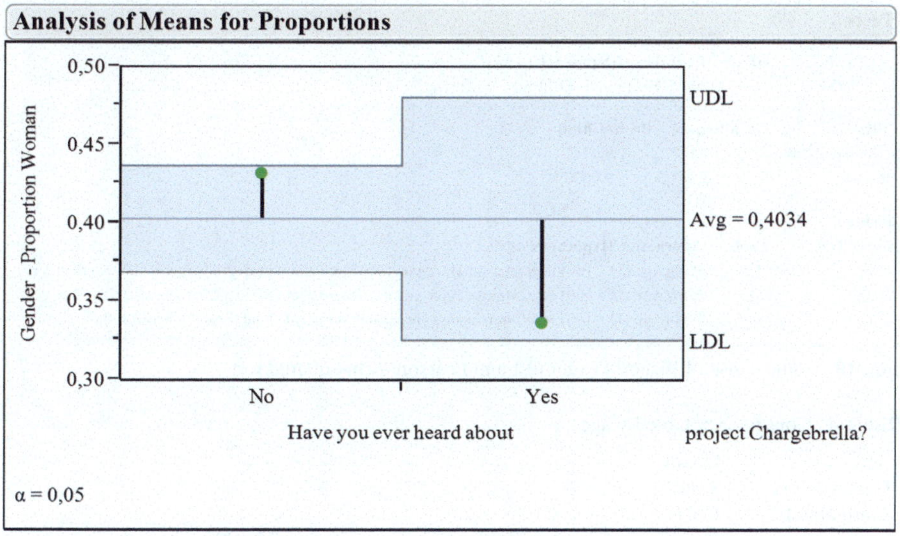

Fig. 8 Analysis of means for proportions – woman

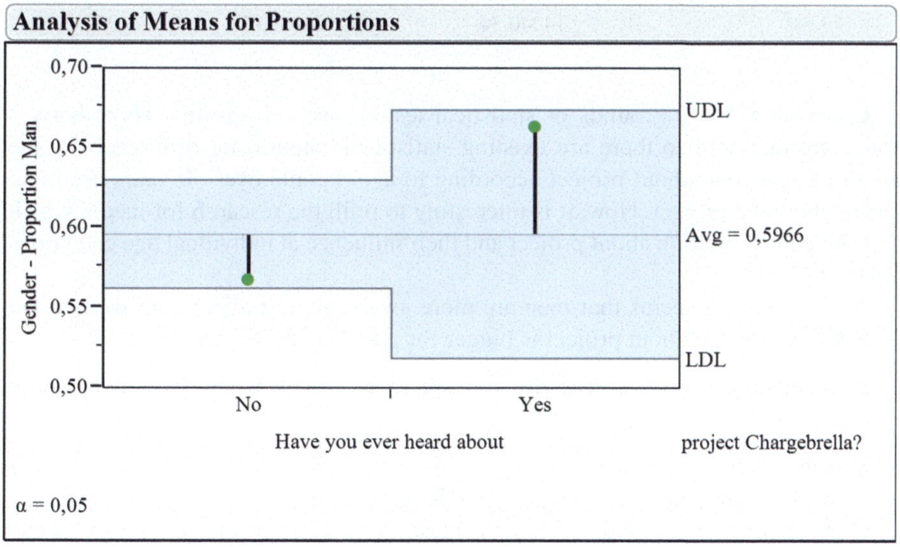

Fig. 9 Analysis of means for proportions – man

Like in previous hypothesis (Hypothesis 2), the results have confirmed a difference in the area of knowledge about project. Although, at the level of significance alfa 0,05, it was confirmed only by Fisher's exact test. At the level of significance alfa 0,1, it was confirmed by both tests (Pearson, Fisher) (Fig. 10, Table 4).

Tests

N	DF	-LogLike	RSquare (U)
354	1	1,5886580	0,0118

Test	ChiSquare	Prob>ChiSq
Likelihood Ratio	3,177	0,0747
Pearson	3,337	0,0678

Fisher's Exact Test	Prob	Alternative Hypothesis
Left	0,9751	Prob(Age=31 - 100) is greater for Have you ever heard about project Chargebrella?=No than Yes
Right	0,0515	Prob(Age=31 - 100) is greater for Have you ever heard about project Chargebrella?=Yes than No
2-Tail	0,0827	Prob(Age=31 - 100) is different across Have you ever heard about project Chargebrella?

Fig. 10 Comparison of Fisher's exact test and Pearson's chi-squared test

Table 4 Contingency table by age

Have you ever heard about project?	Count Total % Col % Row %	Age 0-30	Age 31-100	
	No	22062,1589,43214,729	267,3457,7810,57	24669,49
	Yes	3610,2325,3533,64	7120,1733,8166,36	10730,40
		14240,34	21059,65	352

Conclusion 3: On grounds of statistical testing, we can say that Hypothesis 3 was confirmed and so there are existing statistically significant differences in the area of knowledge about project according to age. People over 30 years are more aware about the project. Now, it is interesting to refill the research for media which distributed information about project and their influence at individual age categories (Fig. 11).

At first sight, it seems that men are more awake about project than women but actually knowledge about project is bigger for people over 30 years (Fig. 12).

2. How many days in a year do you spend at places where you don't have electricity?

   Which device do you use there?

   How much are you willing to pay for Chargebrella?

Conclusion 4: On grounds of statistical testing, we can say that people who spend less than 12 days at places without power supplying are willing to pay about 201–350 € for Chargebrella. It is much more than people who spend at these places longer time (see Fig. 13).

MEN

Have you ever heard about project Chargebrella?

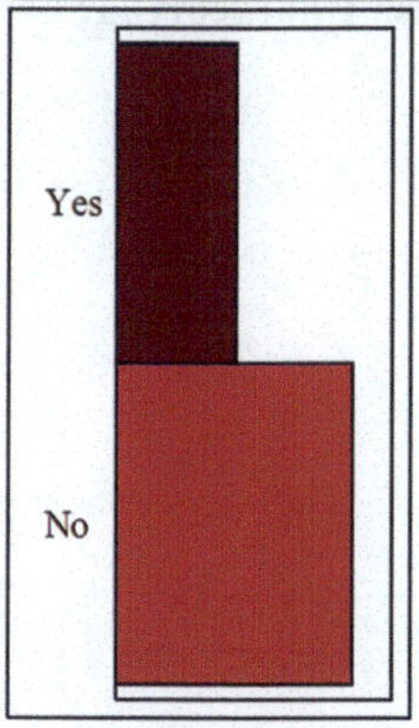

Frequencies

Level	Count	Prob
No	139	0,66190
Yes	71	0,33810
Total	210	1,00000
N Missing	3	

2 Levels

Test Probabilities

Level	Estim Prob	Hypoth Prob
No	0,66190	0,70000
Yes	0,33810	0,30000

Test	ChiSquare	DF	Prob>Chisq
Likelihood Ratio	1,4189	1	0,2336
Pearson	1,4512	1	0,2283

Method: Fix hypothesized values, rescale omitted

Confidence Intervals

Level	Count	Prob	Lower CI	Upper CI	1-Alpha
No	139	0,66190	0,595525	0,722467	0,950
Yes	71	0,33810	0,277533	0,404475	0,950
Total	210				

Note: Computed using score confidence intervals.

Age

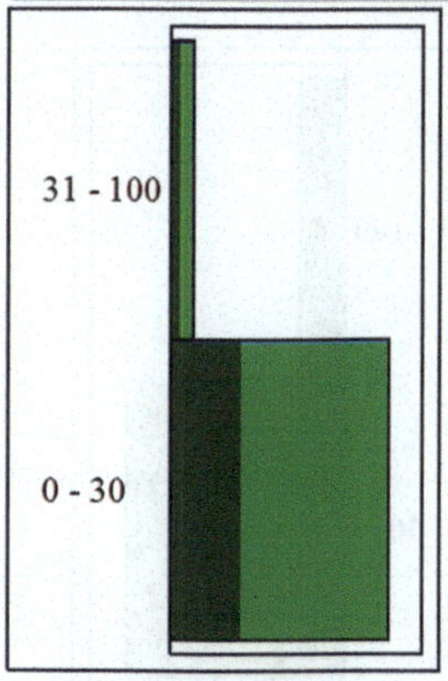

Frequencies

Level	Count	Prob
0 - 30	192	0,90566
31 - 100	20	0,09434
Total	212	1,00000
N Missing	1	

2 Levels

Fig. 11 Graphical comparison questions of "Have you ever heard about project Chargebrella?" and "How old are you?" for men

WOMEN

Age

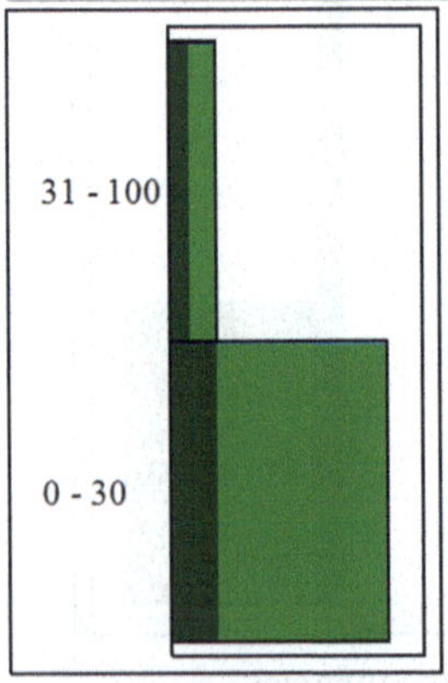

Frequencies

Level	Count	Prob
0 - 30	117	0,82394
31 - 100	25	0,17606
Total	142	1,00000
N Missing	0	

2 Levels

Have you ever heard about project Chargebrella?

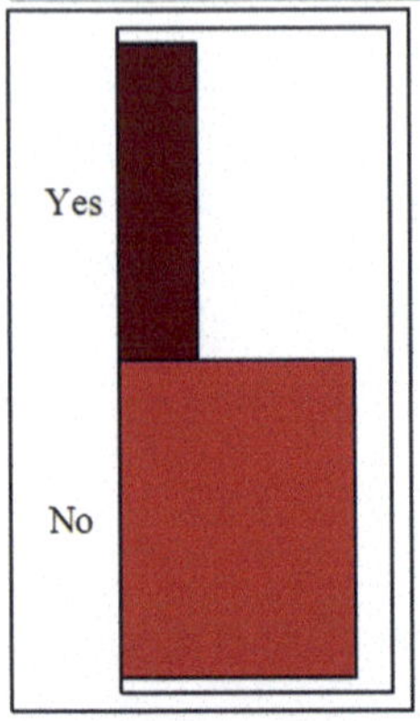

Frequencies

Level	Count	Prob
No	106	0,74648
Yes	36	0,25352
Total	142	1,00000
N Missing	0	

2 Levels

Test Probabilities

Level	Estim Prob	Hypoth Prob
No	0,74648	0,70000
Yes	0,25352	0,30000

Test	ChiSquare	DF	Prob>Chisq
Likelihood Ratio	1,5087	1	0,2193
Pearson	1,4608	1	0,2268

Method:Fix hypothesized values, rescale omitted

Confidence Intervals

Level	Count	Prob	Lower CI	Upper CI	1-Alpha
No	106	0,74648	0,669086	0,810888	0,950
Yes	36	0,25352	0,189112	0,330914	0,950
Total	142				

Note: Computed using score confidence intervals.

Fig. 12 Graphical comparison questions of "Have you ever heard about project Chargebrella?" and "How old are you?" for women

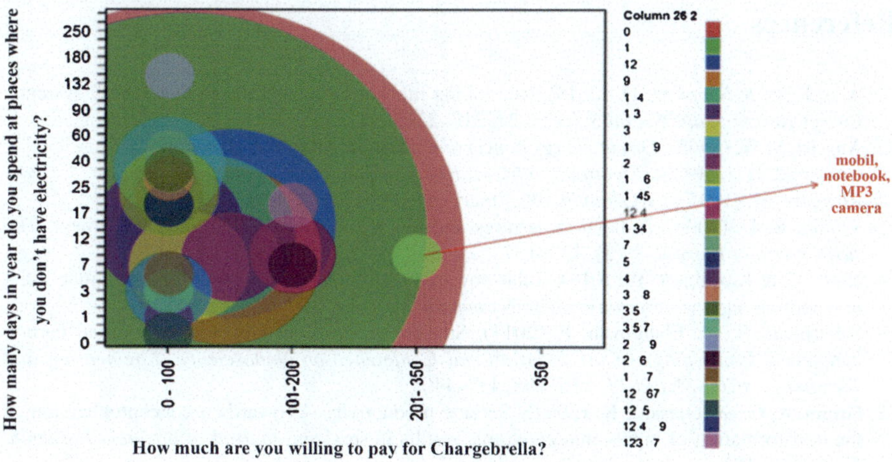

Fig. 13 Graphical comparison questions of "How many days in year do you spend at places where you don't have electricity?" and "How much are you willing to pay for Chargebrella?"

6 Conclusion

In this article, we dealt with a marketing research for a new start-up project called Chargebrella. The research was done through the questionnaire method, which we evaluated by the statistical program JMP.

On the ground of the research, we can say that Chargebrella is mostly applicable for people who spend their time at places where electric power is not available from 7 to 12 days of a month. We found out that these people are willing to pay for the product about 200 €, it is a very interesting sum of money to cover whole costs and also for attractive profit besides similar products, which are available in the market.

We can confirm that the research was very important for the determination of target group, what we did at the end. Because of that, we suggest to orientate to the group of people from 18 to 30 years old, what the research showed, too.

In conclusion, we can say that nowadays start-up projects could be successful only if they will be presented in an interesting way not only for ordinary people but also for investors.

Acknowledgments Authors would like to thank anonymous reviewers for their constructive comments and suggestions that have made a significant contribution to improving this work. This study was supported by the Scientific Grant Agency – VEGA 1/0754/18 – Evaluation of Geometric and Topological Properties of Buildings and the Grant Agency for Culture and Education – Project KEGA No. 002TUKE-4/2017, "Innovative didactic methods of the educational process at the university and their importance in raising teachers' educational competence and developing student competences."

References

1. Alic, J. A., & Sarewitz, D. (2016). Rethinking innovation for decarbonizing energy systems. *Energy Research and Social Science, 21*, 212–221.
2. Angrist, S. W. (1965). *Direct energy conversion*. Boston: Allyn and Bacon, Inc..
3. Goldsmid, H. J. (1986). *Electronic refrigeration*. London: Pion.
4. Rowe, D. M. (1995). *CRC handbook of thermoelectrics*. Boca Raton: CRC Press.
5. Vining, R. L. (2009). Judicial departures and the introduction of qualified retirement. *The Justice System Journal, 30*(2), 139–157.
6. Kato, T., & Kawahara, Y. (2013). Solar-powered cell phone access point for cell phone users in emerging regions. *International Telecommunication Union, 6533792*, 157–164.
7. Sabarinath, B., & Elangovan, P. (2011). Storage of sun's heat by using modified carbon nanotubes. *Proceedings of the International Conference on Nanoscience, Engineering and Technology, ICONSET 2011, 6167998*, 437–440.
8. Grönroos, C., & Ojasalo, K. (2004). Service productivity – Towards a conceptualization of the transformation of inputs into economic results in services. *Journal of Business Research, 57*(4), 414–423.
9. Wiersema, F. (2013). The B2B Agenda: The current state of B2B marketing and a look ahead. *Industrial Marketing Management, 42*(4), 470–488.
10. Barney, J. B., Wright, M., & Ketchen, D. J. (2011). The future of resource-based theory. *Journal of Management, 37*(5), 1299–1315.
11. Samli, A. C., Wirth, G. P., & Wills, J. R. (1994). High-tech firms must get more out of their international sales efforts: An exploration in developing a competitive edge. *Industrial Marketing Management, 23*(4), 333–342.
12. Thai, M. T. T., & Chong, L. C. (2013). Dynamic experimental internationalization: Strategy of SMEs from a transition economy. *Journal of International Entrepreneurship*. Online first. https://doi.org/10.1007/s10843-013-0110-4.
13. Srivastava, R. K., Fahey, L., & Christensen, H. K. (2001). The resource-based view and marketing: The role of market-based assets in gaining competitive advantage. *Journal of Management, 27*(6), 777–802.
14. Kozlenkova, I. V., Samaha, S., & Palmatier, R. (2014). Resource-based theory in marketing. *Journal of the Academy of Marketing Science, 42*(1), 1–21.
15. Vorhies, D. W., Morgan, N. A., & Mason, C. H. (2009). Market orientation, marketing capabilities, and firm performance. *Strategic Management Journal, 30*, 909–920.
16. Khouri, S., Rosova, A., Straka, M., & Behun, M. (2018). Logistics performance and corporate logistic costs, their interconnections and consequences. *Transformations in Business & Economics: International Journal of Scholarly Papers*. – Kaunas (Litva): Kauno fakultetas, ISSN 1648-4460, *17*(2), 426–446.
17. Behunova, A., Behun, M., & Knapcikova, L. (2018). Simulation software support of manufacturing processes in engineering industry. *TEM Journal: Technology, Education, Management, Informatics*. – Novi Pazar (Srbsko): Association for Information Communication Technology Education and Science, *7*, 849–853.
18. Švec, S., et al. (1988). *Metodológia vied o výchove*. Bratislava: IRIS. ISBN 8088779735, p. 300.

Adaptive Assembly Approach for E-Axles

Muaaz Abdul Hadi, Markus Brillinger, and Franz Haas

1 Introduction

In the last century, researches were focussed on low-cost products and achieving them with mass production with highly efficient Dedicated Manufacturing Systems (DMS). These systems are used for manufacturing high quantities of a similar product at low throughput times. Hence, DMS are fixed and have a monotonous sequence of steps. If an additional process step for one part is required, the efficiency and quality of this system decrease significantly [1].

Now given the shift in recent years, researches are focussing on Flexible Manufacturing Systems (FMSs) to keep pace with the ongoing mass customization. Flexible Manufacturing Systems are versatile and adaptive to variety of products. But, the complexity of FMS and costs of implementing such a system are quite high. Also, FMS has a lower productivity compared to DMS, as the production steps are not conducted simultaneously [2]. The quality is of concern as the variety increases. Thus, an advantage of FMS would be that it has high degree of flexible automation. It is also noted that majority of users in the automobile industry are not satisfied with the FMSs because of a variety of problems including lack of reconfigurability as a result of their fixed capacities and functionalities [3]. These are the two opposed types of manufacturing systems.

M. Abdul Hadi (✉) · M. Brillinger
Pro2Future GmbH, Graz, Austria
e-mail: muaaz.abdul-hadi@pro2future.at; markus.brillinger@pro2future.at

F. Haas
Pro2Future GmbH, Graz, Austria

Institute of Production Engineering, Graz, Austria
e-mail: franz.haas@tugraz.at

L. Knapcikova et al. (eds.), *4th EAI International Conference on Management of Manufacturing Systems*, EAI/Springer Innovations in Communication and Computing, https://doi.org/10.1007/978-3-030-34272-2_23

One of the challenges of the twenty-first century is the dynamic interaction between the distinct manufacturing processes and adaptability machines developed by engineers keeping the process quality of the same standards standards [4]. The variety in vehicle types of electromobility (e-mobility) is high and the batch size is low, which in turn makes the manufacturing and assembling costs higher [5], along with the efforts to increase quality. Thus, the manufacturers are dependent to match this high flexibility, quality and variety. To match the high flexibility of an FMS system while considering qualitative efforts, a use case of an e-axle assembly is being done. E-axle is considered as the market maturity of the electric vehicle sector is low [6]. Hence, the goal would be to implement a set of adaptive technologies for an adaptive assembly system with respect to e-mobility. Also, the focus is to achieve the right balance between the machines and humans to make the assembly process simpler, faster and less expensive by combining the proven methodologies.

This article outlines the planned research in terms of investigating how the aforementioned adaptivity and quality can be achieved in an e-axle assembly. To do so, the existing process design of the assembly is analysed to identify the technological gaps. Further, to bridge this disparity, requirements of adaptive assembly system are described. Additionally, the research gap is presented by combining the benefits of these concepts and presenting the various technologies. Finally, with the help of verification models, the article draws an outline of expected results.

2 State of the Art

2.1 Product and Process Analysis

In the current e-axle assembly layout, a single variety of e-axle (type-B) is being assembled. However, the number of e-axles being assembled must be increased, as shown in Table 1 (three types). But the number of types can further increase over a period. Hence, our focus is to make the assembly line adaptive and error free from the current manual process. The axles can be of pure BEV (battery electric vehicle) or an Hybrid vehicular type. For a BEV, both the front and rear sides are equipped with electric motor (e-motor). They are primarily differentiated by the number of stages and speed. Table 1 describes the differentiation of these axles in detail.

To develop an adaptive assembly system and to get a better understanding of the e-axle assembly, the assembly sequence of an existing e-axle (type-B) (as illustrated in Fig. 1) is analysed first (short overview).

The assembly sequence is an ideal case of flow or series assembly. All the parts are pre-assembled in a separate station and then moved to the assembly stations. All the tasks are performed manually. There is an end-of-line testing station which does the leakage test as well as the full-function test. Since the assembly process is completely manual, the time taken is higher. Thus, the aim is to make the assembly line adaptive thereby reducing the assembly time for this specific e-axle.

Table 1 Types of e-axles

e-axle type	A	B	C
Type	BEV	Hybrid	BEV
Front	e-motor for each wheel	e-motor for each wheel	e-motor (differential used)
Rear	e-motor for each wheel	Internal Combustion Engine (ICE)	e-motor (differential used)
Speed/ Gear	1 Speed	1 Speed	1 Speed
Stage	2 Stage	1 Stage	2 Stage
Gear ratio	14–15	10	10–15
Differential	No	No	Yes
	2 motors in each axle (one for each wheel) Independent torque for each wheel		Bevel gear differential
Gearing	Helical gearing – offset design (can be coaxial design as well)	Planetary gearing –coaxial design	Helical gearing – offset design
Cooling system	Water cooling in stator	Oil cooling	Oil cooling
Parking lock	Yes	No	Yes
Quantity required	4000/year for 3 years	3333/year for 9 years	Not specified

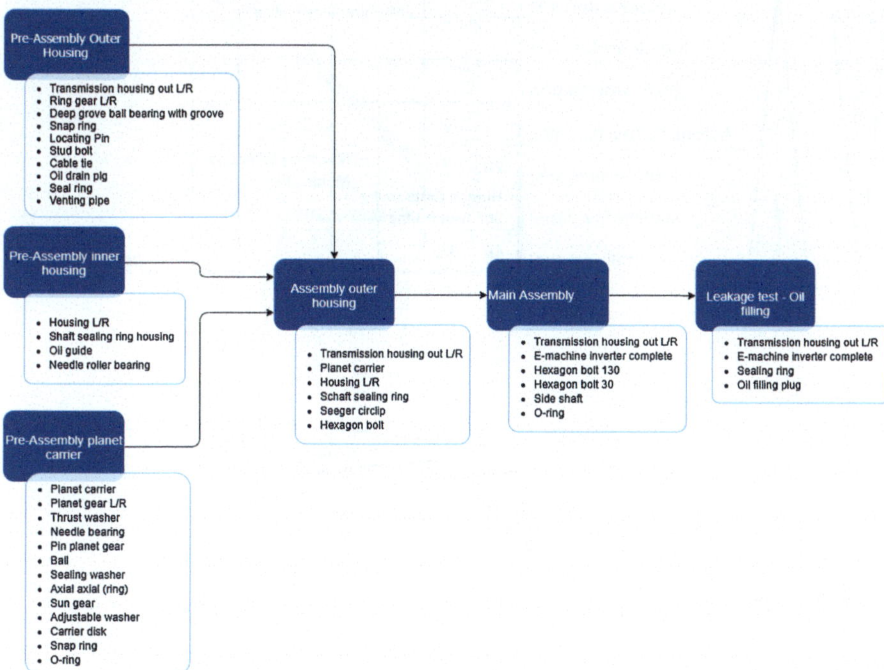

Fig. 1 Structure of the e-axle assembly (type-B)

Considering the yearly output is low, which is not more than 10,000 axles/year, implementing a fully automatic assembly would not be feasible and cost-effective. As described in a case study done in [7], implementing an automated system for lower throughput per day can be expensive. Hence, a combination of completely automated system and manual tasks must be found to make the system cost-effective and adaptive [7].

For implementing these adaptive technologies, the initial Level of Automation (LoA) must be defined. The scientific approach is to perform a DYNAMO++ methodology which further classifies into 12 steps including LoA (Level of Automation) Matrix [8–10]. This methodology helps in increasing the Level of Automation (LoA) [8]. The initial steps have been completed and the current LoA for the above e-axle assembly (type-B) has been determined, as shown in Fig. 2. In the current assembly process, there are 92 tasks which are distributed in the matrix as shown. The implementation of the adaptive technologies must be followed which increases the LoA in the directions shown by the arrows. Finally, this improvement in LoA is measured to determine the increase in cognitive and physical automation.

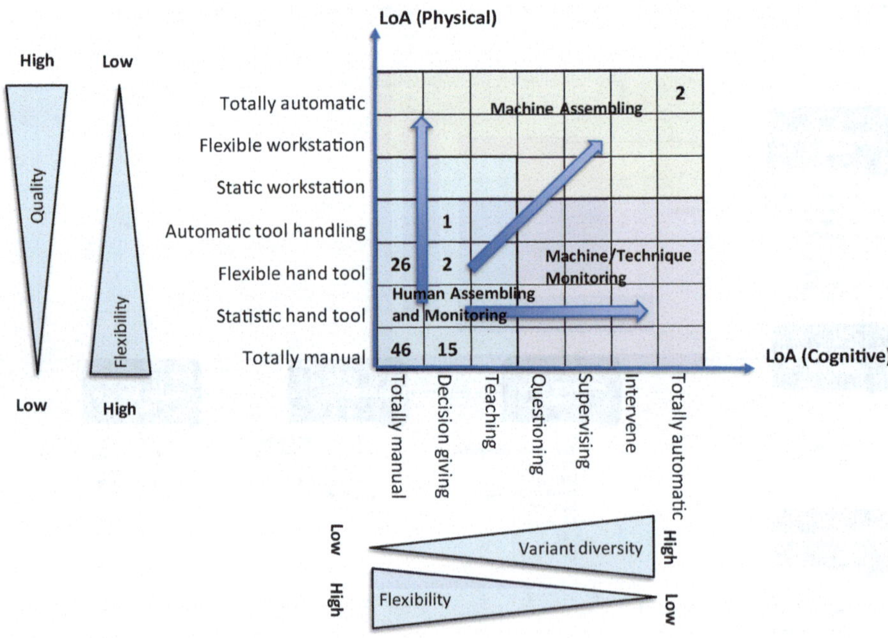

Fig. 2 LoA matrix

2.2 Adaptive Assembly Concepts

After the thorough literature review phase, four main concepts were derived. In this section, we will briefly elaborate the concepts being used to make a system adaptive. These concepts are selected, as they are most suitable for ramp-up of high variety, low batch size assembly by achieving adaptability. The technologies that must be implemented must be a well-balanced solution of automated and manual tasks. With this as the focus, the various concepts have been described below in a nutshell.

Migration Manufacturing The number of variants of each e-axle is increasing considerably with slight variations. Migration manufacturing helps with a method that can manufacture these different parts on the same assembly line by implementing internal loop lines [11]. The use case of migration manufacturing with meandering technique has been explained in [11].

Holonic Manufacturing System Holonic Manufacturing System (HMS) is a concept used for increasing the flexibility, agility and reconfigurability of the manufacturing process [12]. Each unit of HMS is represented by an autonomously working unit called holon [13]. A holon is defined in the holonic paradigm as a unit that advocates the use of autonomous and cooperative manufacturing units [12]. These holons can interact and communicate with other holons and build a hierarchy, which in HMS is termed as holarchy [13]. If any assembly station breaks down, a multi-function (MF) station can be utilized to continue the process. The docking station (DS holon) decides whether (and when) to divert the part from the main line in case of a bottleneck and sends a signal to AGV (Autonomous Guided Vehicle) that transports the picked-up part. Hence, there is coordination between these holons. However, the assembly stations can still be manually operated [12].

Reconfigurable Manufacturing System Reconfigurable Manufacturing System (RMS) can be defined as an intermediate between DMS and FMS [14]. However, the concept of reconfigurability is applicable for a specific part family of products [2] and customized flexibility [15]. It bridges the gap between the high flexibility and high cost of totally flexible machines and the low flexibility and low cost of fully dedicated machines [16, 17]. Reconfigurability at lower levels, such as machines, cells and shop floors, is achieved by changing the hardware resources [14]. The throughput of RMS is higher than the throughput of FMS, but is lower than that of DMS for the same investment cost [18]. There are six core characteristics and principles that an RMS system can achieve: scalability, convertibility, diagnosability, customization, modularity and integrability [18]. Reconfiguration technologies can be implemented on various aspects of an assembly station such as machine, inspection, system [18] and small assembly stations. This system can also be called as a hybrid system where one can obtain volume flexibility with low investment shown in [7]. The system is economical because the movements are reduced to minimum [7]. Also, the output can be increased as the time required by the worker decreases.

Cognitive Factory – HMI (Human Machine Interface) Being cognitive is about flexibility and faster adaption to change. The easy interaction between humans and machine is the key success of a cognitive factory. These are assistance systems that actively support the worker. This also allows automatic knowledge transfer and collaboration between experts and unskilled workers [19]. Sensors and actuators form the main basis of the basic interaction between the assistance systems and humans [20]. As described in [19], this sensor network can be based on initial measurement units (IMUs), cameras and processing units. To simplify the understanding, the two chosen functional cases that can be derived include:

- Input/observation techniques: Hand gesture recognition – the movement of hands (such as grasping) can be tracked by the sensors or a camera [21] and this can be integrated with pick-to-light system for high product and component variety [8].
- Output modalities: Head Mounting Devices (HMDs), such as retina display or augmented reality (AR) [22], are suitable. Visual screen at a static position shows the next assembly steps [21]. Text-to-speech system can also be considered as the assembly is noise free [21].

Achieving the right balance between the two dedicated and flexible manufacturing systems by combining the different concepts explained above would be an ideal way of achieving the right flexibility. Each approach that has been studied would ideally fit the assembly line of low volume and high variety of batch production. Hence, an ideal direction is to implement the best aspects of each concept to achieve this flexibility, adaptability and low costs. Table 2 summarizes the benefits of each concepts.

These concepts focus on one or more core areas of an assembly plant along with their KPIs (Key Performance Indicators). For example, if "throughput" is considered, the concept of HMS, RMS, and Cognitive Factory achieves this KPI. Likewise, the concept of Migration Manufacturing focusses on the KPI: "Area" and so on. These KPIs form the basis as a requirement of an adaptive assembly for high variety. As shown in Table 3, the concepts focus on four main core areas of the assembly plant which in turn has several KPIs. Marked "x" indicates that the concept targets a specific core area.

3 Research Direction

With the concepts and their focussing area and KPIs explained, the technologies to execute these concepts must be described. These technologies are clustered based on the features in an assembly plant. These are then categorized by their level of practicality which is explained in the further section. Furthermore, they are evaluated subsequently to depict the adaptability achieved.

Table 2 Benefits of each concept

Parameters	Migration manufacturing	Holonic manufacturing system (HMS)	Reconfigurable manufacturing system (RMS)	Cognitive factory – HMI
Ideal for	Increasing variants [11]	Flexible and dynamic allocation of resources [13]	Quick and easy adjustments to new products [2]	Increasing productivity [8]
Initial investment	10–30% less than FMS [11]	Higher than DMS, but lower than FMS	Lower than automated system [7]	High initial equipment cost
Overall efforts for implementing	50–80% lesser compared to individual lines [11]	Higher initial efforts than DMS	Depends on the level of reconfigurability	Comparatively lower than RMS and HMS
Other advantages	Faster break-even point than an additional line; 5–14% lesser variable cost [11]	Increase in productivity and throughput [12]	High responsiveness to fluctuating markets [18]; movements of operator are reduced to minimum [7]	Pick-to-light can be used for variety of tools [8]

Table 3 Concepts, core areas, and their KPIs

		Concepts			
Core area	KPIs	Migration manufacturing	Holonic manufacturing system (HMS)	Reconfigurable manufacturing system (RMS)	Cognitive factory – HMI
Layout	Area (m^2), design of layout	×			
Process	Throughput, Overall Equipment Efficiency (OEE), quality		×	×	×
Machine	Cost, throughput, quality, performance		×	×	×
Logistics	Time, inventory		×		

3.1 Derived Technologies

Achieving the maximum adaptability in the assembly process with a high variety of e-axles is the goal of these concepts. Also, another goal is maintaining the right balance between the automated systems and manual work, keeping the small volumes, high variety, and finally costs in mind. To enable this adaptability, as shown in Fig. 3, the derived morphological matrix has various technologies based on their Level of Practical Application (LoPA). These technologies can also be classified individually on their Technology Readiness Level (TRL) [23]. This matrix can be served as a building frame for adaptability. The aspects or features tagged with an asterisk ($_*$) are the aspects that are being focussed on for the current assembly type and these aspects have higher practical implications.

LoPA (Level of Practical Application)	Low	Medium	High
Focussing Area	- Research phase - Simulation approach *	- Evaluation phase - Prototype implementation	- Implementation phase - Applicable in all industries
Level of Risk	- Very high risks	- Medium to high risks *	- Zero to Low risks *
Communication	- Real-time Wireless communication, 5G - Self-organised wireless networks, Neuronal network	- Wireless communication * - Holonic communication *	- Real-time bus interfaces * - Mobile networks * - EMUX and High perf. communication *
Sensors	- Miniaturized sensors - Smart sensors - Vibration device *	- Multi sensor fusion - Networked sensors - Innovative safety sensors (fail-safe system) *	- Motion sensors * - Temperature sensors - Pressure sensors - Acceleration sensors
Actuators		- Intelligent actuators - Networked actuators - Safety actuators	- Pneumatic grippers in robot *
Human Machine Interfaces	- Human behaviour model - Object recognition (YOLO) - Semantics visualization - Exoskeleton	- Voice controls - Gesture controls - Augmented reality * - Virtual reality *	- Intuitive controls * - Pick-to-light systems * - Digital watch/static screen *
Layout/Logistics	- Matrix manufacturing - Bionic layout structure	- Meandering technique * - Matrix manufacturing * - Logistics: Agent based communication *	- Series/flow/parallel production, etc. - Automatic Guided Vehicles (AGV), AG cart *
Machine	- Hybrid Machine - Additive Manufacturing	- Reconfigurable Machine System, RM Tools (RMT) * - Collaborative Robots (Cobot) *	- Mass Production by dedicated machines, High variety & low volume flexible manufacturing machines - Robots, smart tools *
Embedded Systems/ Software	- Industry 4.0 simulation - Miniaturized/smart sensors - Multicriteria situation awareness - Artificial Intelligence	- Energy harvesting - Machine learning - Digital Twin *	- AutoID technologies - Big-Data, cloud-computing
Probable TRL Level	1-3	4-6	7-9

Fig. 3 LoPA morphological matrix

3.2 Evaluation Techniques

The various technologies specified previously are to be implemented and verified with the help of verification models. To help implementing and testing the reliability of the adaptive systems, the recent approaches, which are explained further, can be enforced. These techniques help in implementation and verification of the mentioned adaptive concepts.

FMEA The Failure Mode and Effect Analysis (FMEA) is done on the current assembly layout (based on e-axle type-B) to better understand the errors that are occurring during the process [24]. With 92 tasks in the current assembly, the human-led errors during the assembly are studied and a thorough analysis of the severity, occurrence and detection is done. The values of these parameters are provided from the plant managers, as they have an overview of the manifestation of these errors. Figure 4 describes some of the general errors that occur along with their RPNs (Risk Priority Numbers). After the implementation of adaptive technologies, the FMEA is done to determine the reduced RPN. Therefore, it indicates the reduction in errors by implementing adaptive technologies.

Simulation Model Currently, the assembly process is completely manual with high process times and this has been implemented in the assembly model in Plant Simulation tool. The simulation model helps in determining the bottleneck and the process steps. A bottleneck is defined as a workstation limiting the production efficiency of the entire process [25]. The simulation model allows to calculate the effectiveness of various methods and processes (such as HMI, RMS, etc.) and for a variety of e-axles. The creation of simulation model is done by using a seven-step approach as described by Law [26]. The computer simulation models can be freely improved and further simulations to the improved processes can also be applied freely [27].

The implementation was done as per the layout and station timings. The bottlenecks were clearly seen from the statistic graphs derived from the plant simulation. Furthermore, the changes (adaptive technologies) are also implemented in the simulation tool to determine the increase in throughput and efficiency of the system. Also, with the help of simulation tool, the errors during the ramp-up production are considerably reduced [27]. Further, a simulation model can be used to visualize in real time and focus on the affecting parameters [27]. This approach can also be linked to the concept of digital twin [28].

Other Approaches There are various other methodologies that are being done to determine the priority of each concept. Also, further steps of DYNAMO++ methodology, as described in [8], will be done. The cost–benefit analysis is done to improve the process performance. For example, a pairwise comparison will be done for various adaptive technologies, an FMEA depicting the benefits of each adaptive concept, and finally a cost–benefit analysis.

Process Step/Input	Potential Failure Mode	SEVERITY (1 - 10)	Potential Causes	OCCURRENCE	DETECTION (1 - 10)	RPN
Information/ Assembly operations	Poor or missing documentation - Process description	9	Huge rework and additional times that causes unknown expenses	9	2	162
Assembly operations/ Manufacturing operations	Operating errors	7	Negligence, unskilled or unqualified, or no presence of mind	8	6	336
Assembly operations/ Information	Longer assembly times than allocated	9	Missing or unclear workflow structure	10	6	540
Person related/ Assistive operations/ Information	Distraction of the worker	7	No presence of mind	7	6	294
	Wrong decisions for basic worker operations	9	Due to Lack of Management experience and management skills or improper recruiting	7	3	189
Ergonomics/ Assistive operations/ Information	Lack of prevention measures	8	No fixed monitoring techniques	3	5	120
Maintenance	Equipment failure	9	Due to Equipment aging or Lack of testing and Maintenance	8	5	360
	Wrong tools, broken tools, missing aids	7	Increase in assembly times	6	6	252
Quality	Error in quality inspection	8	No fixed inspection methods	7	6	336

Fig. 4 FMEA analysis of current process

4 Expected Results

The adaptive technologies characterized in the research gap would be implemented to the current manual assembly, thereby making it adaptive. The described verification models would focus on examining the adaptability of the assembly process. It could also lead to the integration of reconfigurable assembly machines for high variety with human machine interfaces. Another benefit of building an adaptive assembly is the increase in quality or decrease in the current manual errors. This has been illustrated by the FMEA. Furthermore, by implementing these techniques, the costs of complex machines are relinquished. Hence, this would form a basis for achieving a high variety error-free production.

Acknowledgments The authors gratefully acknowledge the support from Pro^{2}Future GmbH. Pro^{2}Future is funded as part of the Austrian COMET Program – Competence Centers for Excellent Technologies – under the auspices of the Austrian Federal Ministry of Transport, Innovation and

Technology, the Austrian Federal Ministry for Digital and Economic Affairs, and the Provinces of Upper Austria and Styria. COMET is managed by the Austrian Research Promotion Agency FFG.

References

1. Ko, J., Hu, S. J., & Huang, T. (2005). Reusability assessment for manufacturing systems. *CIRP Annals. – Manufacturing Technology*. https://doi.org/10.1016/S0007-8506(07)60062-6.
2. Abou-El-Hossein, K. A., Theron, N. J., & Ghobashy, S. (2015). Design of machine tool based on reconfigurability principles. *Applied Mechanics and Materials*. https://doi.org/10.4028/www.scientific.net/amm.789-790.213.
3. Mehrabi, M. G., Ulsoy, A. G., Koren, Y., & Heytler, P. (2002). Trends and perspectives in flexible and reconfigurable manufacturing systems. *Journal of Intelligent Manufacturing*. https://doi.org/10.1023/A:1014536330551.
4. Sugiarto, I., Axenie, C., & Conradt, J. (2016). From adaptive reasoning to cognitive factory: Bringing cognitive intelligence to manufacturing technology. *International Journal of Industrial Research and Applied Engineering*. https://doi.org/10.9744/jirae.1.1.1-10.
5. Marcel Schwartz, M. S., Dipl.-W irt.-Ing., Dominik Kolz, M. S., & Katharina Heeg, M. A. (2016). Dienstleistungsinnovationen für Elektromobilität – Förderung von Innovation und Nutzerorientierung. Amsterdam.
6. Electric Car (Market) Data. https://evobsession.com/electric-car-sales/.
7. Wiendahl, H. P., ElMaraghy, H. A., Nyhuis, P., Zäh, M. F., Wiendahl, H. H., Duffie, N., & Brieke, M. (2007). Changeable manufacturing – Classification, design and operation. *CIRP Annals. – Manufacturing Technology*. https://doi.org/10.1016/j.cirp.2007.10.003.
8. Fasth-Berglund, Å., & Stahre, J. (2013). Cognitive automation strategy for reconfigurable and sustainable assembly systems. *Assembly Automation*. https://doi.org/10.1108/AA-12-2013-036.
9. Dencker, K., Fasth, Å., Stahre, J., Mårtensson, L., Lundholm, T., & Akillioglu, H. (2009). Designing proactive assembly systems (ProAct) – Criteria and interaction between automation, information, and competence. *Annual Reviews in Control, 33*(2), 230–237.
10. Lotter, B., & Wiendahl, H.-P. (2008). Changeable and reconfigurable assembly systems. In *Changeable and reconfigurable manufacturing systems*. London: Springer.
11. Meichsner, T. P. (2009). Migration manufacturing – A new concept for automotive body production. In *Changeable and reconfigurable manufacturing systems*. London: Springer.
12. Bussmann, S., & Sieverding, J. (2001, October). Holonic control of an engine assembly plant: An industrial evaluation. In 2001 IEEE international conference on systems, man and cybernetics. E- systems and e-man for cybernetics in cyberspace (cat. No. 01CH37236) (Vol. 1, pp. 169–174). IEEE.
13. Gräßler, I., & Pöhler, A. (2017). Implementation of an adapted holonic production architecture. *Procedia CIRP, 63*, 138–143.
14. Bi, Z. M., Lang, S. Y. T., Shen, W., & Wang, L. (2008). Reconfigurable manufacturing systems: The state of the art. *International Journal of Production Research*. https://doi.org/10.1080/00207540600905646.
15. Koren, Y., & Shpitalni, M. (2010). Design of reconfigurable manufacturing systems. *Journal of Manufacturing Systems*. https://doi.org/10.1016/j.jmsy.2011.01.001.
16. Katz, R. (2007). Design principles of reconfigurable machines. *International Journal of Advanced Manufacturing Technology*. https://doi.org/10.1007/s00170-006-0615-2.

17. Abele, E., Liebeck, T., & Wörn, A. (2006). Measuring flexibility in investment decisions for manufacturing systems. *CIRP Annals. – Manufacturing Technology*. https://doi.org/10.1016/S0007-8506(07)60452-1.
18. Koren, Y., Gu, X., & Guo, W. (2018). Reconfigurable manufacturing systems: Principles, design, and future trends. *Frontiers of Mechanical Engineering, 13*(2), 121–136.
19. Gorecky, D., Worgan, S. F., & Meixner, G. (2011). *COGNITO: A cognitive assistance and training system for manual tasks in industry*. In Proceedings of the 29th annual European conference on cognitive ergonomics.
20. ElMaraghy, H. A. (2008). *Changeable and reconfigurable manufacturing systems*. New York: Springer.
21. Wallhoff, F., AblaBmeier, M., Bannat, A., Buchta, S., Rauschert, A., Rigoll, G., & Wiesbeck, M. (2007, July). Adaptive human-machine interfaces in cognitive production environments. In 2007 IEEE international conference on multimedia and expo (pp. 2246–2249). IEEE.
22. Funk, M., & Schmidt, A. (2015). Cognitive assistance in the workplace. *IEEE Pervasive Computing*. https://doi.org/10.1109/MPRV.2015.53.
23. Böckenkamp, A., Mertens, C., Prasse, C., Stenzel, J., & Weichert, F. (2017). *A versatile and scalable production planning and control system for small batch series* (In industrial internet of things (pp. 541–559)). Cham: Springer.
24. Pascu, C. I., & Paraschiv, D. (2016). Study about improving the quality process performance for a steel structures components assembly using FMEA method. *Applied Mechanics and Materials*. https://doi.org/10.4028/www.scientific.net/amm.822.429.
25. Betterton, C. E., & Silver, S. J. (2012). Detecting bottlenecks in serial production lines – A focus on interdeparture time variance. *International Journal of Production Research*. https://doi.org/10.1080/00207543.2011.596847.
26. Law, A. M. (2009). *How to build valid and credible simulation models*. In Proceedings – Winter Simulation Conference.
27. Kikolski, M. (2016). Identification of production bottlenecks with the use of Plant Simulation software. *Engineering Management in Production and Services*. https://doi.org/10.1515/emj-2016-0038.
28. Zhuang, C., Liu, J., & Xiong, H. (2018). Digital twin-based smart production management and control framework for the complex product assembly shop-floor. *International Journal of Advanced Manufacturing Technology*. https://doi.org/10.1007/s00170-018-1617-6.

Parameter Identification of Technological Equipment for Ensuring the Reliability of the Vibration Separation Process

Ivan Pavlenko, Vitalii Ivanov, Oleksandr Gusak, Oleksandr Liaposhchenko, and Vsevolod Sklabinskyi

1 Introduction

The separation and purification of products are an expensive but a crucial step in chemical technology and production [1]. In this case, when developing the corresponding technological equipment, it is necessary to operate a large amount of information that is not always available for reliable design.

As a result of the competitive global market, there are a few generic forms of separator equipment which have been developed, manufactured, and implemented into enterprises. Considering this fact, in 2007, S. Tarleton and R. Wakeman attempted to give a descriptive overview of the equipment types and the main alternatives available to the design engineer. Particularly, the advantages and disadvantages of equipment are highlighted, and an effort has been made to provide quantitative values.

In 2015, L. Sorsamaki and M. Nappa proposed an idea for developing the rapid methodology and the related "short-cut" model for estimating parameters of various separation equipment for solid–liquid, liquid–liquid, gas–liquid, and gas–solid systems. The proposed approach is successfully used to provide the separation, filtration, and solid drying of heterogeneous systems [2].

I. Pavlenko · V. Ivanov (✉) · O. Gusak · O. Liaposhchenko · V. Sklabinskyi
Sumy State University, Sumy, Ukraine
e-mail: i.pavlenko@omdm.sumdu.edu.ua; ivanov@tmvi.sumdu.edu.ua

L. Knapcikova et al. (eds.), *4th EAI International Conference on Management of Manufacturing Systems*, EAI/Springer Innovations in Communication and Computing, https://doi.org/10.1007/978-3-030-34272-2_24

Due to the above-mentioned, this paper is aimed at solving an urgent problem of ensuring the reliability of the vibration separation process and designing the related technological equipment. This problem can be solved using both the parameter identification and probabilistic approach.

2 Literature Review

The importance of the stated problem is highlighted by a row of up-to-date researches on the development of high-performance separation equipment. Particularly, the articles [3, 4] present new methods for ensuring the separation process. The paper [5] is devoted to the separation of gas–liquid two-phase flow through an independent component analysis. A new method to estimate surface-separator optimum operating pressures for the oil and gas industry is proposed in the research [6].

In the article [7], the effect of vibration energy on the separation of fine particles in a vibrated medium is considered. The research work [8] is aimed at developing the multistage separation processes in a vibrated gas-fluidized bed.

Ways to improve the reliability of compressor equipment are presented in the research papers [9, 10] using the inertial gas-dynamic separation process and a new method for the improvement of assembly's reliability. The methodology for providing numerical simulations of the separation process in SPR-separator is proposed in the paper [11].

Interdisciplinary research in the field of separation technology is presented in the article [12]. Particularly, the stationary aeroelasticity problem for dynamic deflection elements of separation devices has been solved using the parameter identification procedure. Additionally, problems of kinematics and dynamics of particles on the oscillating surface are analytically solved in the paper [13].

The physical and mathematical models of crossed movement and gas–liquid flow interaction with captured liquid film in the inertial-filtering separation channels are developed in the article [14]. Analytical and numerical approaches for modeling the distribution and migration of a liquid in the fibrous filter layer in the process of inertial-filtering separation are proposed in the paper [15]. Additionally, mathematical modeling of gas-cleaning equipment with a highly developed phase contact surface is presented in the article [16].

The problem of increasing the quality of technological equipment is solved practically by means of up-to-date technologies, inventions, and ideas [17]. Particularly, the improvements using automation in companies based on lean production with maximum productivity and efficiency are identified and implemented. However, the fundamental approach based on the operating process in separation equipment needs to be developed. For this purpose, the mathematical model and the related analytical approach for parameter identification of the separation channel are presented below.

3 Research Methodology

3.1 The Mathematical Model of the Disperse Phase Motion in a Separation Channel

To study the process of vibration separation for a moving or suspended layer, the mechanism of the motion for dropping liquid should be considered. For this purpose, different equations are used depending on the operating features of the physical process. Particularly, in a vibration separator with laminar or turbulent modes for a dispersed phase in a gas flow, the interaction of forces acting on single particles can be estimated for the case of relatively small concentrations of the dropping liquid.

The motion of a dropping liquid in a plane channel is considered. The flow is bounded by two walls along with the flow. The first wall is stationary, and the second one oscillates monoharmonically with the amplitude a of vibration velocity, frequency ω_0, and wavelength $2\pi/\lambda$. The corresponding design scheme is presented in Fig. 1.

The velocity field is described by its components by the following equations obtained previously:

$$\begin{cases} u = u_0 + \frac{6\kappa_1 a}{\lambda h}\left[\sin\omega_0 t - \frac{1}{2}\sin\left(\omega_0 - \lambda u_0\right)t - \frac{1}{2}\sin\left(\omega_0 + \lambda u_0\right)t\right] \\ v = \frac{\kappa_2 a}{2}\left[\cos\left(\omega_0 - \lambda u_0\right)t - \cos\left(\omega_0 + \lambda u_0\right)t\right], \end{cases} \quad (1)$$

where κ_1, κ_2 – dimensionless parameters determined according to the methodology presented in the article [18].

The unsteady motion of dropping fluid is complicated since the corresponding differential equations, in general, do not have an analytical solution, but can be

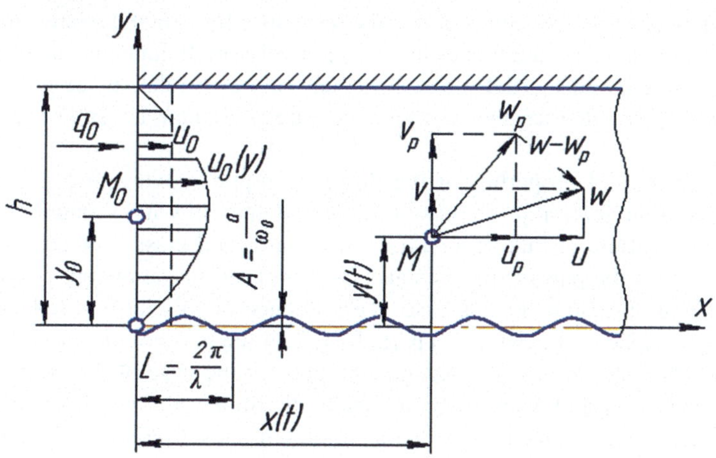

Fig. 1 The design scheme of the operating channel for separation of a gas–liquid mixture

solved numerically for specific conditions only. Otherwise, it is necessary to apply approximate numerical methods for solving the stated problem. In this case, the following simplifications are considered: droplets are spherical, and the motion is mainly laminar with small Reynolds numbers. In this case, the resistance force is conditionally proportional to the droplet velocity [19].

For a relatively slow motion of a liquid phase in a gas stream, the Basset–Boussinesq–Oseen equation is considered for a stationary environment [20–22]. However, the generalized Tchen's equations [23, 24] are used for the unsteady motion of the gas–liquid flow with the correction of Corrsin–Lumley [25, 26].

In the two-dimensional formulation, the general system of differential equations of the steady motion of a droplet fluid has the following form [27]:

$$\begin{cases} \left(1+\frac{\gamma}{3}\right)\frac{du_p}{dt}+\beta u_p=\beta u+\gamma\frac{du}{dt}+\frac{9}{\rho_p d_p}\sqrt{\frac{\mu\rho}{\pi}}\int_0^t\frac{1}{\sqrt{t-\theta}}\left(\frac{du}{dt}-\frac{du_p}{dt}\right)d\tau; \\ \left(1+\frac{\gamma}{3}\right)\frac{dv_p}{dt}+\beta v_p=\beta v+\gamma\frac{dv}{dt}+\frac{9}{\rho_p d_p}\sqrt{\frac{\mu\rho}{\pi}}\int_0^t\frac{1}{\sqrt{t-\theta}}\left(\frac{dv}{dt}-\frac{dv_p}{dt}\right)d\tau-(1-\gamma)g+\frac{F_y}{m_p}, \end{cases} \tag{2}$$

where u_p, v_p is the velocity components of the dropping liquid; d_p, $V_p = \pi d_p{}^3/6$ is the diameter and volume of a particle; ρ, ρ_p is the densities of gas-liquid flow and droplets, respectively; $m_p = \rho V_p$ is the mass of a particle; μ is the dynamic viscosity of a medium; g is the gravitational acceleration; F_y is the other volume force; $\beta = 1/\tau$ is the factor of inertia, inversely proportional to the time constant τ; $\gamma = 2\rho/(3\rho_p)$ is the density factor.

Recent equations contain components of local and convective forces of inertia [28]. However, since in previous researches, the hydrodynamics of the gas–liquid flow was considered without the convective forces of inertia, the effect of these forces should be studied separately by the determination of the volume force F_y and its further substitution into Eq. (2).

The equations also contain resistance forces for the case of a uniform medium. In this case, it is necessary to consider that the resistance force in the relative motion of particles is described by the Stokes law for relatively small diameters d_p of droplets [29]. The influence of inertial forces for small values of relative velocities can be neglected [30]. Another term is due to the pressure gradient as a result of the particle acceleration.

Since the inertial properties of the dropping liquid moving in a gas flow are different from the inertial properties of the particle moving in a vacuum, a reaction force of the medium occurs as the effect of the attached mass. The corresponding term determines the part of the resistance force related to the energy consumption for moving the medium. In this case, the influence of this term is reduced to an imaginary increase in the mass of a particle by half the mass of displaced medium.

The penultimate term of the second equation for the case of the vertical axis y contains the gravity and Archimedes forces. Furthermore, it is necessary to consider the force that slows the fluctuations of the droplet in a gas–liquid flow. The deviation of the flow mode from the steady one is due to the Basset force [31], which increases the instantaneous resistance of particles in the flow. This force is determined by the integral term. Its impact on the droplet motion should be studied separately. However, it should be noted that for the case of relatively high acceleration of a particle, the Basset force becomes essential. Consequently, the instantaneous resistance is multiplied with respect to the value of the steady motion.

The system of Eq. (2) well describes the case of a small concentration of the liquid phase, and every drop can be considered as a single in the gas flow. For small concentrations of the liquid phase, droplets under different speeds increase the dissipation of kinetic energy. As a result, a delay of particles from the gas flow is observed.

Finally, under a significant concentration of liquid phase, liquid droplets interact as a result of their collision and distortion of streamlines near them. Consequently, additional kinetic energy is required to move particles under conditions of the compressed motion.

3.2 Trajectories of Droplets in the Gas–Liquid Mixture

A system of nonlinear integral differential Eq. (2), considering the velocity components of the gas flow (1), can be solved both numerically and analytically using the Caputo generalized Hukuhara derivative [32, 33].

However, in this research, the study will be aimed at finding the precise analytical solution under the following assumptions, which allow simplifying the system of Eq. (2):

1. Density factor is relatively small ($\gamma \ll 1$). This allows neglecting the term $\gamma/3$, the force of inertia for the attached mass proportional to $\gamma du/dt$, and the term $-\gamma$ determining the Archimedes force.
2. Considering the features of the Basset force and convective forces of inertia, they should be studied separately.

Finally, the system of equations describing the motion of liquid droplets takes the following form:

$$\begin{cases} \frac{du_p}{dt} + \beta u_p = \beta u; \\ \frac{dv_p}{dt} + \beta v_p = \beta v - g. \end{cases} \tag{3}$$

This system has been previously solved for the case of the steady velocity field. However, in the case of the oscillating wall, expressions (1) are needed to be considered.

It should be noted that in the case of vertical walls, the term "$-g$" is needed to be removed from this equation. Additionally, in the case of changing the relative position of the stationary and nonstationary walls, the term "$-g$" is needed to be replaced by "g."

The system (3) shows that the motion of dropping liquid in a plane gas–liquid flow with an oscillating wall is a response to a polyharmonic impact with frequencies ω_0, $(\omega_0 - \lambda u_0)$, and $(\omega_0 + \lambda u_0)$. The general solution of this system is obtained as the following superpositions:

$$\begin{cases} u_p = \frac{dx_p}{dt} = u_0 + A_2 \sin(\omega_0 t - \varphi_2) - A_3 \sin[(\omega_0 - \lambda u_0)t - \varphi_3] - \\ \qquad - A_4 \sin[(\omega_0 + \lambda u_0)t - \varphi_4]; \\ v_p = \frac{dy_p}{dt} = B_3 \cos[(\omega_0 - \lambda u_0)t - \varphi_3] - B_4 \cos[(\omega_0 + \lambda u_0)t - \varphi_4] - g\tau, \end{cases} \tag{4}$$

where x_p, y_p – coordinated, which determine the position of the dropping liquid in time.

Integration of the system (4) with respect to time allows obtaining the following parametric equations of trajectories for the dropping liquid:

$$\begin{cases} x_p(t) = u_0 t - \frac{A_2}{\omega_0}\cos(\omega_0 t - \varphi_2) + \frac{A_3}{\omega_0 - \lambda u_0}\cos[(\omega_0 - \lambda u_0)t - \varphi_3] + \\ + \frac{A_4}{\omega_0 + \lambda u_0}\cos[(\omega_0 + \lambda u_0)t - \varphi_4] + \frac{A_2}{\omega_0}\cos\varphi_2 - \frac{A_3}{\omega_0 - \lambda u_0}\cos\varphi_3 + \frac{A_4}{\omega_0 + \lambda u_0}\cos\varphi_4; \\ y_p(t) = y_0 + \frac{B_3}{\omega_0 - \lambda u_0}\sin[(\omega_0 - \lambda u_0)t - \varphi_3] - \frac{B_4}{\omega_0 + \lambda u_0}\sin[(\omega_0 + \lambda u_0)t - \varphi_4] + \\ \qquad + \frac{B_3}{\omega_0 - \lambda u_0}\sin\varphi_3 - \frac{B_4}{\omega_0 + \lambda u_0}\sin\varphi_4 - g\tau t. \end{cases} \tag{5}$$

In general, it can be proved that the absolute trajectories of the mass center for droplets are Lissajous curves [34]. However, separate cases should be investigated for different values of the dimensionless criteria $\omega_0\tau$ and $\lambda u_0/\omega_0$.

4 Results

4.1 *Time to Capture Dropping Liquid*

Time T_s to capture dropping liquid in the separation channel is determined under the condition of achieving the y-coordinate width h of the channel, i.e., $y(T_s) = h$:

$$\begin{aligned} &\frac{B_3}{\omega_0 - \lambda u_0}\{\sin[(\omega_0 - \lambda u_0)T_s - \varphi_3] + \sin\varphi_3\} - \\ &- \frac{B_4}{\omega_0 + \lambda u_0}\{\sin[(\omega_0 + \lambda u_0)T_s - \varphi_4] + \sin\varphi_4\} - g\tau T_s = h - y_0. \end{aligned} \tag{6}$$

This time can be reduced by locating the oscillating wall in the upper part of the channel. In this case, the design scheme is changed symmetrically opposite, and the term "$-g\tau T_s$" is needed to be replaced to "$g\tau T_s$."

In a separate case, when $\lambda u_0/\omega_0 \ll 1$, time to capture the dropping liquid in the separation channel is determined by the following formula:

$$T_s = \frac{18\mu\,(h - y_0)}{\rho_p g d_p^2}. \tag{7}$$

4.2 Area of the Effective Capture of Dropping Liquid

The length of the separation channel, which passes dropping liquid, can be determined using dependencies (5) and (7) as follows:

$$L_s\,(y_0) = x_p\,(T_S) = \frac{u_0\,(h - y_0)}{g\tau} = \frac{6u_0}{g\tau}\frac{y_0}{h}\left(1 - \frac{y_0}{h}\right)^2. \tag{8}$$

The particles most distant from the stationary wall have a significantly lower initial velocity at the inlet, but a greater transverse velocity component under the impact of the oscillating wall. In the flow core, the situation is the opposite: particles have the greatest value of the initial velocity while the impact of the oscillating wall is relatively insignificant. Finally, particles near the stationary wall have a relatively low initial velocity, and the impact of the oscillating wall is neglected.

Due to the above-mentioned, Eq. (8) determines the distribution of droplets along the length of their capture. Thus, the distribution of the value L_s, depending on the initial position of the droplet at the inlet, is shown in Fig. 2.

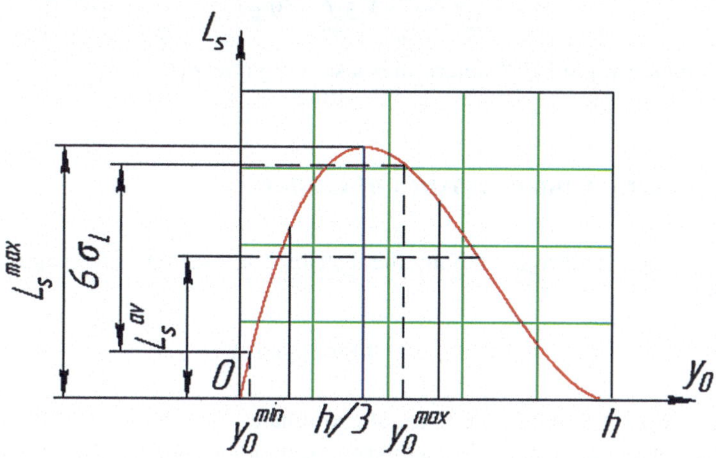

Fig. 2 Area of the effective capture of dropping liquid on the stationary wall of the separation channel

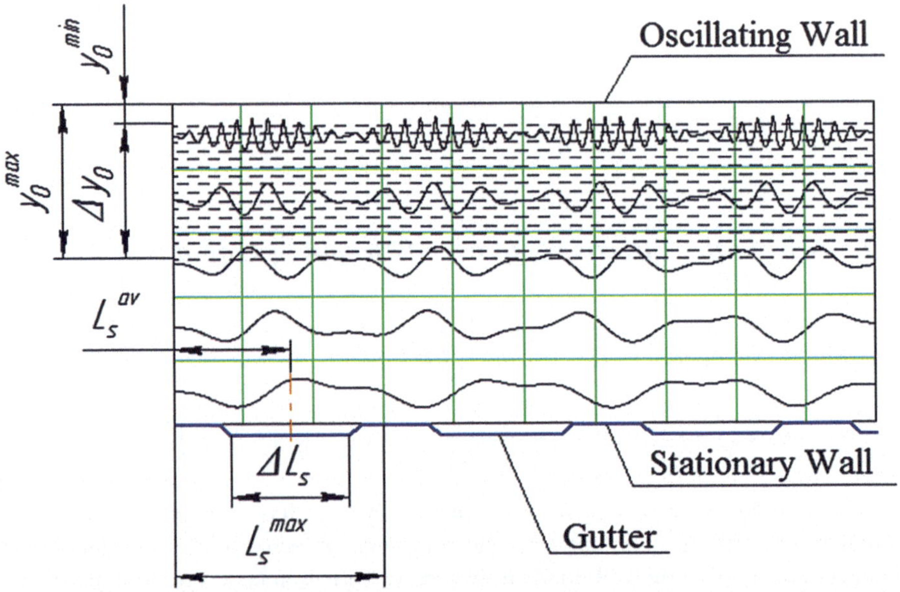

Fig. 3 Geometric characteristics of the separation channel

Research on the extremum of the dependence (8) indicates that the maximum value L_s^{max} of the length L_s is reached for the value $y_0 = h/3$. In this case,

$$L_s^{max} = L_s\left(\frac{h}{3}\right) = \frac{8}{9}\frac{q_0}{g\tau}. \tag{9}$$

The estimated value L_s^{max} determines a step of gutters (Fig. 3).

4.3 Geometric Characteristics of Gutters

Considering Eq. (8) allows obtaining the distribution function of the capture length as follows:

$$L_s(\overline{y}_0) = 6.75 L_s^{max} y_0 (1 - \overline{y}_0)^2, \tag{10}$$

where $\overline{y}_0 = y_0/h$ – dimensionless initial coordinate.

For the intensification of the vibration separation process for a gas–liquid mixture, it is necessary to locate most of the droplets in an effective area of the vibration impact.

Applying a probabilistic approach to determine the effective area of capturing of the droplets capture from a gas–liquid flow, the mathematical expectation, as the average value of the distribution function (10) in a range from 0 to $L_s{}^{max}$, is determined as follows:

$$L_s^{av} = \frac{1}{L_S^{max}} \int\limits_0^{L_s^{max}} L_s\left(\overline{y}_0\right) d\overline{y}_0 = 0.563 L_s^{max} = \frac{q_0}{2g\tau}. \quad (11)$$

The estimated value $L_s{}^{av}$ determines the location of the center of a gutter (Fig. 3). The ratio of its value to the maximum ($L_s{}^{av}/L_s{}^{max} = 0.56$) indicates the shift $0.5L_s{}^{max}$ of the gutter's center relative to the half-step.

The mean square deviation σ_L allows evaluating the effective width of gutters. Considering for the engineering purposes the coefficient of variation 0.2, the value of mean square deviation is equal to $\sigma_L = 0.2L_s^{av} = 0.113L_s^{max}$.

Applying the rule of "3σ" allows evaluating the confidence interval to capture droplets: $\left[L_s^{av} - 3\sigma_L, L_s^{av} + 3\sigma_L\right] = [0.4, 1.6]\, L_s^{av} = [0.225,\ \ 0.900]\, L_s^{max}$. In this case, the width of this interval determines the effective width of a gutter: $\Delta L_S = 6\sigma_L = 1.2L_s^{av} = 0.675L_s^{max}$. This value provides the capture of most of the droplets from the gas–liquid flow.

To determine the effective area of the vibration impact, it is necessary to determine the values of the initial coordinates y_0 providing the values of borders for the confidence interval as positive real roots of the cubic equations:

$$\left[\begin{array}{l} \overline{y}_0^{min}\left(1 - \overline{y}_0^{min}\right)^2 = \frac{0.225}{0.675} = 0.033; \\ \overline{y}_0^{max}\left(1 - \overline{y}_0^{max}\right)^2 = \frac{0.900}{0.675} = 0.133. \end{array}\right. \quad (12)$$

Applying a numerical method for solving these equations, the following values are obtained: $y_0^{min} = 0.036L_s^{max}$; $y_0^{max} = 0.464L_s^{max}$. As a result, the width of the effective area of the vibration impact is determined by the following formula:

$$\Delta y_0 = y_0^{max} - y_0^{min} = 0.428L_s^{max}. \quad (13)$$

Thus, the above-mentioned methodology allows evaluating the geometrical parameters and effective area for locating gutters on the stationary wall in the separation channel.

5 Conclusions

Thus, in this paper, the motion of dropping liquid particles of a gas–liquid mixture in a separation channel with the oscillating wall was considered. As a result of the mathematical modeling and parameter identification, the analytical expressions for velocity components of particles were obtained. Trajectories of liquid droplets were determined analytically. Particularly, the velocity components were determined as a superposition of harmonic components with frequencies depending on the flow rate and vibration characteristics of the oscillating wall were determined for the case of the nonstationary upper wall considering gravity forces.

Using the probabilistic approach, the geometric characteristics of gutters of the stationary wall were determined considering the physical features of transferring particles from the flow to the stationary wall. It was determined that the step of gutters is equal to the maximum estimated length of the liquid capture, as well as their centers are shifted relative to the center of this step in the direction of the gas flow. Finally, the effective width of gutters was evaluated, and the effective zone of vibration impact was detected.

The next research will be aimed at the study of the dispersed phase motion in a turbulized flow, as well as at the impact of convective forces of inertia on the kinematic characteristics of the dropping liquid. Particularly, it should be substantiated that considering these forces leads to opposite effects depending on the relative position of the stationary and nonstationary walls in the separation channel.

Acknowledgments The achieved results were obtained within the research project "Development and Implementation of Energy Efficient Modular Separation Devices for Oil and Gas Purification Equipment" (State Reg. No. 0117U003931) funded by the Ministry of Education and Science of Ukraine.

References

1. Tarleton, E. S., & Wakeman, R. J. (2007). *Solid/liquid separation: equipment selection and process design*. Netherlands: Elsevier Ltd.
2. Kaijaluoto, S., Sorsamaki, L., Aaltonen, O., & Nakari-Setala, T. (2010). Bark-based biorefinery – From pilot experiments to process model. In *AIChE Annual Meeting, Conference Proceedings* (Vol. 83459).
3. Gulsoy, O. Y., & Gulcan, E. (2019). A new method for gravity separation: Vibrating table gravity concentrator. *Separation and Purification Technology, 211*, 124–134. https://doi.org/10.1016/j.seppur.2018.09.074.
4. Zhang, G., Willemin, A. S., Brion, A., Piet, M. H., Moby, V., Bianchi, A., Mainard, D., Galois, L., Gillet, P., & Rousseau, M. (2016). A new method for the separation and purification of the osteogenic compounds of nacre ethanol soluble matrix. *Journal of Structural Biology, 196*(2), 127–137. https://doi.org/10.1016/j.jsb.2016.03.016.
5. Xu, Y., Wang, X., Cui, Z., Dong, F., & Yan, Y. (2010). Separation of gas-liquid two-phase flow through independent component analysis. *IEEE Transactions on Instrumentation and Measurement, 59*(5), 1294–1302.
6. Ling, K., Wu, X., Guo, B., & He, J. (2013). New method to estimate surface-separator optimum operating pressures. *Oil and Gas Facilities, 2*(3), 65–76. https://doi.org/10.2118/163111-PA.
7. Zhou, E., Zhang, Y., Zhao, Y., Luo, Z., Yang, X., Duan, C., Dong, L., & Fu, Z. (2018). Effect of vibration energy on fluidization and 1–6 mm coal separation in a vibrated dense medium fluidized bed. *Separation Science and Technology, 53*(14), 2297–2313. https://doi.org/10.1080/01496395.2018.1445757.
8. Wang, S., Yang, Y., Yang, X., Zhang, Y., & Zhao, Y. (2019). Dry beneficiation of fine coal deploying multistage separation processes in a vibrated gas-fluidized bed. *Separation Science and Technology, 54*(4), 655–664. https://doi.org/10.1080/01496395.2018.1505917.

9. Liaposhchenko, O. O., Sklabinskyi, V. I., Zavialov, V. L., Pavlenko, I. V., Nastenko, O. V., & Demianenko, M. M. (2017). Appliance of inertial gas-dynamic separation of gas dispersion flaws in the curvilinear convergent-divergent channels for compressor equipment reliability improvement. In *IOP Conference Series: Materials Science and Engineering* (Vol. 233, p. 012025). https://doi.org/10.1088/1757-899X/233/1/012025.
10. Tarelnyk, V., Martsynkovskyy, V., & Dziuba, A. (2014). New method of friction assemblies reliability and endurance improvement. *Applied Mechanics and Materials, 630*, 388–396. https://doi.org/10.4028/www.scientific.net/AMM.630.388.
11. Liaposhchenko, O., Pavlenko, I., Demianenko, M., Starynskyi, O., & Pitel, J. (2019). The methodology of numerical simulations of separation process in SPR-separator. In *CEUR Workshop Proceedings* (Vol. 2353, pp. 822–832).
12. Pavlenko, I., Liaposhchenko, A., Ochowiak, M., & Demyanenko, M. (2018). Solving the stationary hydroaeroelasticity problem for dynamic deflection elements of separation devices. *Vibrations in Physical Systems, 29*, 1–7.
13. Pylypaka, S., Klendiy, M., & Zaharova, T. (2019). Movement of the particle on the external surface of the cylinder, which makes the translational oscillations in horizontal planes. In *Advances in Design, Simulation, Manufacturing. DSMIE-2018. Lecture Notes in Mechanical Engineering* (pp. 336–345), https://doi.org/10.1007/978-3-319-93587-4_35
14. Liaposchenko, O., Pavlenko, I., & Nastenko, O. (2017). The model of crossed movement and gas-liquid flow interaction with captured liquid film in the inertial-filtering separation channels. *Separation and Purification Technology, 173*, 240–243. https://doi.org/10.1016/j.seppur.2016.08.042.
15. Sklabinskyi, V., Liaposhchenko, O., Pavlenko, I., Lytvynenko, O., Demianenko, M. (2019). Modelling of liquid's distribution and migration in the fibrous filter layer in the process of inertial-filtering separation. In *Advances in Design, Simulation, Manufacturing. DSMIE-2018. Lecture Notes in Mechanical Engineering* (pp. 489–497), https://doi.org/10.1007/978-3-319-93587-4_51
16. Plyatsuk, L. D., Ablieieva, I. Y., Vaskin, R. A., Yeskendirov, M., & Hurets, L. L. (2018). Mathematical modeling of gas-cleaning equipment with a highly developed phase contact surface. *Journal of Engineering Sciences, 5*(2), F19–F24. https://doi.org/10.21272/jes.2018.5(2).f4.
17. Araujo, A. F., Varela, M. L. R., Gomes, M. S., Barreto, R. C. C., & Trojanowska, J. (2018). Development of an intelligent and automated system for lean industrial production, adding maximum productivity and efficiency in the production process. In *Advances in Manufacturing. Lecture Notes in Mechanical Engineering* (pp. 131–140), https://doi.org/10.1007/978-3-319-68619-6_13.
18. Pavlenko, I., Liaposhchenko, O., Sklabinskyi, V., Ivanov, V., & Gusak, O. (2018). Hydrodynamic features of gas-liquid flow movement in a separation device plane channel with an oscillating wall. *Problemele Energeticii Regionale, 3*(38), 62–70. https://doi.org/10.5281/zenodo.2222360.
19. Feng, J. (2010). A deformable liquid drop falling through a quiescent gas at terminal velocity. *Journal of Fluid Mechanics, 658*, 438–462. https://doi.org/10.1017/S0022112010001825.
20. Yadav, M. P., & Agarwal, R. (2019). Numerical investigation of fractional-fractal Boussinesq equation. *Chaos, 29*, 013109. https://doi.org/10.1063/1.5080139.
21. Kambe, T. (2017). New scenario of turbulence theory and wall-bounded turbulence: Theoretical significance. *Geophysical and Astrophysical Fluid Dynamics, 111*(6), 448–507. https://doi.org/10.1080/03091929.2017.1376055.
22. John, V. (2016). *The Oseen equations. Finite element methods for incompressible flow problems* (Vol. 51, pp. 243–300). Cham: Springer Series in Computational Mathematics. https://doi.org/10.1007/978-3-319-45750-5_5.
23. Parmar, M., Haselbacher, A., & Balachandar, S. (2011). Generalized Basset–Boussinesq–Oseen equation for unsteady forces on a sphere in a compressible flow. *Physical Review Letters, 106*, 084501. https://doi.org/10.1103/PhysRevLett.106.084501.

24. Chouippe, A., & Uhlmann, M. (2019). On the influence of forced homogeneous-isotropic turbulence on the settling and clustering of finite-size particles. *Acta Mechanica, 230*(2), 387–412. https://doi.org/10.1007/s00707-018-2271-7.
25. Petersen, A. J., Baker, L., & Coletti, F. (2019). Experimental study of inertial particles clustering and settling in homogeneous turbulence. *Journal of Fluid Mechanics, 864*, 925–970. https://doi.org/10.1017/jfm.2019.31.
26. Chun, J., Koch, D., Rani, S., Ahluwalia, A., & Collins, L. (2005). Clustering of aerosol particles in isotropic turbulence. *Journal of Fluid Mechanics, 536*, 219–251. https://doi.org/10.1017/S0022112005004568.
27. Kelbaliev, G. I., Ibragimov, Z. I., & Kasimova, R. K. (2010). Deposition of aerosol particles in vertical channels from an isotropic turbulent flow. *Journal of Engineering Physics and Thermophysics, 83*(5), 908–916. https://doi.org/10.1007/s10891-010-0413-4.
28. Cencini, M., Bec, J., Biferale, L., Boffetta, G., Celani, A., Lanotte, A. S., Musacchino, S., & Toschi, F. (2019). Dynamics and statistics of heavy particles in turbulent flows. *Journal of Turbulence, 7*, 36. https://doi.org/10.1080/14685240600675727.
29. Arakawa, K. (2016). Dynamic sliding friction and similarity with Stokes' law. *Tribology International, 94*, 77–81. https://doi.org/10.1016/j.triboint.2015.08.007.
30. Rosen, T. (2014). *The influence of inertia on the rotational dynamics of spheroidal particles suspended in shear flow*. Stockholm: Royal Institute of Technology.
31. Daitche, A. (2015). On the role of the history force for inertial particles in turbulence. *Journal of Fluid Mechanics, 782*, 567–593. https://doi.org/10.1017/jfm.2015.551.
32. Armand, A., Allahviranloo, T., & Gouyandeh, Z. (2016). General solution of Basset equation with Caputo generalized Hukuhara derivative. *Journal of Applied Analysis and Computation, 6*(1), 119–130. https://doi.org/10.11948/2016010.
33. Edwards, J. T., Ford, N. J., & Simpson, A. C. (2002). The numerical solution of linear multi-term fractional differential equations: Systems of equations. *Journal of Computational and Applied Mathematics, 148*(2), 401–418. https://doi.org/10.1016/S0377-0427(02)00558-7.
34. Gomez-Lopez, A., Ferrer, V. H., Rincon, E., Aguayo, J. P., Chavez, A. E., & Vargas, R. O. (2019). Large-amplitude oscillatory shear flow simulation for a FENE fluid. *Rheologica Acta, 58*(1), 241–260.

Optimizing Component Production with Multi-axis Turning Technology

Peter Michalik, Michal Hatala, Luboslav Straka, Michal Petrus, Jozef Macej, Jozef Jusko, and Peter Tirpak

1 Introduction

Machining technologies, as well as technologies related to the completion of work, have developed significantly in recent years. The demand for surface quality as well as the precision requirements for engineering products is increasing. The development of technology that is used today is increasingly associated with the development and use of new metal materials. These materials have high strength parameters, chemical and physical properties as well as high wear resistance. However, machining technology is concerned not only with better processing of these materials, but also with optimization of the control and production process. The prediction of machining thin-walled parts was by Kuram et al. [1].

Metalworking is one of the most common and important technological processes of today, where more and more emphasis is placed on working time and product creation with the best and highest quality properties [2, 3]. Metalworking is a method without which aviation, automobile, engineering, and other technically oriented industries would not be able to deal with today. Metal processing is carried out by various processes. Such processes include, for example, turning, milling, drilling, slotting, ... [4, 5]. All of these processes can be done either individually on separate machines, such as lathes, milling machines, drills, shaping machines, or we can optimize production and perform multiple processes on one machine [6, 7]. Machining is one of the most important methods of removing materials in manufacturing technology. Simply put, it is a set of material workflows that include other processes, such as drilling, shaping, sawing, grinding, and more. Machining is

P. Michalik (✉) · M. Hatala · L. Straka · M. Petrus · J. Macej · J. Jusko · P. Tirpak
TU Kosice, FVT with a Seat in Presov, Bayerova, Slovakia
e-mail: peter.michalik@tuke.sk; michal.hatala@tuke.sk; luboslav.straka@tuke.sk; michal.petrus@tuke.sk; jozef.macej@tuke.sk; jozo.jusko@gmail.com; peter.tirpak@tuke.sk

L. Knapcikova et al. (eds.), *4th EAI International Conference on Management of Manufacturing Systems*, EAI/Springer Innovations in Communication and Computing, https://doi.org/10.1007/978-3-030-34272-2_25

practically part of the production of all metals and other materials, e.g., wood. The second most used technology in machining is turning, where the outer and inner cylindrical surfaces are most often machined Fig. 1 [8, 9].

These operations are most often performed on a lathe. These include thread cutting, groove turning, grinding, grooving, drilling, and more Fig. 2.

Turning workpieces with a length of several meters can make it difficult to provide balanced cutting. In this case, it makes sense to cut using two towers and use a CNC machine that supports additional weight compensation functions.

High volume manufacturers, who have to produce different parts from time to time, have the ability to program machines in-house with great benefits in terms of both time and flexibility and possibility machining pipe conveyor components [10–13]. Other advantages are the following options: program on machine, creating machining cycles for applications such as hole cutting, grooving, drilling, or contouring, program measurement cycles, zero point setting, part/program time

Fig. 1 Typical rotation components made by technology turning

Fig. 2 Multi-axis double spindle turning machining CNC center

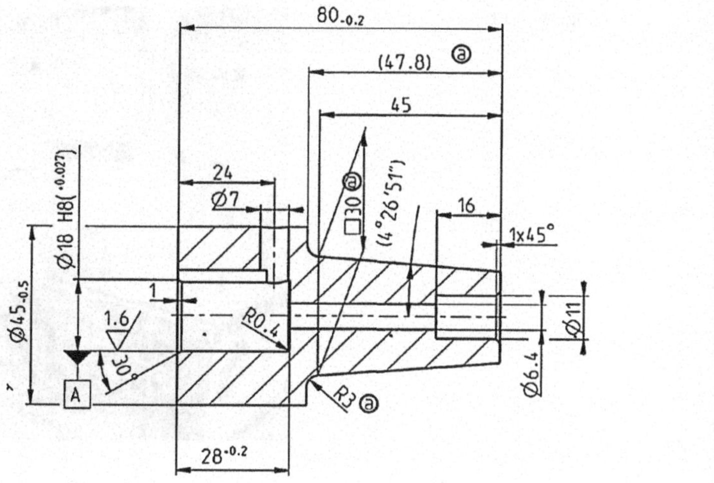

Fig. 3 Component dimension

estimation, workpiece, or tool measurement [14, 15]. To avoid the risk of collisions on CNC multi-axis lathes and protect valuable equipment, tools, and workpieces from damage, some manufacturers provide intelligent collision control software [16]. By modeling the machine path and stopping it when a problem is detected, it also reduces downtime and costly repairs. Good CNC lathes can often also help in milling [17, 18]. Investing in a machine with milling capabilities makes sense because it gives you the freedom to order milling without the need to invest in a second machine. Advanced features on selected CNC devices allow you to cut up to four workpieces at the same time on a standard lathe. The flexible route and axis assignment functions on the CNC enable synchronized motion [19, 20]. Using another CNC function, two workpieces can also be modeled simultaneously. By its name, the Swiss type of turning is suitable for making small parts with high precision in watches, medical and dental technology. Universal CNC machines should support this type of machining and be capable of producing multiple parts in one cycle. Vibration is a problem that adversely affects metal cutting, such as turning. This phenomenon has several serious consequences for threading on a lathe. The solution is to have a good CNC machine with software support that automatically controls the lathe speed to minimize the impact of vibration on the final thread quality [21, 22].

2 Current Component Production Technology

The current component is made of C45 steel, dimensions, and three-dimensional (3D) view Fig. 3. Manufacturing process according to geometric shape and dimensions prescribes technological operations such as turning, shaping, and milling.

Fig. 4 Lathes SPT 16NC

From the company's machinery, according to the manufacturing process for turning operations, an NC lathe SPT 16 NC (Fig. 4) is used, on which the outer and inner cylindrical surfaces will be produced. The machine has the following parameters. Maximum turning diameter 200 mm. Maximum swing diameter 340 mm. Maximum machined length 120 mm. Maximum bar material diameter 40 mm. Maximum length of bar material 1500 mm. Spindle speed range 40–4000 rpm. Total machine input 45,0 kVA. Main motor power 16 kW. Machine weight 4800 kg.

Slotting machine HOV 25 is a single-purpose machine tool Fig. 5. It serves to produce grooves for pens and wedges on the outer or inner cylindrical and planar surfaces. According to the manufacturing process on this machine, a 6P9 tongue groove will be produced on an inner diameter of 6 mm. Technical parameters of slotting machine HOV 25 are: Slotting height 160 mm. Table diameter 320 mm. Cross table adjustment 320 mm. Longitudinal adjustment of table 200 mm. Main motor power 1.5 kW. Dimensions 930 × 1200 mm. Machine weight 1050 kg.

For milling of the needle surfaces, a planar milling machine FD40V PFF Fig. 6. with the following technical parameters is prescribed in the manufacturing process. Table clamping surface 560 × 1800 mm. Table clamping grooves four pieces. Width/spacing 22 × 100 mm. Table movement X: 1250 mm, Y: 430 mm, Z: 475 mm. Spindle cone 50. Spindle diameter 120 mm. Spindle speed – number of steps 16 – range 35.5–1120 min^{-1}. Spindle distance from table clamping surface 75–550 mm. Spindle axis distance from stand guide 450 mm. Rapid traverse longitudinal and transverse 3200 mm/min. Rapid traverse vertical 800 mm/min. Electric motor for 3 kW feeds. Electromotor 25 kW. Power input 38 kVA. Maximum spindle torque 3200 Nm. Machine surface 2800 × 3900 mm. Weight 7040 kg.

Fig. 5 Slotting machine HOV 25

Fig. 6 Classical milling machine FD40V PFF

3 Designed-Optimized Component Manufacturing Technology

When optimizing production, errors were avoided, many machines were eliminated when machining the part. Only one machine was used, instead of lathe, drill, shaping and milling machine, turning double spindle center CTX alpha 500 Fig. 7.

The technical data of the multi-axis two-spindle CNC turning center are: Maximum turning diameter 240 mm. Maximum pendulum diameter 500 mm. Maximum workpiece length with tailstock (can be machined) 500 mm. Maximum workpiece length with opposite spindle (can be machined) 470 mm. Maximum chuck size 225 mm.

Main Spindle:
Spindle motor maximum speed 6000 rpm. Inverter power (100% DC) 13 kW (AC). Torque (100% DC) 172 Nm. Spindle diameter in front bearing 100 mm. Maximum rod permeability diameter 66 mm counterspindle:
Spindle motor maximum speed 6000 rpm. Inverter power (100% DC) 12 kW (AC).
Torque (100% DC) 172 Nm. Spindle diameter in front bearing 90 mm.
Tool Head (Standard):
Tool holder according to DIN 69880 12. Number of driven tools/maximum speed 12/5000 n/rpm. Inverter power (100% DC) 5.4 kW (AC). Torque (100% DC) 18 Nm. Feed rate X/Y/Z 30/22.5/30 m/min. Machine weight 5000 kg. Maximum thermal stability due to water cooling, integrated spindle motor.

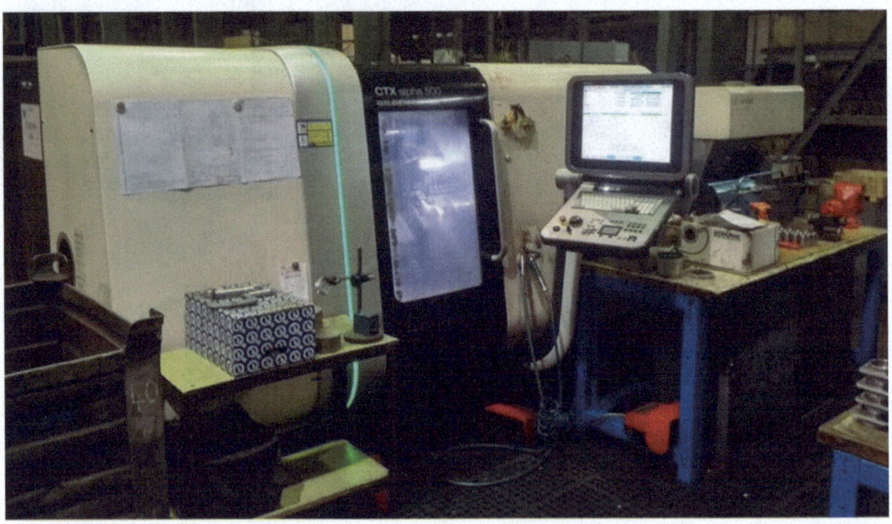

Fig. 7 Multi-axis double spindle turning CNC machining center CTX alpha 500

When choosing the tools used in the optimized manufacturing process on the CTX alpha 500 turning machine, the company's tools are taken into account. Compared to the current production technology, turning tools remain unchanged, the only change to one slotting tool being HORN SH117.1725.1.10 and the use of three driven tools, i. roughing cutter DORMER Product Selector S944 Fig. 8. This mill cuts a 0.1 mm additional pyramid to each side of the needle that is used for subsequent smoothing by the next tool.

The last tool used in the optimized technological process is the DORMER S717 Ø6 Fig. 9, which is used to create the radius R3 and then smooth the surface of the part to the desired size pyramid 23/30.

A sorting device is used to speed up machine alignment DMG VIO 210 MICROSET Fig. 10, which shortens side-by-side working hours, eliminating downtime multi-axis double spindle CNC machining center CTX alpha 500.

On the left side of the component, the technological hole Ø7 is drilled from the front of the component 24 mm. This technological opening serves for subsequent grooving of the groove. HORN SH117.1725.1.10 cuts 6P9 width to tolerance (0; −0.01) to 24 mm and to 20.50 (+0.15; +0.25). The CTX alpha 500 two-spindle turning center automatically clamps the part to the left chuck Fig. 11 and turns the right side.

Fig. 8 Roughing milling tool DORMER S944 16.00 HM titanium aluminum nitride (TiAIN)

Fig. 9 Fig. 9. Finishing milling DORMER Product Selector S717 Ø6

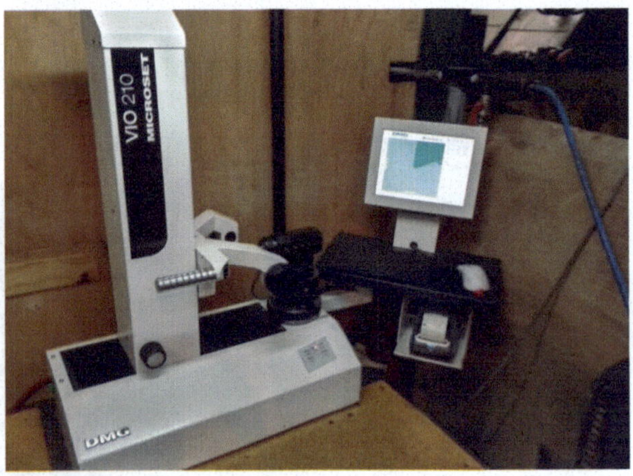

Fig. 10 The adjusting device DMG VIO 210 MICROSET

The first step of turning the right side of the part is to align the face to a total length of l = 80–0.2. Subsequently, the cone is roughened to Ø32.5/Ø42.4 along the length l = 47.8–0.1 at an angle of $4^\circ\ 26'51''$. Drill a Ø6.4 hole into the Ø18 hole connection. With the help of the bore, the hole Ø11 is deepened to the depth $l = 16 + -0.2$ from the forehead. The next step is milling with a radially driven tool Ø16. Roughing cutter Ø16 cuts square to dimension pyramid 23.2/ 30.2 to length 45mm with R3 mm run. After roughing with a roughing cutter. Milling cutter with diameter Ø6 mm is made to dimension pyramid 23/ 30 to a total length of 47.8 mm pyramid.

Fig. 11 Clamping component in the left chuck

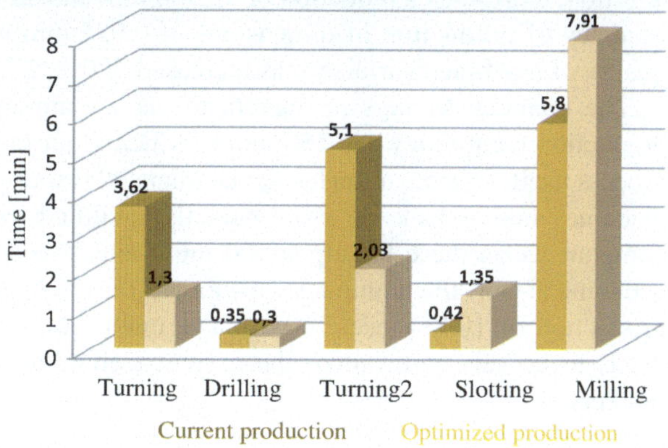

Fig. 12 Evaluation of current and optimized component production

4 Evaluation of Current and Optimized Component Production

If one standard work costs 2.75 euros, we will show the difference at 1000 pieces. Savings in the production of 1000 pieces, with a time saving of one part of 12 min, represent 200 h Fig. 12. The financial savings in the production of a component with a pyramidal surface is 550 euros.

5 Discussion

The article deals with the optimization and assessment of production using turning, milling, drilling, and slotting technology. Four machine tools, milling, turning, drilling, and slotting, have been replaced by multi-axis double spindle turning CNC center machine. New milling tools have been designed with diameters 16 and 6 mm.

The technological process, set-up and clamping of tools in the production of a pyramid part have been selected and implemented to achieve accuracy tolerance of 7.

6 Conclusion

The optimization of the manufacturing process of the cylindrical and pyramid-shaped part was carried out by means of a two-spindle, multi-axis CNC turning center CTX alpha 500, thus reducing the number of clamps of the manufactured part from five to two. A reduction in the number of machine tools from four to one has been achieved. At the same time, the machining tools were changed and the DMG VIO 210 MICROSET tools were added. It has been achieved a reduction in production times, which, with the annual volume of the type of 1000 pieces in question, represents a reduction of 12,500 min and, at the same time, increasing the accuracy of production of dimensions of 0.025 mm has been increased. Also, the quality of machined surfaces was increased $200 \times 2{,}75 = 550$ euros.

The financial savings in manufacturing a component with pyramid-optimized production technology are 550 euros, which is an improvement in the production process itself. Optimization has saved financial resources, operating on conventional machines and a lack of professionally qualified operators. The production of components on the CTX alpha 500 multi-axis two-spindle turning center is more economical for the company. Although the company will save € 550 in the production of 1000 pieces, it will save mainly on employment, as one employee needs to be serviced, unlike the current technology, which requires four skilled workers.

Acknowledgments This work is a part of the research project VEGA 1/0045/18.

References

1. Kuram, E., et al. (2013). Multi-objective optimization using Taguchi based grey relational analysis for micro-milling of Al 7075 material with ball nose end mill. *Measurement, 46*, 1849–1864.
2. Kulka, J., et al. (2016). Analysis of crane track degradation due to operation. *Engineering Failure Analysis, 59*, 384–395.
3. Miko, E., et al. (2012). Analysis and verification of surface roughness constitution model after machining process. *Procedia Engineering, 39*, 395–404.
4. Michalik, P., et al. (2013). Programming CNC machines using computer-aided manufacturing software. *Advanced Science Letters, 19*, 369–373.
5. Michalik, P., et al. (2010). Intelligently programming of holes machining. *Manufacturing Engineering, 9*, 63–65.
6. Michalik, P., et al. (2011). CAM software products for creation of programs for CNC machining. *Lecture Notes in Electrical Engineering, 141*, 421–425.

7. Fabian, M., et al. (2014). Influence of the CAM parameters and selection of end-mill cutter when assessing the resultant surface quality in 3D milling. *Applied Mechanics and Materials, 474*, 267–272.
8. Olejárová, Š., et al. (2017). Measurements and evaluation of measurements of vibrations in steel milling process. *Measurement, 106*, 18–25.
9. Balog, M., et al. (2016). Methodical framework of flexibility production evaluation in terms of manufacturing plant. *Key Engineering Materials, 669*, 568–577.
10. Fedorko, G., et al. (2015). Failure analysis of irreversible changes in the construction of the damaged rubber hoses. *Engineering Failure Analysis, 58*, 31–43.
11. Molnár, V., et al. (2013). Statistical approach for evaluation of pipe conveyor's belt contact forces on guide idlers. *Measurement, 46*, 3127–3135.
12. Molnár, V., et al. (2015). Monitoring of dependences and ratios of normal contact forces on hexagonal idler housings of the pipe conveyor. *Measurement, 64*, 168–176.
13. Fedorko, G., et al. (2013). Failure analysis of textile rubber conveyor belt damaged by dynamic wear. *Engineering Failure Analysis, 28*, 103–114.
14. Knapčikova, L., et al. (2016). Material recycling of some automobile plastics waste. *Przemysł Chemiczny, 95*, 1716–1720.
15. Novotný, L. (2018). Imperfection generation in finite element models of welding. *Science and Technology of Welding and Joining, 23*(2), 148–157.
16. Novak-Marcincin, J., et al. (2014). Use of alternative scanning devices for creation of 3D models of machine parts. *Tehnicki Vjesnik, 21*(1), 177–181.
17. Murčinková, Z., et al. (2017). Research and analysis of stress distribution in multilayers of coated tools. *International Journal of Materials Research, 108*, 495–506.
18. Lehocka, D., et al. (2017). Comparison of the influence of acoustically enhanced pulsating water jet on selected surface integrity characteristics of CW004A copper and CW614N brass. *Measurement, 110*, 230–238.
19. Duplakova, D., et al. (2018). Determination of optimal production process using scheduling and simulation software. *International Journal of Simulation Modelling (IJSIMM), 17*, 609–622.
20. Michalik, P., et al. (2014). Comparison measurement of the distance between axes of holes with the Roundtest RA-120 and Thome Präzision-Rapid. *Applied Mechanics and Materials, 616*, 284–291.
21. Duplak, J., et al. (2018). Comprehensive analysis and study of the machinability of a high strength aluminum alloy (EN AW-AlZn5.5MgCu) in the high-feed milling. *Advances in Production Engineering And Management, 13*, 455–465.
22. Duplakova, D., et al. (2019). Illumination simulation of working environment during the testing of cutting materials durability. *Ain Shams Engineering Journal, 10*, 161–169.

Development of Chosen Social and Economic Indicators of Using Raw Materials in the Context of Sustainability in Slovakia

Katarina Culkova, Andrea Rosova, Michal Cehlar, and Henrieta Pavolova

1 Introduction

The base of raw materials provides, in the frame of a certain region, the value that is possible to evaluate with optimal usage complexly, and by this way profit for the owner, state and mining company, as well as a single region is created [1]. Profit for the state can be made possible by reasonable and rational usage of deposits, as defined in the legislation of raw materials usage. Profit for the mining organization can be profit creation during provision of its sustainability. Profit for the region is in the sense of administration's creation of wealth for its region's inhabitants, which is viewed through the support of new working posts, support of regional tax creation, or taxes, orientated directly to the region and the availability of raw material sources that are necessary for the region's development. The owner of reserved deposits of raw materials in conditions of Slovakia is single state, but the state does not have definite obligation to evaluate economically all recorded deposits of reserved raw materials [2]. It means that problems with using raw materials are actual and necessary for the business sector, as well as for the private sector [3].

K. Culkova (✉) · M. Cehlar
Technical University Košice, FBERG, Institute of Earth Sources, Košice, Slovakia
e-mail: katarina.culkova@tuke.sk; michal.cehlar@tuke.sk

A. Rosova
Technical University Košice, FBERG, Institute of Logistics, Košice, Slovakia
e-mail: andrea.rosova@tuke.sk

H. Pavolova
Technical University Košice, Košice, Slovakia
e-mail: henrieta.pavolova@tuke.sk

L. Knapcikova et al. (eds.), *4th EAI International Conference on Management of Manufacturing Systems*, EAI/Springer Innovations in Communication and Computing, https://doi.org/10.1007/978-3-030-34272-2_26

The goal of the contribution is therefore to search for the development of the most important social and economic indicators that influence the usage of raw materials, mostly from the viewpoint of sustainable development of the Slovakian industry.

2 State of the Problem

Presently, raw materials are being mined in all counties of Slovakia. The majority of these mined raw materials are being consumed in the regions of Slovakia. Raw materials, which present surplus in Slovakia, are exported to the surrounding countries. Raw materials in shortage are imported to Slovakia. Attention is necessary to be given mainly to social and economic indexes, to which employment, gross domestic product (GDP), average monthly wage, migration and criminality belong. Such indexes characterize changes, connected with inhabitants of the country [4].

The availability of raw materials is perceived from physical and economic side [5]. According to the Constitution of the Slovak Republic (SR), the raw material wealth must be protected and effectively used in connection with the long-term needs of economic and social development of the society with regard to environmental aspects of sustainable development, beginning with geological research and using the verified stocks of raw materials.

A process of using raw materials is subjected presently to legislative norms that divide the deposits into two categories: reserved and non-exclusive. Non-exclusive deposits are characterized by an attribute of ownership – owner of the land. Deposits reserved are part of raw materials wealth of the state and in the sense of the Constitution of SR, they are in ownership of the state and this ownership is unchangeable. Effectiveness of using the state ownership is interpreted through rationality. Since the raw materials are irrecoverable, the state must take care of using their savings with the goal of providing protection and rational usage of raw materials wealth as the ownership of Slovakia, together with regarding the sustainability principles [6].

One of the most important tasks in the future should be to increase the processing level and products' finalization on the base of non-metallic raw materials. Stocks and quality of ore raw materials, as the determining factor of their usage, prove their decreasing economic importance in the structure of raw materials wealth of the country. Similarly, domestic energetic sources (except brown coal and lignite) have, due to the volume of imported commodities, only small importance. The total rate of value of raw materials mining on GDP is negligible (0.5%), but this statistical data does not include the value of consequently processed and adjusted commodities on mineral base that is multiply higher and they present a significant element in the economy and foreign trade of Slovakia [7].

Earth sources have an impact on the economic growth. Their prices and the precise channels through which oil price shocks affect economic activity are only partially known, it is necessary to study the area more. The overall effect of energy prices on growth is negative. Arshad et al. (2016) found evidence that high-energy price decreases the real interest rate, investment and stock prices [8]. It also decreases the real value of local currency. High-energy price puts pressure on government expenditures and increases unemployment in the country. Also, Brown and Yücel (2002) surveyed linking fluctuations in energy prices to aggregate economic activity [9]. Cunado and Perez de Gracia (2005) studied the impact of oil price shocks on both economic activity and consumer price indexes, suggesting that oil prices have a significant effect on both economic activity and price indexes, although the impact is limited to the short run and becomes more significant when oil price shocks are defined in local currencies [10]. Many societies today express their desire to participate in the decision-finding on the development of their economic environment. In this area, a sustained and sustainable mine development requires the collaboration with the host communities concerned, which means that it has to be developed in a process commonly termed social licensing. Falck (2016) argues to make social licensing an integral element of economic (business risk) management for mining companies [11]. The last decade and a half witnessed a dramatic growth in mining activity in many developing countries. The relationships between mining and political economic change had been studied as well by Bebbington et al. (2008) [12].

Presently, energy conservation and pollution reduction have become two of the most significant factors in economic development and social circulate [13]. Economics–Energy–Environment (3E) system is necessary. The number of countries is struggling to understand, plan and realize the energy–environment–economic nexus [14]. Especially, coal has been an important commodity and it is one of the oldest mining enterprises in the country. The energy generated, especially the electricity and synthetic fuels, impacts on all sectors of the economy and the society. It is therefore necessary to determine the key players in the coal mining industry who are the beneficiaries of the process that underpins the socio-economic attributes of the industry [15].

3 Methodology

Due to the evaluation of the present state of raw materials usage in Slovakia, it is necessary to limit the position of the mining sector according to classification in the conditions of Slovakia. According to SK NACE Rev. 2, published by the Statistic Office of Slovakia, official indication of the mining industry is mentioned in section B – Mining and quarrying. The sector includes the mining of minerals, appearing naturally as solid minerals (coal and ores), liquid (petroleum) or gas (earth gas). In the frame of mentioned section, there are divisions 05–06: mining and quarrying of fossile fuels (coal, lignite, petroleum and gas), division 07–08, including: mining and quarrying of metal ores, various minerals and stone (Statistic Office SR, 2011). In spite of the historic tradition in mining and processing of ore raw materials, presently Slovakia does not belong to states with developed mining industry. According to United nations for developmend and business (UNCTAD), states, in which rate of mining and processing of raw materials on GDP is higher then 25%, belong to category of states with developed mining industry.

According to the national strategy for permanently sustainable development of Slovakia, the position of the natural environment and usage of raw materials in Slovakia are not sustainable from the long-term view. The present situation of the raw materials base in Slovakia is almost characterized by total exhaustion of ore raw materials stocks, big stocks, but different measures of non-metallic and construction materials usage, as well as a total limitation by state control over mining. Influences on the country and living environment, caused by mining, are vast. Such influences present only one of the most serious environmental problems of Slovakia (Government Resolution, SR, 2016) [16]. Sector B – Mining and quarrying, due to the reporting from the side of Statistical Office SR, is mentioned in the analysis as part of aggregated sector Industry in total [17].

From the view of social and economic impacts of mining, it is important to follow up mainly employment and wages in mining. Also, it is necessary to search the development and placement of mining companies in Slovakia. This way we analyzed individual regions of Slovakia. With the help of the social and economic indexes, we made an analysis of the macro environment and an analysis of the mining industry. Due to the evaluation of development, we followed up the influence of the mining industry on the GDP, employment and wage (due to the extended analysis, we will mention only the results and conclusions of the analysis).

4 Results

The usage of raw materials base directly and indirectly influences not only the environmental area of sustainable development, mainly by negative impacts by the way of environmental risk, presenting sources of sustainability instability, but also its social and economic area. Using raw materials influences the employment of inhabitants, and in this way it contributes to the increase of household incomes and also positively influences their purchasing power and living standard. At the same time, it also contributes to the creation of incomes to state budget and budget of administrations by payment for mining areas, mined mineral or stocking of gases, as well as local taxes and fees, or obligatory payment from dependent activity, etc.

The employment in mining organizations recorded a more or less decreasing trend of development, which connected logically with the development of raw material (RM) mining and usage of technical and technological equipment during RM mining in Slovakia. During the whole analyzed period, a higher employment was recorded for employees, working at surface mining, which is again logically connected with the prevailing surface way of RM exploitation in Slovakia. The highest employment was recorded in 2000, when employment achieved level 13,968 employees (6231 at deep mining and 7737 at surface mining) and on the other hand, the lowest level of employment was achieved in 2016, when the employment decreased to 5476 employees (from which 1666 employees worked in underground mining and 3180 at the surface mining). In the following period, the total employment decreased to 45.5% of employment in 2000, with an average number of employees amounting to 8273 employees per year. In underground mining, we recorded in the analyzed period an average employment of 3169 employees per year with its decrease in 2017 against 2000 by 62.3% (which means 3884 persons) from original employment. At the surface mining, decrease of employment was rather smooth, when recording a decrease by 48.2% (which means 3731 persons) from original employment in 2000, while averagely at surface mining 5105 employees per year were employed (Fig. 1).

Considering development of employment in Slovakia that was characterized by a variable tendency in 2000–2017 (Fig. 2), and a higher described development of employment in the mining sector, the rate of the mining sector also recorded an analogical trend of development, which means a fluctuating tendency (Fig. 3).

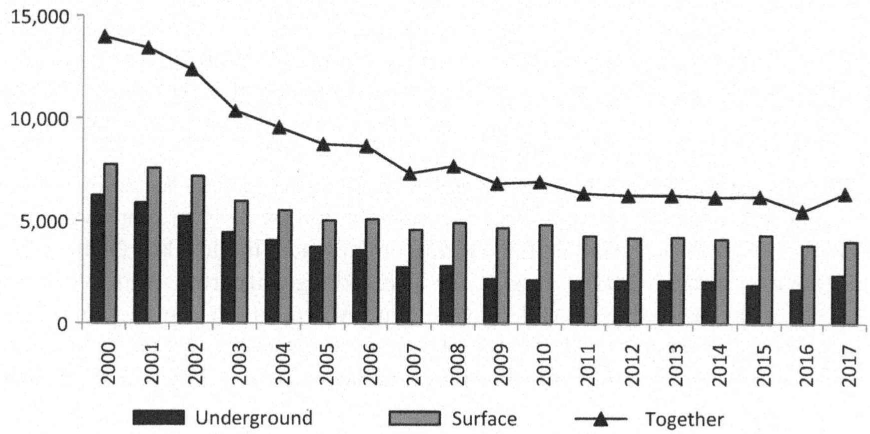

Fig. 1 Development of employment in the mining organization of Slovakia. (Source: Processing according to [18])

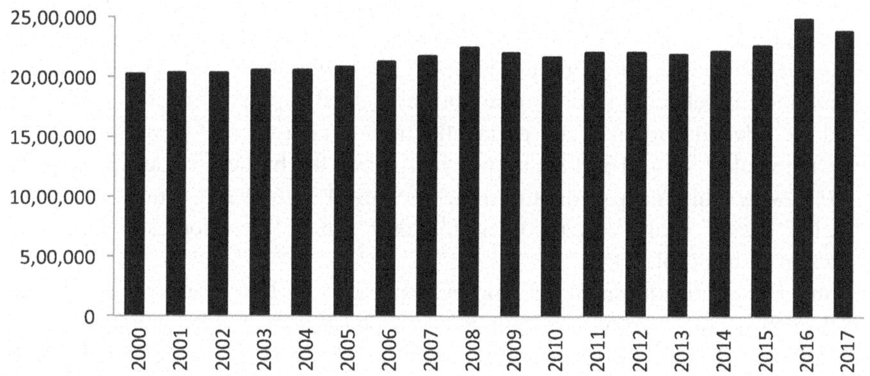

Fig. 2 Development of employment in Slovakia. (Source: Processing according to [19])

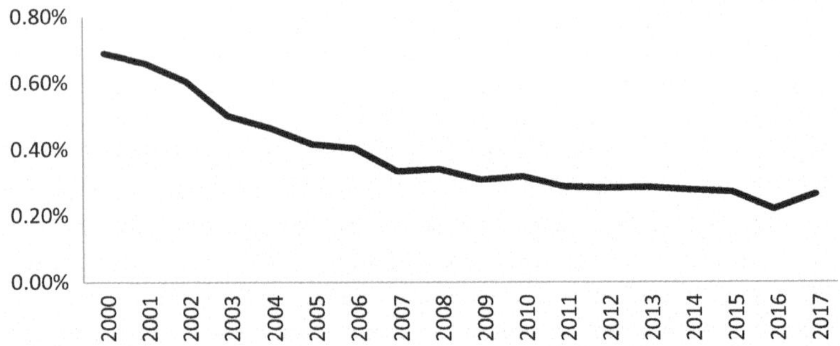

Fig. 3 Development of mining rate on employment of Slovakia. (Source: Processing according to [18, 19])

The average employment recorded in the analyzed period in Slovakia amounted to 2,178,705 employees per year, with the highest measure of employment in 2016 (2,390,000 employed) and with the lowest measure in 2000 (2,024,848 employed), which in comparison with development of employment in the exploitation of raw material base means that the average rate of mining on Slovakian employment achieved a level 0.39% in the analyzed period.

The highest rate of mining was recorded in 2000, when 0.69% rate of mining was achieved, which connected logically with the lowest level of employment in mining and on the other hand with the lowest employment in Slovakia. On the other hand, the lowest rate in 2016 – level 0.22% resulted logically from the highest level of employment in Slovakia and the lowest employment in the exploitation of raw material base. In general level, the rate of mining on total employment of Slovakia was characterized by a fluctuating tendency with decrease by 0.42% in 2017 against 2000 (Fig. 3).

The exploitation of raw material base contributes similarly to the creation of value added, which can be characterized as the difference between the production of commodities in basic prices and intermediate consumption of production factors in purchase prices. Production consists of products, produced in a common accounting period and intermediate consumption consists of the value of products and services, consumed in the production process as inputs, excluding investment property, whose consumption is reflected in the consumption of fixed capital. Also, the gross value added recorded in the analyzed period fluctuating character of development with an average value – 56,616 mil. Eur. per year, while the highest value was recorded in 2017, when this macroeconomic index achieved 73,602 mil. Eur; on the other hand the lowest value was recorded in 2000, when the value added recorded 37,698 mil. Eur (Table 1).

Similarly as development of employment, also in the frame of gross value added creation in the mining sector we recorded the fluctuating tendency in the analyzed period (Fig. 4), while mining recorded the average value added at level 29,581 mil. Eur. per year, which means 0.53% average rate. Generally, we can

Table 1 Development of gross value added in Slovakia

year	SR [mil. Eur]	Mining [mil. Eur]
2000	37 698.76	186.65
2001	39 486.60	187.48
2002	41 118.53	246.22
2003	42 973.05	275.94
2004	44 973.02	236.28
2005	47 492.91	249.22
2006	52 266.60	204,73
2007	58 021.92	328.49
2008	61 691.94	368.03
2009	58 314.00	372.83
2010	61 368.25	320.89
2011	62 898.59	344.24
2012	64 433.97	336.65
2013	65 167.74	337.71
2014	66 708.76	318.14
2015	69 295.49	352.96
2016	71 585.46	326.86
2017	73 602.75	331.21

Source: Processing according to [19]

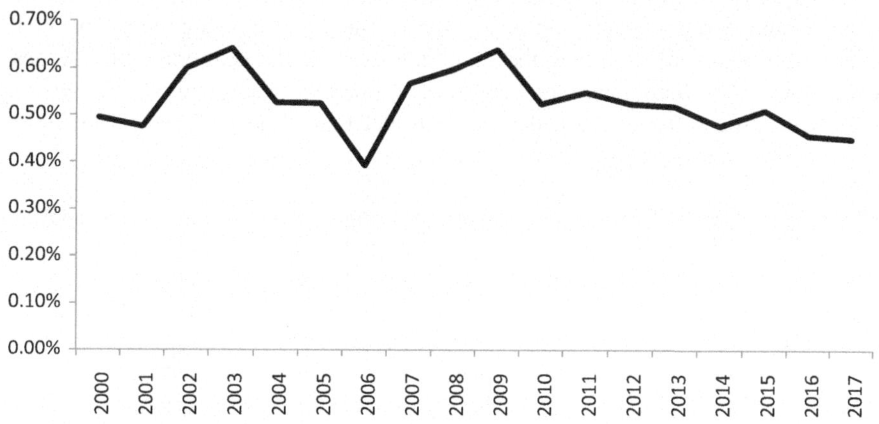

Fig. 4 Development of rate of mining on gross value added in Slovakia. (Source: Processing according to [19])

state that the highest value added in mining was recorded in 2009 with level 37,283 mil. Eur, on the other hand the lowest value was recorded in 2000, when the mining recorded value added at level 18,665 mil. Eur. Comparing 2000 and 2017, we found gross value added, produced by mining, decreased in 2017 against 2000 by 14,456 mil. Eur. The highest rate of mining on value added was recorded in 2003 with level 0.64%.

On the other hand, the lowest rate of mining on value added decreased in 2017 against 2000 by 0.05% (Fig. 4).

The development of mining and connected single exploitation of raw material base is also determined by domestic and foreign investments that support increasing economic effectiveness. Mainly, foreign direct investment (FDI) in Slovakia influences positively sustainable development, since it increases employment by creation of new working posts, GDP creation; it contributes to the implementation of innovative technologies or increase of qualification of domestic human capital. The FDI can be generally characterized as investment of money or other monetary values or rights with the aim to base, obtain or extend permanent economic relations of domestic investor to business abroad, resp. foreign investor investing business in Slovakia. This is mainly due to the mentioned FDI and rate of mining on FDI being considered as direct indicators of dynamic development of Slovakia in interaction with investment attractiveness of Slovakia for foreign investors. The FDI development in Slovakia is characterized more or less by an increasing trend with a slight decrease against 2013 by 2.62%. According to the available data, we can state that the average FDI level in the analyzed periods was approximately 27,140,447 tis. Eur. per year, while exploitation of raw material base recorded an average FDI level 232,371 tis. Eur. per year (Table 2), with average rate on total FDI in the analyzed period at level 0.88%.

The exploitation of raw material base participates similarly as mentioned indexes of social and economic development, also in the GDP development. The GDP development was ranked among the basic indicators of the economic effectiveness of Slovakia. Also this index, similarly as mentioned earlier, recorded a fluctuating trend. According to the available data, we can state that in the analyzed period 2000–

Table 2 FDI development in Slovakia and mining

Year	SR [thousand Eur]	Mining [thousand Eur]
2000	4 097 365.44	43 508.43
2001	5 376 306.25	65 094.73
2002	7 270 627.89	63 859.35
2003	12 490 727.63	71 948.77
2004	15 141 732.45	86 089.53
2005	19 700 181.02	102 308.47
2006	23 338 906.57	312 402.28
2007	29 075 292.32	300 263.44
2008	35 846 869.24	350 298.28
2009	36 469 023.00	396 317.98
2010	37 665 095.00	426 960.00
2011	40 173 448.00	476 802.00
2012	42 304 005.00	497 996.00
2013	42 071 874.94	211 726.00
2014	40 969 239.26	209 928.00
2015	42 256 452.00	102 437.00

Source: processing according [20]

2017 the GDP recorded averagely 583,135 mil. Eur. per year, while the highest value was recorded in 2015 at level 7,889,644 mil. Eur and on the other hand the lowest value was recorded in 2000 at level 3,160,129 mil. Eur. Exploitation of the raw material base in Slovakia participated at the GDP creation in the analyzed period averagely by 583,135 mil. Eur. per year, while the highest level of raw material base usage on GDP in Slovakia was recorded in 2008 (3713 mil. Eur) and the lowest value was recorded in 2002 (1543 mil. Eur) (Table 3). During evaluation of the rate of the raw material base on GDP creation, we can state that this was directly determined by the development of single GDP in interaction with development of the mining sector that in the analyzed period recorded average rate on single GDP creation in Slovakia 0.54% per year.

Also, the rate of the raw material base usage on GDP creation, similarly as the above-mentioned analyzed indexes, recorded a fluctuating trend, determined by the GDP development in Slovakia, as well as by demand and offer development of single raw material base. The highest rate of mining, which means usage of the raw material base in Slovakia, was recorded in 2000, when the rate of mining on GDP creation in Slovakia achieved 0.79% and the lowest rate was recorded in 2014, when the rate achieved only 0.42% (Fig. 5).

According to the above-mentioned development of employment in mining organization and development of average wage in Slovakia, we can state that the income of state from payment from employees and employer by social security contributions presented averagely 278 mil. Eur. per year, when the highest incomes were recorded in 2007, presenting approximately 256 mil. Eur, which connected logically with the development of average wage and number of employees, since in 2007 the average annual wage in Slovakia was 71,891 Eur and the number of

Table 3 GDP development in Slovakia and mining

year	SR [mil. Eur]	Mining [mil. Eur]
2001	34 310.622	158.0
2002	37 279.842	154.3
2003	41 404.346	207.0
2004	46 101.527	160.7
2005	50 415.092	177.7
2006	56 272.653	204.2
2007	63 053.882	263.6
2008	68 491.623	371.3
2009	64 023.061	329.6
2010	67 577.288	358.1
2011	70 627.205	337.9
2012	72 703.513	330.3
2013	74 169.873	334.6
2014	76 087.789	319.8
2015	78 896.443	354.8

Source: Processing according to [19]

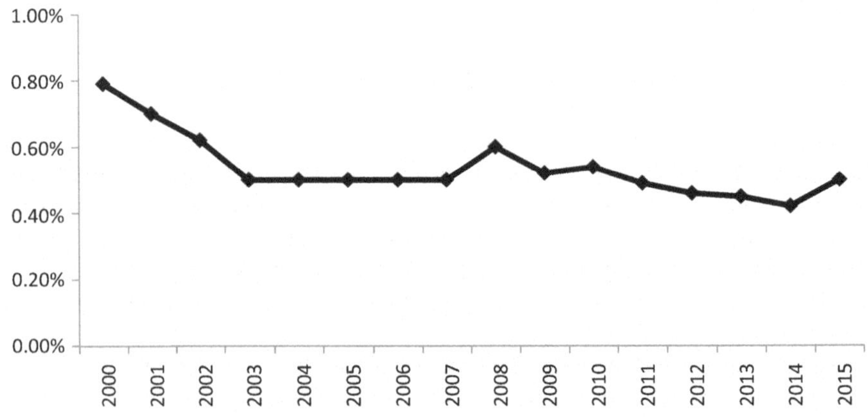

Fig. 5 Rate of mining on GDP in Slovakia. (Source: processing according [18])

Fig. 6 State incomes from payment to social security of employees. (Source: Processing according to [18])

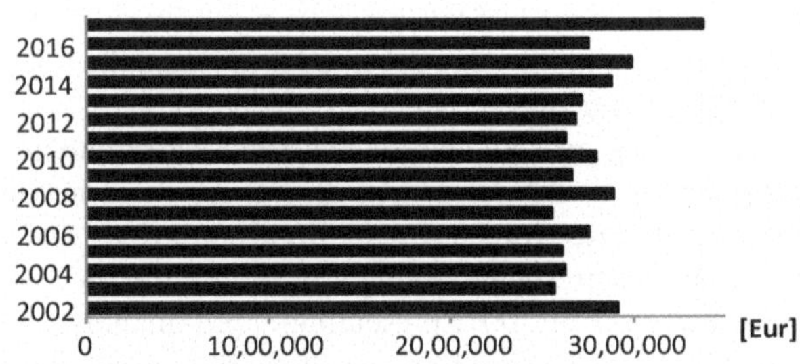

employees was 7312. In 2017 the average annual wages achieved 1095 Eur and the number of employees in mining organizations was 6353 (Fig. 6).

Income of the state budget, budget of administrations and environmental fund also presented payment for mining area, mined minerals and stocking gases. In the analyzed period, the highest incomes were created in 2007, achieving approximately 5.56 mil. Eur, from which only 2.3% presented payment for the mining area (approximately 12,655 tis. Eur), 20.0% payment for stocking gases (approximately 112 mil. Eur) and yet 77.8% (approximately 432 mil. Eur) presented payment for mined minerals (Fig. 7). During the analyzed period, the highest rate of state budget income and administration incomes presented payment for mined minerals, when its level was quantified according to the conditions amended by Government Decree No. 50/2002 Z. z. about payment for mining area, mined minerals and stocking of gases or liquids, as amended lately. The payment for the mining area that presented the lowest incomes of state and administrations is paid at the level 80% to administration, on whose cadastral area the mining space is situated, and 20% to the state budget.

Due to the above-mentioned reasons, we can state that the usage of raw materials is characterized by positive impacts on the social and economic area of sustainable development in Slovakia, which is different from the negative impacts on the environmental area of sustainability.

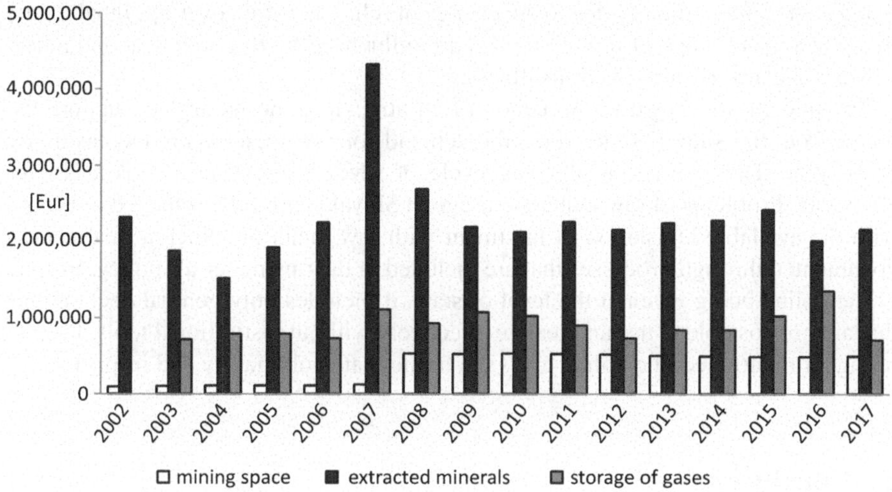

Fig. 7 Development of payment from RM exploitation in SR. (Source: Processing according to [18])

5 Discussion

The effectiveness of raw material used is changing in accord with internal and external factors. The raw material policy must respect the principles of social and ecologically orientated market economy during mining and processing of raw materials and to also regard effective usage of the natural resources. The mining activity is characterized by technological severity and high rate of human work, and in this way with regard to the cost structure it presents an important source of employment.

In spite of Slovakia being a small country, it has its own rich sources of raw materials. Mining is participating at GDP in Slovakia. The majority of mined raw materials are consumed in the domestic market, as, for example, building stone, dolomite, gravel and sand. Mostly, petroleum, earth gas, iron ores and black coal are imported to Slovakia [21]. Among the positive impacts of mining belong increasing average monthly wage in the sector. People automatically with increasing wages have a need to buy more and more products and services for provision of their needs, which can be met by using raw materials.

Decrease of employment in mining can be considered as negative social and economic impact of mining activity in Slovakia. New companies purchase technologies, by which they increase the volume of mining, sales of products, and at the same time the number of employees is decreasing. Further reason for the employment decreasing in mining can be due to the disinterest of people to take up such a physically demanding job. Mining activity influences mostly living environment and inhabitants of the country. Mining of raw materials influences living environment mainly due to the change of relief and soil covering. Inhabitants, living in surroundings of mining works, are influenced by too high dust and noise, which can cause rising of various illnesses.

In spite of the negative influence of mining, it is necessary to support the sector from the side of state, since the demand for raw materials is increasing by every year. The estimation of a life cycle of several stocks in Slovakia is over 200 years. Problems of raw materials usage in Slovakia are fully reflected in accord with the available legislation of treatment with raw materials. Such a treatment is documented through processes that are included in the raw materials policy. In spite of the policy being given at the level of state, it includes only general declarations that are not possible to use and realize effectively without institutional tools. In spite of higher mentioned, the policy gives the frame that is obligatory and inspiring.

6 Conclusion

The worldwide importance of raw materials is obvious from the interpretation of their influence on the world economy, state economy and regions' economy. Costs of the raw materials availability are also reflected through the availability of energetic raw materials. Raw material safety is mainly about availability, and raw material base is more important in the present time than in the past. Also, the European Union (EU) put raw material safety among basic priorities. This cannot be limited or disadvantaged by aspect of no effective or improper decisions of member states, especially states that do not have a direct impact on their raw material base and that deform the raw material base by tendentious opinions to the raw material base and its usage. The problems of raw material base using in Slovakia are fully reflected in the available legislation. The legislation to threat with raw materials is recorded through processes of raw material policy. The raw material policy of the state consists only of general declarations that through institutional tools cannot be effectively used and realized. In spite of higher mentioned, it gives the frame that is obligatory and inspiring. The transformation of competences downwards to the autonomy brought new quality of management, when not state, but autonomy participates in the public affairs. Competences takeover brought also the responsibility of decision about using raw materials base. The results can serve as the basis for the management of mining organizations towards sustainability, which will demand research of other indicators in all regions and industries, contributing to national economy.

Acknowledgements The contribution presents partial results of project solving VEGA No. 1/0515/18, VEGA No. 1/0651/2019 and KEGA 006TUKE-4/2019.

References

1. Ciruelos, J., & Duchene, M. (1983). Modele sur micro-ordinateur pour l'analyse de la faisabilite financiere d'un project minier. Industrie Minérale – Les techniquies.
2. Zuberec, J., Tréger, M., Lexa, J., & Baláž, P. (2005). Raw material of Slovakia. ŠGÚ DŠ, Bratislava, 351 p.
3. Cehlár, M., & Maras, M. (2001). Economic evaluation of mining projects. TU, FBERG, Košice.
4. Merciu, G. L., Merciu, F. C., & Cercleux, A. L. (2016). The assessment of social and economic impacts associated to an abandoned mining site. Case study: Ciudanovita (Romania). International Conference – Environment at a Crossroads: SMART approaches for a sustainable future. *Procedia Environmental Sciences, 32*, 420–430.
5. Rybár, P., Cehlár, M., Engel, J., & Mihok, J. (2005). Evaluation of mineral deposits. *Acta Montanistica Slovaca, 1*, 136.
6. Švarda, T. Sustainable development as possible medicament for problems (in Slovak). Available at: http://svarda.blog.sme.sk/c/205978/Trvalo-Udrzatelny-rozvoj-ako-mozny-liek-na-problemy-ludstva-Snad.html.
7. Peter, B. (2013). Perspective of domestic energetic raw materials using (in Slovak). ŠGÚ DŠ. *Mineralia Slovaca, 45*, 225–232.
8. Arshad, A., Zakaria, M., & Junyang, X. (2016). Energy prices and economic growth in Pakistan: A macro-economic analysis. *Renewable and Sustainable Energy Reviews., 55*, 25–33.
9. Brown, S. P. A., & Yücel, M. K. (2002). Energy prices and aggregate economic activity: An interpretative survey. *Quarterly Review of Economics and Finance, 42*(2), 193–208.
10. Cunado, J., & Perez de Gracia, F. (2005). Oil prices, economic activity and inflation: Evidence for some Asian countries. *Quarterly Review of Economics and Finance, 45*(1), 65–83.
11. Falck, W. E. (2016). Social licencing in mining – Between ethical dilemmas and economic risk management. *Mineral Economics, 29*, 97–104.
12. Bebbington, A., Hinojosa, L., Bebbington, D. H., Burneo, M. L., & Warnaars, X. (2008). Contention and ambiguity: Mining and the possibilities of development. *Development and Change, 39*(6), 887–914.
13. Liu, D., Yang, X., Tian, X., Wu, R., & Wang, L. (2011). Study on integrated simulation model of economic, energy and environment safety system under the low-carbon policy in Beijing. *Procedia Environmental Sciences, 5*, 120–130.
14. Ur Rehman, S. A., Cai, Y., Mirjat, N. H., Walasai, G. D., & Nafees, M. (2019). Energy-environment-economy nexus in Pakistan: Lessons from a PAK-TIMES model. *Energy Policy, 1*, 200–211.
15. Mathu, K., & Chinomona, R. (2013). South African coal mining industry: Socio-economic attributes. *Mediterranean Journal of Social Sciences, 4*(14), 347–358.
16. Government resolution. Available at: http://www.vlada.gov.sk/programove-vyhlasenie-vlady-sr-na-roky-2016-2020/
17. Klinda, J., & Lieskovská, Z. (2011). Report on state of living environment in Slovakia in 2010 (in Slovak). MŽP SR, SAŽP SR.
18. HBU SR: Annual report of activity of mining office in SR in 2018. Available at: http://www.hbu.sk/sk/Vyrocna-a-rocna-sprava/Rocne-spravy.alej
19. Statistic office of Slovakia. (2017). Available at: http://www.statistics.sk/pls/elisw/metainfo.explorer
20. NBS, SR: Inward FDI financial flows. (2018). Available at: https://nbs.sk/sk/vyhladavanie-na-strankach-nbs
21. Raw material of Slovakia. (2014). Available at: http://www.geology.sk/new/sites/default/files/media/doc/rocenky_NS/rocenka_%20NS_10.pdf.

Index

L. Knapcikova et al. (eds.), *4th EAI International Conference on Management of Manufacturing Systems*, EAI/Springer Innovations in Communication and Computing, https://doi.org/10.1007/978-3-030-34272-2

K

L

M

T

The manufacturer's authorised representative in the EU is Springer Nature Customer Service Centre GmbH, Europaplatz 3, 69115 Heidelberg, Germany. If you have any concerns regarding our products, please contact ProductSafety@springernature.com

Printed and bound by CPI Group (UK) Ltd, Croydon, CR0 4YY
25/09/2026
02222901-0008